Biodiversity and Protected Areas

Biodiversity and Protected Areas

Special Issue Editors

Karen Beazley
Robert Baldwin

MDPI • Basel • Beijing • Wuhan • Barcelona • Belgrade

MDPI

Special Issue Editors

Karen Beazley
Dalhousie University
Canada

Robert Baldwin
Clemson University
USA

Editorial Office
MDPI
St. Alban-Anlage 66
4052 Basel, Switzerland

This is a reprint of articles from the Special Issue published online in the open access journal *Land* (ISSN 2073-445X) from 2018 to 2019 (available at: https://www.mdpi.com/journal/land/special_issues/biodiversity_protectedareas)

For citation purposes, cite each article independently as indicated on the article page online and as indicated below:

LastName, A.A.; LastName, B.B.; LastName, C.C. Article Title. *Journal Name* **Year**, *Article Number*, Page Range.

ISBN 978-3-03897-732-2 (Pbk)
ISBN 978-3-03897-733-9 (PDF)

Cover image courtesy of Robert Baldwin.

Contents

About the Special Issue Editors

Karen Beazley School for Resource and Environmental Studies, Dalhousie University, 6100 University Ave., P.O. Box 15000, Halifax, Nova Scotia, B3H 4R2, Canada. Interests: biodiversity conservation; protected area system design; conservation biology; landscape ecology; road ecology; Indigenous perspectives; environmental justice.

Robert Baldwin Forestry and Environmental Conservation Department, 261 Lehotsky Hall, Clemson University, Clemson, South Carolina 29634, USA. Interests: biodiversity; landscape-scale conservation planning; wetland landscapes; habitat connectivity.

Preface to "Biodiversity and Protected Areas"

Protected areas are key to biodiversity conservation. While the value of protected areas is generally undisputed, challenges remain. Many areas designated as protected were created for objectives other than biodiversity conservation, and those objectives can conflict with biodiversity conservation. Protected area legal status is, in many cases, impermanent. Protected areas are generally too small, isolated, and few to conserve biodiversity on their own, and thus there are calls for connected conservation areas between them, and for their integration into broader landscapes and seascapes [1]. There is a general consensus that the current global suite of protected areas is insufficient to protect biodiversity. Although there is no precise prescription for how much would be enough, systematic conservation planning studies have indicated that 25–75% of a region is required to capture key elements of biodiversity [2]. Studies that address range shifts and movement pathways in response to climate change reveal even more extensive area and connectivity requirements. These and other insights have contributed to recent calls for 'half Earth' [3].

There is an increasing recognition that not all of the area required to maintain biodiversity is likely to be accommodated within protected areas. Other effective area-based measures, connectivity, and management of private lands offer potential complements to protected areas, but may also compete for scarce resources. Increased focus on framing biodiversity and protected area values in terms of ecosystem services and human well-being may not always lead to biodiversity conservation, particularly if narrowly focused on goods and services. There is increasing acknowledgment of the imperative to engage Indigenous communities and recognize their rights to self-governance, territorial lands and resources, including biodiversity, and conservation areas. These and other emergent issues demand transformed approaches to biodiversity and protected areas, which engage diverse communities and boundary spanning collaborations, and may require new conceptual framings.

This Special Issue assembles papers that explore these and other emerging issues around biodiversity and protected areas. We sought papers that examine approaches that show promise or demonstrate success as potential new models and applications that support progress in biodiversity conservation and protected areas in an increasingly challenging and complex context. Papers are from all regions of the world. Our ultimate goal is to identify new ways of moving forward in a context of increasing urgency.

Reference

1. United Nations. Strategic Plan for Biodiversity 2011–2020 including Aichi Biodiversity Targets. 2010. Available online: https://www.cbd.int/sp/.
2. Noss, R.F.; Cooperrider, A. Saving Nature's Legacy: Protecting and Restoring Biodiversity; Island Press: Washington, DC, USA, 1994.
3. Wilson, E.O. Half Earth: Our Planet's Fight for Life; WW Norton & Company: New York, NY, USA, 2016.

Karen Beazley, Robert Baldwin
Special Issue Editors

land

MDPI

Editorial

Emerging Paradigms for Biodiversity and Protected Areas

Robert F. Baldwin [1,*] **and Karen F. Beazley** [2,*]

1   Department of Forestry and Environmental Conservation, Clemson University, Clemson, SC 29634, USA
2   School for Resource and Environmental Studies, Dalhousie University, P.O. BOX 15000, Halifax,
    NS B3H 4R2, Canada
*   Correspondence: baldwi6@clemson.edu (R.F.B.); Karen.Beazley@Dal.Ca (K.F.B.)

Received: 25 February 2019; Accepted: 26 February 2019; Published: 1 March 2019

Despite significant investments in protected areas, biodiversity continues to show the negative influence of human domination of earth's ecosystems with population reductions across many taxa (Dirzo et al. 2014) [1]. Biodiversity loss ("biosphere integrity") is one of two "core planet boundaries", and currently exceeds the "safe operating space for humanity" as an intrinsic biophysical process that regulates the stability of the Earth system: it is at high risk, "beyond the zone of uncertainty" that human perturbations will destabilize the Earth system at a planetary scale (Steffen et al. 2015) [2]. Shifts towards globalization and increased emphasis on ecosystem services pose further challenges to biodiversity and its conservation (Cimon-Morin et al. 2013) [3]. Particularly threatened are organisms that require large, undisturbed areas where natural patterns and processes can occur freely (Laliberte & Ripple 2004) [4]. Such areas are increasingly under-protected. The amount of area under traditional protection varies by country but globally is about 15%, short of the 25–75% required to capture vulnerable biodiversity (Juffe-Bignoli et al. 2014) [5]. Habitat fragmentation is an ongoing process isolating remaining areas of high quality habitat (Haddad et al. 2015) [6]. For example, 50% of the continent of Europe is within 1.5 km of transportation infrastructure (Torres et al. 2016) [7]. A summary of 35 years of studies of habitat fragmentation caused by infrastructural development has shown that it has reduced biodiversity by 13–75% in various regions across the globe (Lawton 2018) [8]. The Global Human Footprint in terrestrial systems increased 9% from 1993 to 2009 with 75% of the surface experiencing measurable pressures (Venter et al. 2016) [9]. Using these same measures, one-third of protected land is influenced by intense human activity (Jones et al. 2018) [10].

The increasing drumbeat of alarm that protected areas, as special and valuable as they are, have not been an adequate answer to the biodiversity crisis is supported by a plethora of studies. Protected areas are often located in the wrong places to protect the greatest diversity (Jenkins et al. 2015) [11], are systematically biased in location (Margules & Pressey 2000) [12], are too small, scattered, and disconnected to protect diversity under changing climatic conditions (McGuire et al. 2016) [13], and internally and at landscape scales are often mismanaged such that biodiversity establishment goals are not achieved (Belote et al. 2016 [14]; Joppa & Pfaff 2009 [15]). Protected areas alone, unless increased in area and landscape-level management practices beyond those currently considered political acceptability, are unlikely to reduce decline let alone stimulate recovery and provide resilience in response to climate changes. On the positive side, there has been a vast increase in research and practical engagement in systematic conservation planning, habitat connectivity, and socioeconomic and cultural mechanisms (Sinclair et al. 2018) [16], and widespread international biodiversity and protected areas initiatives (IUCN 2005 [17], 2017 [18]; UNEP 2010 [19]).

The global map of marine and terrestrial protected areas illustrates the impressive scope and extent of effort that has been invested to create the protected areas estate (Figure 1). There are approximately 200,000 significant terrestrial and 12,000 marine sites globally, covering 15.4% and 8.4%, respectively (World Database on Protected Areas (WDPA)). These numbers are underestimates of the actual amount that is secured from overuse or conversion; for example, in the continental United States (US) alone there are nearly as many as in the WDPA. Yet, in the US there are many multiple use lands that are not specifically managed for biodiversity. On the other hand, the WDPA percentages overestimate the amount of protected area that is permanently secured and effectively managed primarily for biodiversity protection, particularly in the marine realm; only 3.6% of the oceans are formally protected and many of these are not effectively managed (Baillie & Zhang 2018 [20]; Edgar et al. 2014 [21]). Despite inadequate coverage and management, it is evident that humanity has highly valued nature conservation. Much time and resources are dedicated to protected areas but not all of that effort is efficiently spent (Armsworth et al. 2011) [22]. It is clear from databases such as the WDPA and Protected Areas Database of the United States (PADUS), that despite challenges cataloging, classifying, and providing data for protected areas (DellaSala et al. 2001 [23]; Rissman et al. 2017 [24]) a vast amount of international, national, and local effort has been expended for protected areas.

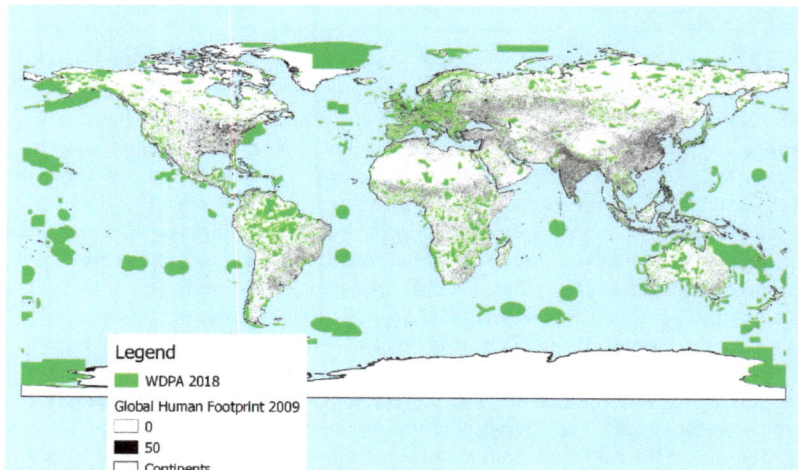

Figure 1. Largest protected areas of the world across marine and terrestrial ecosystems in relation to areas of human land use transformation, the Global Human Footprint. WDPA 2014, Accessed from Protected Planet October 2018. Global terrestrial Human Footprint maps for 2009 from Dryad Digital Repository; HF values stretched from 0 (min) to 50 (max). https://doi.org/10.5061/dryad.052q5.2 (Venter et al. 2016) [9].

The effort has not kept pace with the scale of the task—as much as 50% of the planet needs to be managed for biodiversity (Wilson 2016) [25]. Protected areas alone will likely not accomplish the task. Managing the landscape as a binary problem—protected and not—is too simplistic. A more nuanced view is needed to achieve lasting biodiversity benefits, a view reflected in this special issue. Namely, mechanisms would see the continued establishment of systematically selected protected areas; management of the areas themselves and across political and jurisdictional boundaries improved by meaningful inclusion of the ideas, experience, leadership, and biocultural heritage of local and Indigenous peoples; enhanced quality of life of people; measurements of protected area biodiversity values and ecosystem services provided to society; and, improvements to the habitat quality of the landscape matrix and connectivity made a priority.

The papers in this special issue respond to these unprecedented challenges and imperatives and reflect three broad categories of emerging topics for biodiversity and protected areas. First, there is the topic of database inclusiveness; accuracy, accessibility, and curating is a basic need for assessing protected areas distribution and function. Second, there are issues of social justice and protected areas, in particular the need to reconcile past and on-going 'wrong-doings' that exclude Indigenous and other local peoples from protected areas themselves and from engagement in their establishment, management and benefits despite their long tenure in those areas and development of their cultures based on local biota. By including Indigenous and local people in management, protected area function may be enhanced. Third, there are challenges in protected area planning and management, within their boundaries and at the landscape scale. National and International politics and policies likewise influence function. Papers in this special issue cut across these topics.

1. Protected Areas Databases

Basic tenets of ecological inquiry are to first ask 'where, what, and how much'. Mapping of protected areas and including them in databases so they may be accessed and used in analyses, including those designed to assess biodiversity coverage, has been a vital function of several conservation initiatives. The World Database on Protected Areas, Protected Areas Database of the United States, Canada's Conservation Areas Reporting and Tracking System (Vanderkam 2016) [26], and National Conservation Easement Database are examples of comprehensive attempts to gather, standardize, and serve spatial data and protection attributes. Despite these efforts, standardized data on management interventions and changes in biodiversity inside and outside of protected areas "do not currently exist for any global sample of PAs, but need to be created" if the relationship to biodiversity outcomes is to be understood (Geldmann et al. 2018) [27].

What seems like a simple task of compiling public information into protected areas databases has not been simple for several reasons. First, there is the question of what to include in a protected areas database, i.e., what constitutes 'protection'. Classification systems of protected areas differ, such as the International Union for the Conservation of Nature (IUCN) Global Protected Areas Programme (Dudley et al. 2010) [28], and the GAP system of the United States Geological Survey (Dudley et al. 2010 [28]; Scott et al. 1993 [29]). This is why, for example, the World Database of Protected Areas includes about 200,000 global records, and the Protected Areas Database of the United States includes roughly the same amount (196,000) for the lower 48 states alone, i.e., the PADUS is more inclusive than the WDPA. Even though guidance exists for defining protected areas, such as for international tracking towards numerical targets under the Convention on Biodiversity's Aichi Target 11, it is up to signatory nation states and their sub-national jurisdictions to interpret the guidance for reporting purposes (MacKinnon et al. 2015) [30], and these interpretations may vary significantly, often for political purposes and sometimes triggering perverse consequences such as a proliferation of 'paper parks' that lack demonstrable conservation impact (Barnes et al. 2018) [31]. Second, there is the problem of obtaining accurate polygons or point locations for protected areas. The government and private entities who acquired, map, and manage the protected areas are responsible for providing those data. Metadata of protected area databases describe problems that pertain to lack of accuracy (such as https://gapanalysis.usgs.gov/padus/data/metadata/). In the United States, compilation of digital map data on public protected areas began in earnest in the 1990s (Scott et al. 1993 [29]). Thus it has been an ongoing project. Obtaining private protected area map data encounters reluctance on behalf of providers to share because of privacy concerns (Rissman et al. 2017) [24].

In this special issue, several authors address the issue of databases. Clements et al. [32] identify the emerging phenomenon of private lands conservation as a concern for database development and management. They review reporting procedures from three countries and recommend a process by which data can be reviewed according to 10 principles and subsequently included in the WDPA. Fundamental to their approach is the problem of equity in relation to reporting requirements and management, in particular for private landholders who may not currently receive the benefits of participating in reporting processes. Zurba et al. [33] address these issues in relation to Indigenous Protected and Conserved Areas (IPCAs), and raise the additional, perplexing legal and ethical issue of what may constitute 'Indigenous-led' when it comes to establishing and reporting on IPCAs, and the limitations and hesitations that Indigenous communities may have with reporting and tracking IPCAs within an imposed framework, particularly when they may receive no benefits for doing so.

Furthermore, they may risk being co-opted in the service of national or international quantitative targets for biodiversity protection, and thus inadvertently contribute to the colonial enterprise. Baldwin and Fouch [34] illustrate the role of small protected areas in conserving biodiversity and the special challenges of spatial data at these fine scales. They describe the area distribution of protected areas in the United States and find that they are, on average, very small (median 16 ha, mean 1648 ha). Additionally they identify potential errors of inclusion and exclusion, and hypothesize such errors may disproportionately influence mapping of small areas. Similarly to Clements et al. [32], they note that database errors probably do not accrue to large public ownerships, but rather smaller, private protected areas. In order to accurately assess how well the protected areas estate meets biodiversity goals, a base requirement is accurate, complete, and accessible data on coverage and effectiveness of protected areas.

2. Social Justice and Protected Areas

Protected areas are commonly assumed to be at the 'wild' end of a human domination gradient. This view undoubtedly arises from the history of protected area establishment in that remote 'rocks and ice' areas were systematically selected for conservation due to their low economic value, among other reasons. As human populations and activities proliferate in more remote areas, conservationists have had to grapple with new realities. Many protected areas are now embedded in human dominated landscapes (Figure 2).

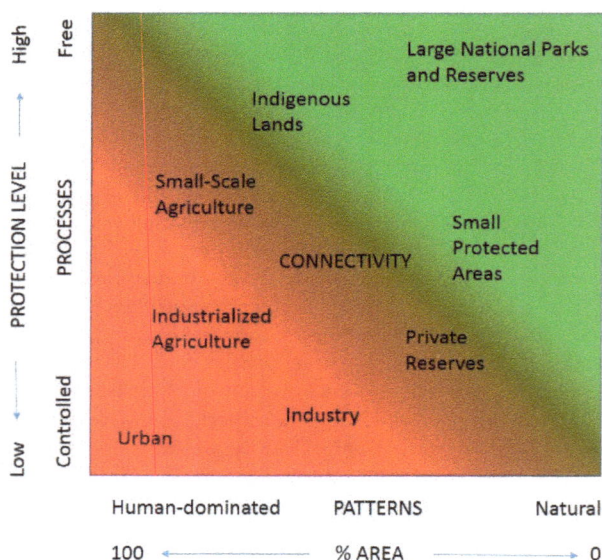

Figure 2. A gradient view of protected areas adapted from a human modification gradient (Theobald 2004) [35]. There are protected areas in virtually every human landscape context. Connectivity is noted in capital letters in the center in the portion of the gradient.

An emerging view is that people living in or near protected areas are not by default a threat to biodiversity and can be management partners. Their wellbeing needs to be part of the conservation equation, especially since poverty near protected areas creates a negative dynamic and can actually be alleviated by progressive management (DeFries et al. 2004) [36]. People have cultures and practices that arose from natural ecosystems, places, and organisms and these histories can become part of the conservation puzzle.

Conservation biology started in the 1980s as a discipline focused on the non-human world and anthropogenic threats to diversity. As realities of conservation have become more apparent, the field has become much more inclusive of research on humans in and around protected areas as more than mere threats but in some ways, mechanisms for conservation (Klein et al. 2008 [37]; Powell et al. 2009 [38]). Thinking about the matrix in which protected areas sit, and how human activities outside protected areas can influence their effectiveness, contributes to a holistic view of landscapes moving from discrete areas of management and protection to working across the entire gradient of land uses (Anderson et al. 2012 [39]; Cushman et al. 2010 [40]). A view that protected areas are at one end of a management spectrum with intervening areas managed across a gradient of human activities has emerged so that a holistic, landscape view of habitat and permeability becomes more in focus. More area needs to be set aside, but the emerging view is that people's livelihoods and wellbeing be considered as part of the conservation equation if biodiversity protection is to be sustainable, and at the same time these matrix activities can be managed and directed to be less harmful and even beneficial for core protected areas.

In their assessment of global biodiversity hotspots, Cunningham and Beazley [41] point out the imperative to attend to both the wellbeing of the people who live there and the establishment of more protected areas. They identify that most hotspots in which protection has not yet reached 17% are located in countries that are struggling economically and also dealing with war, famine, social unrest, and rapid sea level rise. Half of these hotspots have population densities above the global average. They argue that it is unrealistic and unethical in terms of international equity to assume that protected area targets will be met in many of these regions without extensive assistance from the global community, that wealthy nations have a responsibility to address threats to biodiversity from their consumption of trade goods produced in hotspots outside their boundaries, and that novel approaches to biodiversity conservation should support human-nature coexistence in and beyond protected areas.

Special attention to social justice around protected areas is critical in relation to Indigenous communities, many of whom have experienced and continue to be at risk of dispossessions of their lands, livelihoods and wellbeing for conservation and development purposes. Despite international and national efforts around rights and reconciliation, such as the United Nations Declaration on the Rights of Indigenous Peoples, the Truth and Reconciliation Commission of Canada, and the ICCA[1] Consortium (Borrini-Feyerabend & Campese 2017) [42], meaningful engagement and partnership with Indigenous communities in protected area establishment and management is far from the norm. In this issue, McCarthy et al. [43] posit that the knowledge of Indigenous communities is too often ignored in management of protected areas due to cultural and other assumptions. Their study documents high levels of awareness of biodiversity and positive attitudes towards conservation within the local population near a National Park in Mongolia. Improved communication will more meaningfully engage local and Indigenous communities and help to overcome trust issues; however, as they and others point out, it is better to see Indigenous and local communities as collaborators and/or leaders, if possible, rather than solely as recipients or providers of knowledge, and certainly rather than adversaries—a view that is echoed throughout conservation literature (Colchester 2004) [44].

Zurba et al. [33] position Indigenous leadership (sovereignty) and collaborative governance of protected areas as a potential means of conservation through reconciliation. Indigenous-led Protected and Conserved Areas (IPCAs) express traditional values, responsibilities and Indigenous laws and worldviews, through land stewardship that encompasses the understanding of humans and non-humans as one community whose health is intertwined. In a similar vein, Ekblom et al. [45] show that the cultural connection to landscapes for people in Sub-Saharan Africa through ecosystem memories, landscape memories, and place memories is a powerful conservation dimension. Biocultural

---

[1] ICCA is not an acronym. "It is an abbreviation for 'territories and areas conserved by indigenous peoples and local communities' or 'territories of life'." (https://www.iccaconsortium.org/index.php/discover/).

heritage connects identity, social cohesion, and practice with social and political negotiation and is thus, they argue, foundational to promoting stewardship.

IPCAs exemplify the nexus of social justice, protected areas and biodiversity conservation. Through the lens of implementing IPCAs in Canada, Zurba et al. [33] examine wicked problems in relation to Aboriginal and Treaty Rights, Aboriginal title, building a nation-to-nation relationship with Canada, Aichi Target 11, and other international Indigenous rights and biodiversity conservation initiatives. Not the least of these is the exclusionary 'wilderness' paradigm pervasive in parks and protected areas, particularly in the global north and west but increasingly exported around the world, which has thus far proven difficult to supplant despite new more inclusionary paradigms that link Indigenous and other local peoples with conservation lands. Attention to both on-the-ground practices and high-level considerations is critical for equitable and just relations between Indigenous and conservation communities. These relations are necessary for protected areas and other arrangements that serve biodiversity conservation and reconciliation, and thus de-colonize institutions, peoples, and non-human nature. Complementary to the social justice imperative to uphold the rights of Indigenous peoples to their lands and its governance is the recognition that it is essential to meeting conservation goals in a practical sense: Indigenous lands, globally, intersect at least 40% of all protected areas, account for 37% of all remaining ecologically intact landscapes, and encompass >65% of the remotest and least inhabited anthropogenic biomes (Garnett et al. 2018) [46].

3. Protected Areas Planning and Management

Management of protected areas is conceptualized in three spatial categories: (1) management within the boundaries; (2) transboundary or ecosystem management at the landscape scale (Grumbine 1994) [47]; and (3) regional to global management of the establishment, legal status, and distribution of protected areas. Within-boundary management is first priority as a protected area cannot effectively contribute to landscape-level biodiversity processes as a functional core unless populations and communities within its boundaries are intact (Noss et al. 2002) [48]. Source populations for metapopulation dynamics depend on core areas of good habitat, and thus protected areas management should provide propagules for feeding processes at the landscape scale (Hunter & Gibbs 2007) [49]. By the same token, conditions outside of protected areas influence those inside, and landscape-level planning and management should improve within-area conditions. For example, population viability for wide ranging carnivores who have core habitat inside reserves is improved by restoring connectivity (Carroll 2006) [50]. Thinking at the systems level and not just about wildlife, it is impossible to consider landscape-level management and improve matrix conditions without considering human enterprise, as millions of people live and make their livings on the edges of protected areas and within them (DeFries et al. 2007 [51]; DeFries et al. 2010 [52]). Finally, management of the entire protected areas estate through regional, national, and global politics, priorities, and agreements requires a global level of cooperation through entities such as the International Union for the Conservation of Nature. With increasing globalization, attention to transboundary and ecosystem management at biome and continental scales is warranted to attend to large-scale regulating and supporting services and planetary boundaries associated with biodiversity.

In this issue, papers show the complexity of protected areas management and linkages between within-area problems and those external, and how spatial scale of influence extends far beyond the landscape matrix to global governance systems. Belote [53] focuses on the important issue of legal management status of protected areas and the risk of having status diminished, due to changes in national political leadership. His policy-based analysis examines impacts of proposed legislation that would demote 29 Wilderness Study Areas in the United States, and thereby reduce their protections. Since core wilderness areas are the bedrock of effective protected area networks, such policy-based demotions would lead to habitat declines in cores, and degrade overall biodiversity function (Soule & Terborgh 1999) [54]. On the other hand, Pasha et al. [55] describe a system to improve standards for management of wild tigers within areas. They propose that sites be audited against a broadly accepted set of standards and discuss the challenges in implementing such standards across a wide range of protected areas and within a tight community of tiger conservation professionals. Tiger reserves are critically important for the continuance of that species, are scattered across many countries, and have differing management jurisdictions, priorities, and goals (Seidensticker et al. 1999) [56]. Standards and oversight of reserves might help.

Size, distribution, and isolation of protected areas is a central question. While large protected core areas remain the backbone of conservation, a very large number of small areas have been created. Baldwin and Fouch [34] show that there is a very large number of small protected areas that may have been protected for some local value, but whose cumulative biodiversity function is essentially unknown. Small protected areas are very numerous and may provide significant habitat in the matrix, stepping stones for dispersal, and protect localized, rare ecosystems but such spatially scaled research is only now emerging. Ekblom et al. [45] show that there are many small areas in Sub-Saharan Africa protected as sacred sites that also provide habitat for important species diversity. Local conservation is often focused on small natural areas near where people live and work in particular near where conservationists work (Baldwin & Leonard 2015) [57]. While they may not fulfill expectations as large core areas for biodiversity at the ecoregion scale, they may be important for many social, ecological, and economic reasons. More research is needed on small protected areas and how they contribute to overall biodiversity goals, yet it appears they are highly valued.

On the other end of the size spectrum are large protected areas. A small number of countries hold the global legacy of the last of the wild: >70% of the world's remaining wilderness is in five countries—Russia, Canada, Australia, US and Brazil (Watson et al. 2018) [58]. While most of these lands and seas are not officially protected, Wulder et al. [59] in this issue illustrate that large intact areas in Canada's boreal forest region are functioning as de facto protected areas, relatively free from development pressure, and that Canada is in a fairly unique position, globally, to expand the area under its protection, seizing this 'generational opportunity'. Along with formal protections, Wulder et al. [59] elaborate collaborative opportunities through commitments from industry, and provincial and territorial land stewards, and the contributions of First Nations and private protection programs. Consistent with Watson et al., such protections should be established in a way to slow the impacts of industrial activity on large landscapes or seascapes, acknowledge that protecting the livelihoods of Indigenous people can conserve biodiversity just as well as strictly protected areas can, and recognize local Indigenous community rights to land ownership and management. McCarthy et al. [43] examine how Indigenous people living alongside the protected area could improve the management of the protected area itself. They and Zurba et al. [33] show that Indigenous peoples have a good understanding of biodiversity management, and suggest that their systematic inclusion may result in better management outcomes.

Climate and land use change remain the greatest threats to biodiversity and protected areas. Tabor et al. [60] show that combined climate and deforestation risks within the humid tropical biome result in 2 million hectares at extreme risk, which therefore should be prioritized for conservation action. Hamad et al. [61] show that land use changes precipitated by war and economic sanctions have served to fragment forest patches in Iraq, which indicates the tight linkage between social and ecological systems that needs to be observed for the future of biodiversity conservation.

A unique global mechanism for biodiversity conservation based on climate change mitigation is proposed by Githiru and Njambuya [62]. They suggest that REDD+, the United Nations climate change mitigation scheme whose goal is to reduce emissions from deforestation and forest degradation in developing countries, can be harnessed for biodiversity conservation. Other authors caution that globally centralized forest governance should be evaluated in light of desire to give local people, including Indigenous people, more control over natural resources (Phelps et al. 2010) [63]. Githiru and Njambuya [62] address this by urging that bottom-up approaches be included in nested, polycentric schemes, concluding "the power of globalization enables a rural farmer in Kenya to play a role in global climate change mitigation, while a social worker in downtown New York can help conserve Elephants in Africa."

Emphasizing the increasing pressures on protected areas from outside their boundaries, Cunningham and Beazley [41] evaluated conservation threats in global biodiversity hotspots based on changes in human population density and protected area coverage. Over a 20-year timeframe (1995–2015), average population densities in the hotspots increased by 36%, double the global average. The Aichi Target 11 protected area goal of 17% was achieved in only half of the hotspots. In 2015, 15 of 36 global biodiversity hotspots remained in the highest threat category (i.e., population density exceeding global average, and protected area coverage less than 17%). Only two hotspots achieved a target of 50% protection, a scientifically defensible target for the hotspots, which are rich in endemic species and limited to less than 30% of their original habitat extent. They conclude that although conservation progress has been made in most global biodiversity hotspots additional efforts are needed to slow and/or reduce increases in population density and achieve protected area targets, and that such conservation efforts are likely to require support from the global community.

Globalization and externalization of cost are major drivers of biodiversity declines (Weinzettel et al. 2018) [64] as acknowledged by Cunningham and Beazley [41] and by Githiru and Njambuya [62]. There are growing trends and tensions in framing conservation policy in terms of ecosystem services rather than biodiversity (Cimon-Morin et al. 2013 [3]; Kusmanoff et al. 2017 [65]) with implications across boundaries at local to global scales. At the same time, if synergies can be found, these global processes may offer opportunities that support local biodiversity conservation, such as funding for protected area establishment and management, either for internal operations or reducing threats from outside (Angelsen 2008) [66]. Githiru and Njambuya's [62] calls for the polycentric approach of the nested REDD+ process are intended to address some of the globalization-driven biodiversity problems. They argue that such approaches have the potential to harness resources for protected area management and biodiversity conservation by using the appeal of greenhouse gas emissions as a global commodity, and that using carbon to build polycentric policy frameworks and infrastructure could facilitate future development of a similar system for biodiversity. Consistent with Cimon-Morin et al. (2013) [3], while there remain tensions and gaps, site complementarity for ecosystem services and biodiversity through systematic conservation planning could increase the efficiency for both. Global synergies could be derived particularly for large-scale regulating and supporting services such as carbon and climate regulation, which are considered to be at a maximum in intact ecosystems, potentially providing local funding in support of global conservation imperatives.

4. Summary

Protected areas are the gem of biodiversity conservation as they are areas set aside for natural pattern and process to prevail, relatively free from human intervention. Currently at about 15% coverage world-wide, they need to be two- to five-times more extensive if they are to meet science-based estimations of 25–75% protection and more effectively managed to maintain current levels of biodiversity. The scale of biodiversity loss is so great that the protected area estate alone is currently not equal to the task. Human activity dominates the surface of the land, has extensive impacts on the ocean, and dominates many ecosystems (Doney 2010 [67]; Venter et al. 2016 [9]). It is impossible to imagine the future of biodiversity without a profound human component as a major selective force, shaping patterns and processes for millennia, if the earth can bear it.

It may be best to view large core protected areas as one highly concentrated end of a gradient of biodiversity conservation measures with every other part of the landscape managed with a lesser degree of biodiversity benefits in mind. As such it is critical to imagine the losses to biodiversity conservation function should management or designation changes occur within existing protected areas that weaken their ability to meet conservation goals. Adding to the protected areas estate using the tools and techniques of systematic conservation planning is a major goal for governments and conservation organizations throughout the globe (Moilanen et al. 2009) [68]. Even so, the matrix in which protected areas sit contributes to their isolation or connectivity, and thus cannot be ignored (Crooks & Sanjayan 2006) [69].

Human activities in the matrix that would improve connectivity and help meet local conservation goals such as representation of rare endemics can be guided through economic incentives, management partnerships, local-scale conservation planning for small protected areas, agro-ecological programs, wildlife laws and enforcement, and environmental education. Compatible low-intensity human uses that provide complementary bio-cultural conservation and ecosystem services can supplement protected areas and serve to buffer them from external threats. While few examples exist specific to protected areas management outcomes, hundreds of examples around the world support that in human dominated landscapes habitat conditions can be improved by targeted programs and volunteer spirit (Hunter 1990 [70]; Lindenmayer & Hobbs 2004 [71]; Stubbs 2014 [72]; Vandermeer & Perfecto 2007 [73]).

As the planet continues to undergo massive, human-caused changes in climate, biogeography, and ecosystem function, it is important to continue research on mechanisms for biodiversity conservation. The existing protected areas estate is a tribute to the past commitment to conserve biodiversity, yet may reflect a different paradigm than needs to be employed to meet new protection goals. Today, we know more about the 'where' and 'why' for establishment of new protected areas than ever before (Margules & Pressey 2000 [12]; Steffen et al. 2015 [2]). We know less than we should about the 'how' and new scholarship is reflecting a plethora of plausible mechanisms. The papers in this special issue, although not exhaustive by any means, provide a window into emerging topics for biodiversity and protected areas.

References

1. Dirzo, R.; Young, H.S.; Galetti, M.; Ceballos, G.; Isaac, N.J.B.; Collen, B. Defaunation in the Anthropocene. *Science* **2014**, *345*, 401–406. [CrossRef] [PubMed]
2. Steffen, W.; Richardson, K.; Rockström, J.; Cornell, S.E.; Fetzer, I.; Bennett, E.M.; Biggs, R.; Carpenter, S.R.; de Vries, W.; de Wit, C.A.; et al. Planetary boundaries: Guiding human development on a changing planet. *Science* **2015**, *347*, 1259855. [CrossRef] [PubMed]
3. Cimon-Morin, J.; Darveau, M.; Poulin, M. Fostering synergies between ecosystem services and biodiversity in conservation planning: A review. *Biol. Conserv.* **2013**, *166*, 144–154. [CrossRef]
4. Laliberte, A.S.; Ripple, W.J. Range contractions of North American carnivores and ungulates. *Bioscience* **2004**, *54*, 123–138. [CrossRef]
5. Juffe-Bignoli, D.; Burgess, N.D.; Bingham, H.; Belle, E.M.S.; de Lima, M.G.; Deguignet, M.; Bertzky, B.; Milam, A.N.; Martinez-Lopez, J.; Lewis, E.; et al. *Protected Planet Report 2014*; UNEP-WCMC: Cambridge, UK, 2004.
6. Haddad, N.M.; Brudvig, L.A.; Clobert, J.; Davies, K.F.; Gonzalez, A.; Holt, R.D.; Lovejoy, T.E.; Sexton, J.O.; Austin, M.P.; Collins, C.D.; et al. Habitat fragmentation and its lasting impact on Earth's ecosystems. *Sci. Adv.* **2015**, *1*, e1500052. [CrossRef] [PubMed]
7. Torres, A.; Jaeger, J.A.G.; Alonso, J.C. Assessing large-scale wildlife responses to human infrastructure development. *Proc. Natl. Acad. Sci. USA* **2016**, *113*, 8472–8477. [CrossRef] [PubMed]
8. Lawton, G. Roadkill: Biodiversity in crisis. *New Sci.* **2018**, 36–39.
9. Venter, O.; Sanderson, E.W.; Magrach, A.; Allan, J.R.; Beher, J.; Jones, K.R.; Possingham, H.P.; Laurance, W.F.; Wood, P.; Fekete, B.M.; et al. Sixteen years of change in the global terrestrial human footprint and implications for biodiversity conservation. *Nat. Commun.* **2016**, *7*, 12558. [CrossRef] [PubMed]
10. Jones, K.R.; Venter, O.; Fuller, R.A.; Allan, J.R.; Maxwell, S.L.; Negret, P.J.; Watson, J.E.M. One-third of global protected land is under intense human pressure. *Science* **2018**, *360*, 788–791. [CrossRef] [PubMed]
11. Jenkins, C.N.; van Houtan, K.S.; Pimm, S.L.; Sexton, J.O. US protected lands mismatch biodiversity priorities. *Proc. Natl. Acad. Sci. USA* **2015**, *112*, 5081–5086. [CrossRef] [PubMed]
12. Margules, C.R.; Pressey, R.L. Systematic conservation planning. *Nature* **2000**, *405*, 243–253. [CrossRef] [PubMed]

13. McGuire, J.L.; Lawler, J.J.; McRae, B.H.; Nuñez, T.A.; Theobald, D.M. Achieving climate connectivity in a fragmented landscape. *Proc. Natl. Acad. Sci. USA* **2016**, *113*, 7195–7200. [CrossRef] [PubMed]

14. Belote, R.T.; Dietz, M.S.; McRae, B.H.; Theobald, D.M.; McClure, M.L.; Irwin, G.H.; McKinley, P.S.; Gage, J.A.; Aplet, G.H. Identifying Corridors among Large Protected Areas in the United States. *PLoS ONE* **2016**, *11*, e0154223. [CrossRef] [PubMed]

15. Joppa, L.N.; Pfaff, A. High and far: Biases in the location of protected areas. *PLoS ONE* **2009**, *4*, e8273. [CrossRef] [PubMed]

16. Sinclair, S.P.; Milner-Gulland, E.J.; Smith, R.J.; McIntosh, E.J.; Possingham, H.P.; Vercammen, A.; Knight, A.T. The use, and usefulness, of spatial conservation prioritizations. *Conserv. Lett.* **2018**, *11*, e12459. [CrossRef]

17. IUCN. *The Durban Accord: Our Global Commitment for People and Earth's Protected Areas*; IUCN: Gland, Switzerland, 2005.

18. IUCN. *IUCN Green List of Protected and Conserved Areas: Standard, Version 1.1.*; IUCN: Gland, Switzerland, 2017.

19. UNEP. Decision Adopted by the Conference of the Parties to the Convention on Biological Diversity at its Tenth Meeting. In Proceedings of the United Nations Environment Programme (UNEP) Conference of the Parties (COP) to the UN Convention on Biological Diversity (CBD), Nagoya, Japan, 12–29 October 2010.

20. Baillie, J.; Zhang, Y.-P. Space for nature. *Science* **2018**, *361*, 1051. [CrossRef] [PubMed]

21. Edgar, G.J.; Stuart-Smith, R.D.; Willis, T.J.; Kininmonth, S.; Baker, S.C.; Banks, S.; Barrett, N.S.; Becerro, M.A.; Bernard, A.T.F.; Berkhout, J.; et al. Global conservation outcomes depend on marine protected areas with five key features. *Nature* **2014**, *506*, 216–220. [CrossRef] [PubMed]

22. Armsworth, P.R.; Cantú-Salazar, L.; Parnell, M.; Davies, Z.G.; Stoneman, R. Management costs for small protected areas and economies of scale in habitat conservation. *Biol. Conserv.* **2011**, *144*, 423–429. [CrossRef]

23. DellaSala, D.A.; Staus, N.L.; Strittholt, J.R.; Hackman, A.; Iacobelli, A. An updated protected areas database for the United States and Canada. *Nat. Areas J.* **2001**, *21*, 124–135.

24. Rissman, A.R.; Owley, J.; L'Roe, A.W.; Morris, A.W.; Wardropper, C.B. Public access to spatial data on private-land conservation. *Ecol. Soc.* **2017**, *22*. [CrossRef]

25. Wilson, E.O. *Half Earth: Our Planet's Fight for Life*; W.W. Norton & Company: New York, NY, USA, 2016.

26. Vanderkam, R. *Conservation Areas Reporting and Tracking System: Procedures Manual and Schema*; Canadian Council on Ecological Areas: Gatineau, QC, Canada, 2016.

27. Geldmann, J.; Coad, L.; Barnes, M.D.; Craigie, I.D.; Woodley, S.; Balmford, A.; Brooks, T.M.; Hockings, M.; Knights, K.; Mascia, M.B.; et al. A global analysis of management capacity and ecological outcomes in terrestrial protected areas. *Conserv. Lett.* **2018**, *11*, e12434. [CrossRef]

28. Dudley, N.; Parrish, J.D.; Redford, K.H.; Stolton, S. The revised IUCN protected areas management categories: The debate and ways forward. *Oryx* **2010**, *44*, 485–490. [CrossRef]

29. Scott, J.M.; Davis, F.; Csuti, F.; Noss, R.; Butterfield, B.; Groves, C.; Anderson, H.; Caicco, S.; D'Erchia, F.; Edwards, T.C.J.; et al. Gap analysis: A geographic approach to protection of biological diversity. *Wildl. Monogr.* **1993**, *57*, 5–41.

30. MacKinnon, D.; Lemieux, C.J.; Beazley, K.; Woodley, S.; Helie, R.; Perron, J.; Elliott, J.; Haas, C.; Langlois, J.; Lazaruk, H.; et al. Canada and Aichi Biodiversity Target 11: Understanding 'other effective area-based conservation measures' in the context of the broader target. *Biodivers. Conserv.* **2015**, *24*, 3559–3581. [CrossRef]

31. Barnes, M.D.; Glew, L.; Wyborn, C.; Craigie, I.D. Prevent perverse outcomes from global protected area policy. *Nat. Ecol. Evol.* **2018**, *2*, 759–762. [CrossRef] [PubMed]

32. Clements, H.S.; Selinske, M.J.; Archibald, C.L.; Cooke, B.; Fitzsimons, J.A.; Groce, J.E.; Torabi, N.; Hardy, M.J. Fairness and Transparency Are Required for the Inclusion of Privately Protected Areas in Publicly Accessible Conservation Databases. *Land* **2018**, *7*, 96. [CrossRef]

33. Zurba, M.; Beazley, K.F.; English, E.; Buchmann-Duck, J. Indigenous Protected and Conserved Areas (IPCAs), Aichi Target 11 and Canada's Pathway to Target 1: Focusing Conservation on Reconciliation. *Land* **2019**, *8*, 10. [CrossRef]

34. Baldwin, R.F.; Fouch, N.T. Understanding the Biodiversity Contributions of Small Protected Areas Presents Many Challenges. *Land* **2018**, *7*, 123. [CrossRef]

35. Theobald, D.M. Placing exurban land-use change in a human modification framework. *Front. Ecol. Environ.* **2004**, *2*, 139–144. [CrossRef]

36. DeFries, R.S.; Foley, J.A.; Asner, G.P. Land-use choices: Balancing human needs and ecosystem function. *Front. Ecol. Environ.* **2004**, *2*, 249–257. [CrossRef]

37. Klein, C.J.; Chan, A.; Kircher, L.; Cundiff, A.J.; Gardner, N.; Hrovat, Y.; Scholz, A.; Kendall, B.E.; Airame, S. Striking a balance between biodiversity conservation and socioecnomic viability in the design of marine protected areas. *Conserv. Biol.* **2008**, *22*, 691–700. [CrossRef] [PubMed]

38. Powell, R.B.; Cuschnir, A.; Peiris, P. Overcoming governance and institutional barriers to integrated coastal zone, marine protected area, and tourism management in Sri Lanka. *Coast. Manag.* **2009**, *37*, 633–655. [CrossRef]

39. Anderson, M.G.; Clark, M.; Olivero-Sheldon, A. *Resilient Sites for Terrestrial Conservation in the Northeast and Mid-Atlantic Region*; The Nature Conservancy: Boston, MA, USA, 2012; p. 168.

40. Cushman, S.A.; Gutzweiler, K.; Evans, J.S.; McGarigal, K. The Gradient Paradigm: A Conceptual and Analytical Framework for Landscape Ecology. In *Spatial Complexity, Informatics, and Wildlife Conservation*; Cushman, S.A., Huettmann, F., Eds.; Springer: Japan, Tokyo, 2010; pp. 83–108.

41. Cunningham, C.; Beazley, K.F. Changes in Human Population Density and Protected Areas in Terrestrial Global Biodiversity Hotspots, 1995–2015. *Land* **2018**, *7*, 136. [CrossRef]

42. Borrini-Feyerabend, G.; Campese, J. *Self-Strengthening ICCAs—Guidance on a Process and Resources for Custodian Indigenous Peoples and Local Communities*; The ICCA Consortium, 2017; Available online: https://www.iccaconsortium.org/wp-content/uploads/2017/04/ICCA-SSP-Guidance-Document-14-March.pdf (accessed on 28 November 2018).

43. McCarthy, C.; Shinjo, H.; Hoshino, B.; Enkhjargal, E. Assessing Local Indigenous Knowledge and Information Sources on Biodiversity, Conservation and Protected Area Management at Khuvsgol Lake National Park, Mongolia. *Land* **2018**, *7*, 117. [CrossRef]

44. Colchester, M. Conservation policy and indigenous peoples. *Environ. Sci. Policy* **2004**, *7*, 145–153. [CrossRef]

45. Ekblom, A.; Shoemaker, A.; Gillson, L.; Lane, P.; Lindholm, K.J. Conservation through Biocultural Heritage—Examples from Sub-Saharan Africa. *Land* **2019**, *8*, 5. [CrossRef]

46. Garnett, S.T.; Burgess, N.D.; Fa, J.E.; Fernández-Llamazares, Á.; Molnár, Z.; Robinson, C.J.; Watson, J.E.M.; Zander, K.K.; Austin, B.; Brondizio, E.S.; et al. A spatial overview of the global importance of Indigenous lands for conservation. *Nat. Sustain.* **2018**, *1*, 369–374. [CrossRef]

47. Grumbine, R.E. What is ecosystem management? *Conserv. Biol.* **1994**, *8*, 27–38. [CrossRef]

48. Noss, R.; Carroll, C.; Vance-Borland, K.; Wuerthner, G. A multicriteria assessment of the irreplaceability and vulnerability of sites in the Greater Yellowstone Ecosystem. *Conserv. Biol.* **2002**, *16*, 895–908. [CrossRef]

49. Hunter, M.L.; Gibbs, J.P. *Fundamentals of Conservation Biology*; Blackwell Publishing: Oxford, UK, 2007.

50. Carroll, C. Linking connectivity to viability: Insights from spatially explicit population viability models of large carnivores. In *Connectivity Conservation*; Crooks, K.R., Sanjayan, M., Eds.; Cambridge University Press: Cambridge, UK, 2006; pp. 369–389.

51. DeFries, R.; Hanson, A.; Turner, B.L.; Reid, R.; Liu, J. Land use change around protected areas: Management to balance human needs and ecological function. *Ecol. Appl.* **2007**, *17*, 1031–1038. [CrossRef] [PubMed]

52. DeFries, R.; Karanth, K.K.; Pareeth, S. Interactions between protected areas and their surroundings in human-dominated tropical landscapes. *Biol. Conserv.* **2010**, *143*, 2870–2880. [CrossRef]

53. Belote, T.R. Proposed Release of Wilderness Study Areas in Montana (USA) Would Demote the Conservation Status of Nationally-Valuable Wildlands. *Land* **2018**, *7*, 69. [CrossRef]

54. Soule, M.; Terborgh, J. (Eds.) *Continental Conservation: Scientific Foundations of Regional Reserve Networks*; The Wildlands Project; Island Press: Washington, DC, USA, 1999.

55. Pasha, M.K.S.; Dudley, N.; Stolton, S.; Baltzer, M.; Long, B.; Roy, S.; Belecky, M.; Gopal, R.; Yadav, S.P. Setting and Implementing Standards for Management of Wild Tigers. *Land* **2018**, *7*, 93. [CrossRef]

56. Seidensticker, J.; Christie, S.; Jackson, P. (Eds.) *Riding the Tiger: Tiger Conservation in Human Dominated Landscapes*; Cambridge University Press: Cambridge, UK, 1999.

57. Baldwin, R.F.; Leonard, P.B. Interacting social and environmental predictors for the spatial distribution of conservation lands. *PLoS ONE* **2015**, *10*, e0140540. [CrossRef] [PubMed]

58. Watson, J.E.M.; Venter, O.; Lee, J.; Possingham, H.P.; Allan, J.R. Protect the last of the wild. *Nature* **2018**, *563*, 27–30. [CrossRef] [PubMed]

59. Wulder, M.A.; Cardille, J.A.; White, J.C.; Rayfield, B. Context and Opportunities for Expanding Protected Areas in Canada. *Land* **2018**, *7*, 137. [CrossRef]

60. Tabor, K.; Hewson, J.; Tien, H.; González-Roglich, M.; Hole, D.; Williams, J.W. Tropical Protected Areas Under Increasing Threats from Climate Change and Deforestation. *Land* **2018**, *7*, 90. [CrossRef]

61. Hamad, R.; Kolo, K.; Balzter, H. Post-War Land Cover Changes and Fragmentation in Halgurd Sakran National Park (HSNP), Kurdistan Region of Iraq. *Land* **2018**, *7*, 38. [CrossRef]

62. Githiru, M.; Njambuya, J.W. Globalization and Biodiversity Conservation Problems: Polycentric REDD+ Solutions. *Land* **2019**, *8*, 35. [CrossRef]

63. Phelps, J.; Webb, E.L.; Agrawal, A. Does REDD+ Threaten to Recentralize Forest Governance? *Science* **2010**, *328*, 312–313. [CrossRef] [PubMed]

64. Weinzettel, J.; Vačkář, D.; Medková, H. Human footprint in biodiversity hotspots. *Front. Ecol. Environ.* **2018**, *16*, 447–452. [CrossRef]

65. Kusmanoff, A.M.; Fidler, F.; Gordon, A.; Bekessy, S.A. Decline of 'biodiversity' in conservation policy discourse in Australia. *Environ. Sci. Policy* **2017**, *77*, 160–165. [CrossRef]

66. Angelsen, A. (Ed.) *Moving Ahead with REDD: Issues, Options, and Implications*; Center for International Forestry Research: Bogar Barat, Indonesia, 2008.

67. Doney, S.C. The Growing Human Footprint on Coastal and Open-Ocean Biogeochemistry. *Science* **2010**, *328*, 1512–1516. [CrossRef] [PubMed]

68. Moilanen, A.; Wilson, K.A.; Possingham, H.P. (Eds.) *Spatial Conservation Prioritization: Quantitative Methods and Computational Tools*; Oxford University Press: Oxford, UK, 2009.

69. Crooks, K.R.; Sanjayan, M. (Eds.) *Connectivity Conservation*; Cambridge University Press: Cambridge, UK, 2006.

70. Hunter, M.L. *Wildlife, Forests and Forestry: Principles of Managing Forests for Biological Diversity*; Prentice-Hall: Englewood Cliffs, NJ, USA, 1990.

71. Lindenmayer, D.B.; Hobbs, R.J. Fauna conservation in Australian plantation forests—A review. *Biol. Conserv.* **2004**, *119*, 151–168. [CrossRef]

72. Stubbs, M. *Conservation Reserve Program (CRP): Status and Issues*; Library of Congress, Congressional Research Service: Washington, DC, USA, 2014.

73. Vandermeer, J.; Perfecto, I. The agricultural matrix and a future paradigm for conservation. *Conserv. Biol.* **2007**, *21*, 274–277. [CrossRef] [PubMed]

Article

Changes in Human Population Density and Protected Areas in Terrestrial Global Biodiversity Hotspots, 1995–2015

Caitlin Cunningham [1,*] and Karen F. Beazley [2]

1 Interdisciplinary PhD Programme, Dalhousie University, Halifax, NS B3H 4R2, Canada
2 School for Resource and Environmental Studies, Dalhousie University, Halifax, NS B3H 4R2, Canada; karen.beazley@dal.ca
* Correspondence: caitlin.cunningham@dal.ca; Tel.: +1-902-412-2732

Received: 17 October 2018; Accepted: 9 November 2018; Published: 15 November 2018

Abstract: Biodiversity hotspots are rich in endemic species and threatened by anthropogenic influences and, thus, considered priorities for conservation. In this study, conservation achievements in 36 global biodiversity hotspots (25 identified in 1988, 10 added in 2011, and one in 2016) were evaluated in relation to changes in human population density and protected area coverage between 1995 and 2015. Population densities were compared against 1995 global averages, and percentages of protected area coverage were compared against area-based targets outlined in Aichi target 11 of the Convention on Biological Diversity (17% by 2020) and calls for half Earth (50%). The two factors (average population density and percent protected area coverage) for each hotspot were then plotted to evaluate relative levels of threat to biodiversity conservation. Average population densities in biodiversity hotspots increased by 36% over the 20-year period, and were double the global average. The protected area target of 17% is achieved in 19 of the 36 hotspots; the 17 hotspots where this target has not been met are economically disadvantaged areas as defined by Gross Domestic Product. In 2015, there are seven fewer hotspots (22 in 1995; 15 in 2015) in the highest threat category (i.e., population density exceeding global average, and protected area coverage less than 17%). In the lowest threat category (i.e., population density below the global average, and a protected area coverage of 17% or more), there are two additional hotspots in 2015 as compared to 1995, attributable to gains in protected area. Only two hotspots achieve a target of 50% protection. Although conservation progress has been made in most global biodiversity hotspots, additional efforts are needed to slow and/or reduce population density and achieve protected area targets. Such conservation efforts are likely to require more coordinated and collaborative initiatives, attention to biodiversity objectives beyond protected areas, and support from the global community.

Keywords: biodiversity conservation targets; threat assessment; prioritization; biodiversity hotspots; human population density; protected areas

1. Introduction

Anthropogenic activities have generated significant negative impacts on natural environments. No other species has exerted such immense influence on all the other species, and rapid growth of human populations poses additional threats to biodiversity [1–3]. Areas with high human population densities are particularly threatened, through direct and indirect effects [4–7]. To help counter such threats, biodiversity conservation has emerged as a global priority, to be addressed at multiple scales, and including tactics such as the establishment of protected areas and other effective area-based conservation measures [8,9]. Since biodiversity and threats to it are not evenly distributed across the planet, various frameworks for identifying global priorities for conservation have been developed [8].

Conservation priorities are often dictated through frameworks based on vulnerability (relative risk or threat to biodiversity) and irreplaceability (the degree to which options for safeguarding biodiversity values elsewhere are limited or restricted) [8,10–12]. These two measures have been combined in a variety of ways on the global scale to identify conservation priorities, including crisis ecoregions [13], endemic bird areas [14], centers of plant diversity [15], megadiversity countries [16], global 200 ecoregions [17,18], high-biodiversity wilderness areas [19], frontier forests [20], and last of the wild [21]. In a review of global prioritization frameworks, Brooks et al. [8] found that most frameworks prioritized irreplaceability, but that some were 'reactive' (prioritizing high vulnerability) and others 'proactive' (prioritizing low vulnerability). In this study, we focus on global biodiversity hotspots [12–15,17–19]. Biodiversity hotspots meet two key criteria: (i) a minimum of 1500 endemic plant species; and, (ii) a loss of at least 70% of original habitat extent [11]. By this definition they represent a reactive approach to conservation prioritization. They are considered high priorities for conservation attention because of their importance (irreplaceability) in safe-guarding global biodiversity and the threats (vulnerability) they are under.

Global biodiversity hotspots were first identified by Myers [12]. Initially, 10 forest biodiversity hotspots were identified, based on endemic species richness and threats from human activities, with eight hotspots from other ecosystem types added later [22]. An extensive systematic global review subsequently identified seven additional global biodiversity hotspots, for a total of 25 [23,24]. Cincotta et al. [25] published 1995 human population density and growth rates for these 25 biodiversity hotspots, providing an important methodology and baseline study for assessing future levels of these measures in these and other conservation priority areas. The earlier biodiversity hotspot frameworks have been revised several times since their inception. A second systematic global update [26] redefined several hotspots and added others. In 2011, an additional hotspot was identified [27], bringing the total number of global hotspots to 35 [11]. In 2016, the global analysis was updated, bringing the total number to 36 [28]. The 25 hotspots originally identified by Myers, Mittermeier, and colleagues [12,24] remain in the updated set of 36 hotspots, though some have been renamed. Collectively, the global biodiversity hotspots cover a tiny fraction of the planet—just 4.6% of the Earth's total surface area (15.4% of the land area)—but they are important to biodiversity, representing 50% of vascular plants and 42% of terrestrial vertebrates as endemics [11,26].

The Strategic Plan for Biodiversity 2011–2020, signed at COP10, outlines global targets for biodiversity conservation. One of the targets in the plan, Aichi target 11, specifies that "by 2020, at least 17% of terrestrial and inland water areas . . . , especially areas of particular importance for biodiversity and ecosystem services, are conserved through effectively and equitably managed, ecologically representative and well-connected systems of protected areas and other effective area-based conservation measures" [29]. This minimum global target of 17% is politically defined rather than scientifically defensible and, thus, is regarded by many as an interim target [30]. Most scientific studies estimate 25–75% of a region is needed to protect biodiversity, some are calling for 50% (half earth), and new global targets for beyond 2020 are currently being negotiated. Furthermore, the best practice in conservation planning calls for higher than average percentages of protection for rare or geographically-limited remnants of hotspots of diversity and rarity [30–33]. Given that the irreplaceable habitats represented in the global biodiversity hotspots have already been reduced to approximately 14.5% of their original extent, from 23.5 million km^2 to 3.4 million km^2 in 2004 [11,26], strong argument could be made for protecting as much as 100% of their remaining intact ecosystems.

In light of the irreplaceability and vulnerability of these global biodiversity hotspots, we analyze how well their conservation is currently being achieved relative to measures in 1995, in comparison to protected area targets of 17% and 50%, and in the face of one of the biggest threats to biodiversity—increasing human population density (Figure 1), which presents additional challenges for conservation planning [3,25,34–36]. Our assessment follows up on Cincotta et al.'s [25] similar study in 1995, providing an update and comparison, and complements a recent assessment by Weinzettel et al. [37] of conservation priorities among hotspots based on a footprint measured as loss of potential net primary production due to agricultural production and final consumption. Such diverse and on-going analyses are important in moving forward with future conservation strategies, such as new protected area targets beyond 2020. They also serve to illustrate the deficiencies inherent in relying upon discrete surrogate measures of threat no matter how well established, the ineffectual nature of global assessments to direct or inform within site or local conservation initiatives, and the importance of collaborative international and local conservation approaches, including those that extend beyond protected areas.

2. Methods

Quantitative analyses of human population densities (1995, 2015, and projected to 2020) and percentages of protected area coverage (1995 and 2015) were conducted for 36 global biodiversity hotspots identified by Mittermeier and colleagues [11] and the Critical Ecosystem Partnership Fund [28] (Figure 1). Changes in population density and protected area coverage from 1995 to 2015 were calculated. Population densities were compared against global averages, and percentages of protected area coverage were compared against area-based targets outlined in Aichi target 11 (17% by 2020) and half Earth (50%). Average population density and percentage of protected area coverage for each hotspot were then plotted against each other to evaluate relative levels of threat to biodiversity conservation in 1995 and 2015. Changes from 1995 to 2015 in the numbers and suites of global biodiversity hotspots falling within the high threat category (i.e., higher than global average population densities, and lower than target percentages for protected areas) were assessed.

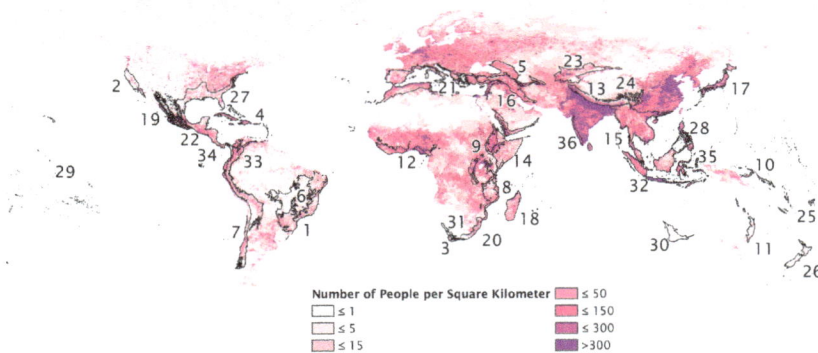

Figure 1. Map showing the human population density (number of people per square kilometer) and the location of the biodiversity hotspots. Hotspots: (1) Atlantic Forest, (2) California Floristic Province, (3) Cape Floristic Region, (4) Caribbean Islands, (5) Caucasus, (6) Cerrado, (7) Chilean Winter Rainfall and Valdivian Forests, (8) Coastal Forests of Eastern Africa, (9) East Melanesian Islands, (10) Eastern Afromontane, (11) Forests of East Australia, (12) Guinean Forests of West Africa, (13) Himalaya, (14) Horn of Africa, (15) Indo-Burma, (16) Irano-Anatolian, (17) Japan, (18) Madagascar and the Indian Ocean Islands, (19) Madrean Pine-Oak Woodlands, (20) Maputaland-Pondoland-Albany, (21) Mediterranean Basin, (22) Mesoamerica, (23) Mountains of Central Asia, (24) Mountains of Southwest China, (25) New Caledonia, (26) New Zealand, (27) North American Coastal Plain, (28) Philippines, (29) Polynesia-Micronesia, (30) Southwest Australia, (31) Succulent Karoo, (32) Sundaland, (33) Tropical Andes, (34) Tumbes-Choco-Magdalena, (35) Wallacea, and (36) Western Ghats and Sri Lanka.

2.1. Population Density

Population density methods followed Cincotta et al. [25]. Average human population density for each biodiversity hotspot was calculated using the *Gridded Population of the World* data for 1995, 2015, and 2020 (predicted) [38]. These data were collated by the Socioeconomic Data and Applications Center at NASA using national censuses and population registers around the world at a 30 arc-second (approximately one kilometer at the equator) resolution. The authors of the data note several limitations of the data, such as the extrapolation of census data to the specific year of the data and the mapping of the geographical boundaries themselves [39]. The data are therefore not appropriate for use at small, localized scales, but are useful for analyses such as ours that focus on broad trends over large areas. Changes in average population density between 1995 and 2015, as well as 2015 and 2020, were calculated via raster calculation. The change in population density for each hotspot was compared to the change in global (across all land, excluding ice or rock covered land; consistent with [25]) average density to identify those biodiversity hotspots in which the average population density is increasing at rates faster than the global average.

2.2. Percent Protected Area

Protected areas data for 1995 and 2015 were obtained from the World Database of Protected Areas (WDPA), United Nations Environment Programme (UNEP) [40]. While most of these data are in polygon form, about 9% of protected areas in this database are recorded as points. Following recommendations in the user manual [41], points were converted to polygons using buffers based on the reported size of the protected area in the database. The dissolve tool was then used to eliminate the double counting of areas of overlap. These processed data were then used to determine the percentage of terrestrial protected area in each biodiversity hotspot in 1995 and 2015. The change in percentage area protected in each hotspot between 1995 and 2015 was also calculated and compared to the global change in percentage of terrestrial protected area over the same time period.

2.3. Combining Population Density and Percent Protected Area to Assess Threat Level

Pearson correlation coefficients were calculated for average population density and percentage of protected area for each biodiversity hotspot. The two factors were then used to determine the level of threat associated with each biodiversity hotspot, specifically: (i) whether average population density in 2015, change in average population density from 1995 to 2015, and predicted change in average population density for 2015 to 2020 were above or below the global averages; and, (ii) whether the 17% and 50% protected areas targets were met. The results for each biodiversity hotspot at each time period were plotted, with population density values on the y-axis, and percent protected area on the x-axis. Results were grouped into quadrants defined by average population densities above or below the global average in 1995 and whether or not the Aichi target 11 goal of 17% terrestrial protection has been met.

3. Results

3.1. Population Density

Over the 20-year time frame, average population density across all biodiversity hotspots increased by 36%, from 76 people per square kilometer (ppl/km^2) to 103 ppl/km2, which was a smaller increase in density than for the world as a whole (47% increase; from 38 to 56 ppl/km^2) (Table A1). However, 11 hotspots saw average population density increases at rates higher than the global average. The highest *raw* change was found in the Philippines (28) (103 ppl/km^2); the highest *rate* of change was found in the Horn of Africa (14) (88.24%). In both 1995 and 2015, the average population density across all biodiversity hotspots was about double the global average in both 1995 (76 ppl/km^2 and 38 ppl/km^2, respectively) and 2015 (103 people/km^2; 56 ppl/km^2) (Figures 2 and 3). In 2015, average population densities ranged from 3 ppl/km^2 in the Succulent Karoo (31) to 345 ppl/km^2 in the Philippines (28); 21 of the 35 biodiversity hotspots had average population densities above the global average.

Overall, the predicted average population density for 2020 in hotspots is 112 ppl/km^2, compared to a predicted global average of 61 ppl/km^2. Average population densities in biodiversity hotspots are predicted to range from 4 ppl/km^2 in the Succulent Karoo (31) to 378 ppl/km^2 in the Philippines (28). In total, 21 of the 35 biodiversity hotspots are expected to have average population density higher than the global average (the same as the 2015 data). This predicted change in average population density represents an 8.74% increase in the biodiversity hotspots, which is on par with the world, which is predicted to see an 8.93% increase. The full results for the population densities in biodiversity hotspots for 1995, 2015, 2020 (predicted), and the changes between them can be found in Table A1.

3.2. Percent Protected Area

In 2015, 23.22% of the terrestrial and inland water areas of biodiversity hotspots was protected, which was above the global total of 14.7% (Table A2). However, there is high variability in the percentage protected of each hotspot, ranging from just 2.19% in the East Melanesian Islands (9) to 57.74% in New Caledonia (25). Overall, 23 hotspots were protecting more than the global average of 14.7% and 19 had met or exceeded the Aichi target 11 goal of 17% protected area. Two hotspots (Cerrado (6) and New Caledonia (25)) exceeded 50% protection. From 1995 to 2015, the protected area within hotspots increased by 12.44% on average, which was higher than the global increase of 4.9% over the same period. In 23 of the 36 biodiversity hotspots, the percentage of land in protected areas increased by more than the global average. The highest increases were seen in Cerrado (6) (36.86%) and the lowest increases were seen in the Horn of Africa (14) (0.46%). The full results for the percentage of area protected in each biodiversity hotspot for 1995 and 2015 and the change between them can be found in Table A2.

3.3. Comparing Population Density and Percent Protected Area

The average human population density and percentage of protected area in the biodiversity hotspots is not statistically correlated for 1995 (correlation coefficient 0.30) or for 2015 (correlation coefficient −0.09) (Table A3). It is worth noting, however, that three (New Caledonia (25), Cerrado (6) and Tropical Andes (33)) of the four hotspots with the highest percentages (>45%) of protected area in 2015 also had relatively low average population densities; among the four, only Cape Floristic Region (3) was above the global average. The Aichi Target 11 aim of 17% protection was met in hotspots with a range of average population densities, from high densities in Japan (17) (29.42% protected; 336 ppl/km^2 average population density) to low densities in Southwest Australia (30) (17.06% protected; 6 ppl/km^2 average population density).

The mean population density and percentage terrestrial area protected were plotted against each other for both 1995 (Figure 2) and 2015 (Figure 3). This placed hotspots into quadrants based on whether or not mean population densities were above the 1995 and 2015 global averages and whether or not the 17% or 50% targets for terrestrial protected area were met. In 1995, there were only three biodiversity hotspots with 17% or more of terrestrial area protected and a mean population density at or below the 1995 global average (Figure 2; Table 1). In 2015, there were five hotspots in this quadrant: the three from 1995 ((Cerrado (6), New Caledonia (25) and New Zealand (26)), plus Southwest Australia (30) and Succulent Karoo (31) (Figure 3; Table 1). If the 2015 global average population density is used, an additional three hotspots would be in this lower-threat group (Tropical Andes (33), Chilean Winter Rainfall and Valdivian Forests (7) and Forests of East Africa (11)). No hotspots met or exceeded a 50% protected area target in 1995, however two (Cerrado (6) and New Caledonia (25)) met the target in 2015, representing the least-threatened hotspots.

In 1995, there were 22 hotspots in the highest-threat quadrant, with mean population densities above the global average and less than 17% protected area (Table 1). In 2015, there were seven fewer hotspots in this quadrant due to increases in protected area, and one due to lower average population density (Mountains of Southwest China (24)), leaving 14 hotspots in the high threat quadrant. If 50% is considered as a protected area target, all but two hotspots (Cerrado (6) and New Caledonia (25)) shift into the higher threat category (mean population densities above the global average and less than 50% protected area).

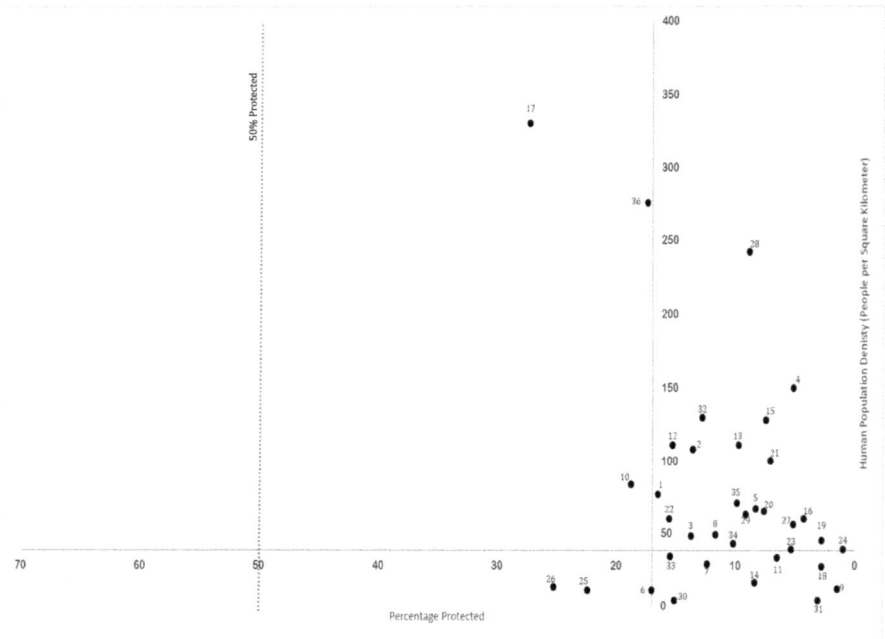

Figure 2. 1995 mean population densities (number of people per square kilometer) and percentage terrestrial area protected in each biodiversity hotspot. The x-axis crosses the y-axis at 38 people per square kilometer (the global average in 1995). The y-axis crosses the x-axis at 17% protected area, the target defined in Aichi target 11; the half-earth target of 50% protection is also shown. The upper-right quadrant represents hotspots that are under the most threat—they have an average human population density above the 1995 global average and less than 17% protected area. Lower-left quadrants represent hotspots under lesser threats—they have average human population densities at or below the global average and at least 17% of the area is under protection, or at least 50% protection (the lowest threat category). Hotspot numbers correspond with Figure 1.

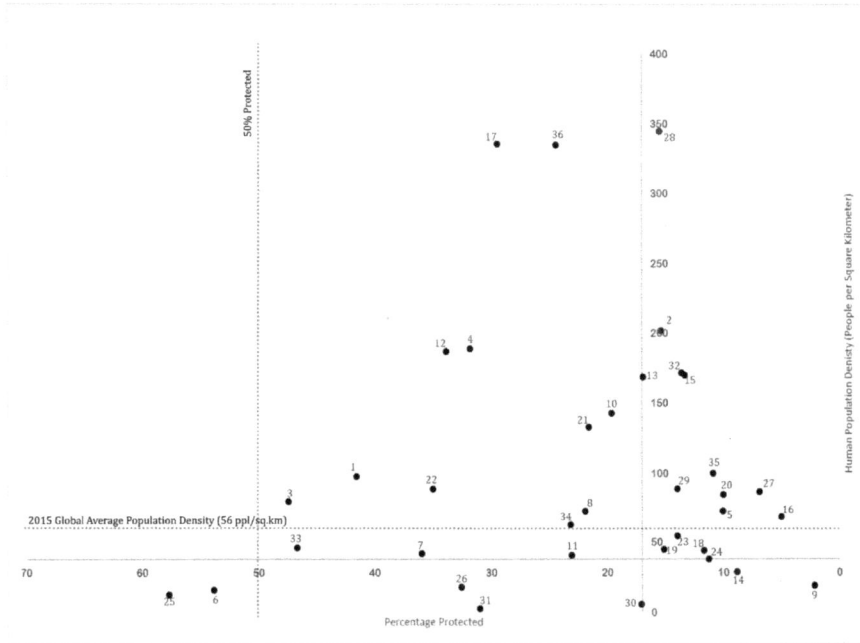

Figure 3. 2015 mean population densities (number of people per square kilometer) and percentage terrestrial area protected in each biodiversity hotspot. The x-axis crosses the y-axis at 38 people per square kilometer (the global average in 1995); the 2015 global average population density (56 people per square kilometer) is also indicated. The y-axis crosses the x-axis at 17% protected area coverage, the target under Aichi target 11; the half-earth target of 50% protection is also indicated. The upper-right quadrant represents hotspots that are under the most threat—they have an average human population density above the 2015 global average and less than 17% protected area. The lower-left quadrant represent hotspots under the least threat—they have an average human population density at or below the 1995 global average and at least 50% protected area. Hotspot numbers correspond with Figure 1.

Table 1. Biodiversity hotspots falling within high- and low-threat categories in analyses based on average human population densities relative to global average and protected area relative to 17% and 50% targets in 1995 and 2015, as shown in Figures 2 and 3.

1995	2015
High Threat Categories	
(Average human population density >1995 global average; <17% protected area)	
Atlantic Forest (1)	Average human population density >2015 and >1995
California Floristic Province (2)	global averages; <17% protected area:
Cape Floristic Region (3)	California Floristic Province (2)
Caribbean Islands (4)	Caucasus (5)
Caucasus (5)	Himalaya (13)
Coastal Forests of Eastern Africa (8)	Indo-Burma (15)
Guinean Forests of West Africa (12)	Irano-Anatolian (16)
Himalaya (13)	Maputaland-Pondoland-Albany (20)
Indo-Burma (15)	North American Coastal Plain (27)
Irano-Anatolian (16)	Philippines (28)
Madrean Pine-Oak Woodlands (19)	Polynesia-Micronesia (29)
Maputaland-Pondoland-Albany (20)	Sundaland (32)
Mediterranean Basin (21)	Wallacea (35)
Mesoamerica (22)	Average human population density >1995–2015 global
Mountains of Central Asia (23)	average; <17% protected area:
Mountains of Southwest China (24)	Madagascar and the Indian Ocean Islands (18)
North American Coastal Plain (27)	Madrean Pine-Oak Woodlands (19)
Philippines (28)	Mountains of Central Asia (23)
Polynesia-Micronesia (29)	
Sundaland (32)	
Tumbes-Choco-Magdalena (34)	
Wallacea (35)	
Low threat categories	
(Average human population density ≤1995 global average; ≥17% protected area)	
(Average human population density ≤1995 global average; ≥17%–<50% protected area:	
	Cerrado (6)
Cerrado (6)	New Caledonia (25)
New Caledonia (25)	New Zealand (26)
New Zealand (26)	Southwest Australia (30)
	Succulent Karoo (31)
(Average human population density ≤1995 global average; ≥50% protected area:	
	Cerrado (6)
	New Caledonia (25)

4. Discussion

4.1. Using Human Population Density as a Measure of Threat to Biodiversity

Cincotta and colleagues [25] examined mean human population densities (based on the same 1995 data used in this paper) and population growth rates between 1995 and 2000 in 25 biodiversity hotspots. For these 25 hotspots, our calculations of 1995 mean human population densities yielded similar results (any differences were non-significant and likely attributable to the different underlying hotspot datasets used). Although Cincotta and colleagues also looked at population growth rates in these hotspots, we analysed actual (1995–2015) changes in population density because population growth rates can be misleading indicators of risk [25]. These different methods yielded different pictures of the changes in human populations in these hotspots. For example, in our analysis, the California Floristic Province (2) was identified as having the fastest growing population density (an 87% increase between 1995 and 2015, as compared to global average of 47%), whereas Cincotta et al. [25] calculated the population growth rate for this hotspot as right at the global average (between 1995 and 2000), which was a relatively low growth rate as compared to other hotspots.

Although human population density is frequently used as a proxy for threat to biodiversity [3,25,42–44], there are caveats to the method. First, it only measures human presence in the discreet analytical unit and does not take into account the indirect effects that localized high densities, such as urban populations, can have on surrounding ecosystems. Cities can alter ecosystems over great distances, potentially even thousands of kilometers away. Second, using average population density throughout a biodiversity hotspot obscures patterns of population distribution within the area. The hotspots are mostly large with population density widely variable within them. For example, the Mountains of Southwest China (24) and Forests of Eastern Australia (11) have similar average population densities (38 and 41 ppl/km^2, respectively) and total areas (262,129 and 254,388 km^2, respectively), but very different patterns of population distribution. In the Mountains of Southwest China (24), the population is spread throughout the hotspot (Figure 4a). Conversely, the human population of the Forests of Eastern Australia (11) is concentrated along the coast in a few, densely populated cities: Sydney, Brisbane, Newcastle and Cairns. The rest of the hotspot is virtually uninhabited (Figure 4b). Obscuring the pattern of population distribution within hotpots also obscures the ecological footprint of the various populations that live throughout biodiversity hotspots. In hotspots where population is distributed throughout, conservation initiatives may be more prevalent throughout the region in a 'land-sharing' approach, in which humans and nature coexist [45–47]. In those where the population is more spatially concentrated, greater opportunities may exists for 'land-sparing', in which larger and more connected protected areas may be set aside. Further, it is well established that land-use activities such as agriculture, timber harvesting and mining can be extensive and intensive in terms of impact on biodiversity, even in areas with low population densities [37].

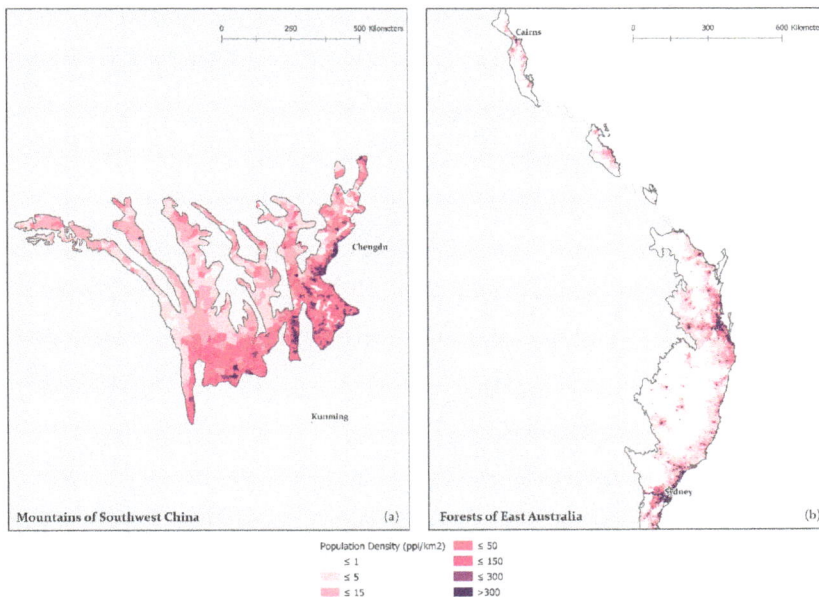

Figure 4. Maps showing the population density of two biodiversity hotspots, the Mountains of Southwest China (11.3% protected area in 2015) (**a**) and the Forests of East Australia (23.1% protected area in 2015) (**b**). Both hotspots have similar average population densities (38 and 41 ppl/km^2, respectively) and total areas (262,129 and 254,388 km^2, respectively), but the patterns of population distribution differ.

Despite these limitations, population density is generally accepted as an indicator for assessing broad trends over large areas and can be considered a better measure of environmental impact than total population or population growth [34]. However, both density and growth rates mask the cultural aspects of the populations they represent and the impacts they have on their surrounding environment (e.g., how affluent they are and what technologies they utilize) [25]. Furthermore, understanding of the human population density-biodiversity relationship is biased towards global, coarse scaled studies [34]. However, given that this study is a global, coarse scaled study, assessing human population density as a factor in biodiversity conservation is appropriate. Although it cannot be said definitively that increasing human population density in biodiversity hotspots directly threatens the biodiversity of these areas, there is a general consensus that the presence of human populations is a key factor contributing to the decline and extirpation of native species [1–3]. Furthermore, an increased human presence can hinder conservation efforts and the establishment of new protected areas in a region [34]. Nonetheless, it is important to measure indicators other than population density alone.

4.2. Wealth Inequality and Conservation in Biodiversity Hotspots

A potentially worrying trend is the wealth inequality between the countries where the biodiversity hotspots are considered critical or threatened in terms of population density and area protected, in comparison to those deemed low or least concern. As noted by Mittermeier and colleagues [11], one of the major outcomes of the hotspot designations was an increase in global conservation funds (in excess of 1 billion dollars, US) being directed at some of the most ecologically sensitive places on Earth. While progress has been made in a number of hotspots (for example, in the Cape Floristic Region (3) protected areas more than tripled in size between 1995 and the present), about half (18 of 35) hotspots do not meet 17% protected area as set out by Aichi target 11, and only two exceed 50%. With the notable exception of the California Floristic Province (2) (one of the wealthiest areas on the planet), hotspots that have not achieved the 17% protected area target are located in countries that are struggling economically [48], many of which are also dealing with other significant challenges such as war (Irano-Anatolian (16) and Horn of Africa (14)—civil wars in Syria, Somalia and Yemen), famine (Horn of Africa (14)—associated with the aforementioned civil wars), social unrest (Indo-Burma (15) and Philippines (28)), and rapid sea level rise (Sundaland (32) and Polynesia-Micronesia (29) [49]. In contrast, despite being one of the wealthiest places on the planet, the California Floristic Province has failed to protect 17%. One contributing factor may be that this hotspot has the highest growth in population density since 1995 and the fourth highest average population density. However, 17% protection has been met in the four other hotspots with equivalent (Guinean Forests of West Africa (12) and Caribbean Islands (4)) or higher (Japan (17) and Western Ghats and Sri Lanka (36)) average population densities. The only other hotspot with higher population density in which the 17% target has not been met is the Philippines (2), which, as noted, is struggling with significant economic and social challenges. Half of the biodiversity hotspots that do not meet the 17% protected area target have average population densities above the global average, a factor that will further complicate conservation efforts in those regions [34].

Wealthy nations, such as the United States, in which the California Floristic Province is located, need to step up protection efforts for global biodiversity hotspots located within their boundaries. Those that are signatory to the Convention on Biodiversity should at a minimum follow through on their commitments to Aichi target 11. Beyond these internal commitments, however, relatively wealthy nations should recognize their responsibilities to other global hotspots in terms of world leadership and stewardship in general, but also in response to their indirect role in the loss of natural habitat in the hotspots, such as through their consumption of agricultural and other products produced in those regions and exported through global trade [37]. It is unrealistic, and unethical in terms of international equity, to assume that conservation targets will be met in economically and socially challenged regions without extensive support from the global community.

4.3. Towards 2020 and Aichi Target 11

Over the 30 years since Myers [12] first introduced the concept of biodiversity hotspots there has been significant progress in conservation within the hotspots, but threats to their conservation have also continued. Looking towards 2020 and beyond, more resources will need to be directed at biodiversity hotspots. It is important that conservation efforts be directed at places that are vulnerable due to high human presence, in addition to regions where fewer people live, though these less-populated places may also be highly threatened by human activities such as resource extraction. As Cincotta et al. [25] note, biodiversity hotspots with low population density may also be highly threatened by extensive and intensive resource extraction, such as through disturbances related to 'over-logging, burning, grazing, mining and commercial hunting that have extracted or degraded natural resources, abetted biological invasion or polluted soil and water resources' [50]. Many hotspots are at risk from land conversions to agriculture, the effects of which are further compounded by international trade and the consumption of the products in wealthy nations [37]. Consideration of these various factors can result in different estimations of threat.

Of the 10 hotspots that Weinzettel et al. [37] identified as facing the greatest threats from agriculture, only four fell into our high threat category (Irano-Anatolian (16), Maputaland-Pondoland-Albany (20), Sundaland (32) and Wallacea (35)), and two into our low threat category (New Zealand (26) and Succulent Karoo (31)). The remaining four (Cape Floristic Region (3), Mediterranean basin (21), Mountains of Southwest China (24) and Western Ghat and SriLank (36)) had human populations densities higher than the 1995 global average, but also had 17% or more in protected area, thus reducing their threat level in our assessment. Such variations point to the importance of considering multiple factors in identifying threats so as to respond with various forms of conservation initiatives to address fundamental and proximate causes, local and global mechanisms, and reactive and proactive approaches. Furthermore, that two different approaches identified four hotspots in common as highly threatened lends confidence to the notion that significant and immediate conservation attention should be brought to these hotspots.

The conservation targets set out in Aichi Target 11 are political in nature, and fall far below the ecologically-based 25–75% protection estimated by scientific studies (for reviews, see for example, [31–33]. The 17% target should therefore be regarded as a stepping stone towards more wide-spread protection. Current levels of protection, even in regions that meet Aichi Target 11, are woefully inadequate. This is especially true in the most biodiverse and/or threatened places on Earth, such as the biodiversity hotspots, where best practices would call for higher than average percentages of protection. Therefore, we echo the calls made by Baillie and Zhang [30], Wilson [51], Locke [52], and others for increased biodiversity conservation targets—that current conservation targets are too low, especially in the most vulnerable regions of the planet. Targets for biodiversity conservation should err on the side of caution, and given the irreplaceable nature of the hotspots, it is not unreasonable to call for 100% of their remaining intact ecosystems to be protected. The consequences of meeting targets that are too low are far greater than meeting those that are too high. On the other hand, a strict focus on percentage-based targets alone can have many unintended and perverse consequences [53], and thus efforts must take into account the needs of local peoples, careful siting and design of protected areas, and provisions for their effective governance and management, to maximize conservation impact.

A factor which was not examined in this study, but which will be important in future conservation planning in the hotspots, is ecological connectivity and permeability, the degree to which landscapes are able to facilitate species movement between them [54,55]. Similar to the different patterns of population density and protected areas in biodiversity hotspots (Figure 5), there are also different degrees to which species movement can be facilitated on the landscape. Although direct connectivity between protected areas is important, the permeability of the matrix within hotspots must also be considered to facilitate the ecosystem processes that keep the hotspots functional [55,56]. Considerations of permeability will be of importance where there is a high degree of human presence on the landscape, as ecological function is especially vulnerable in such landscapes [57].

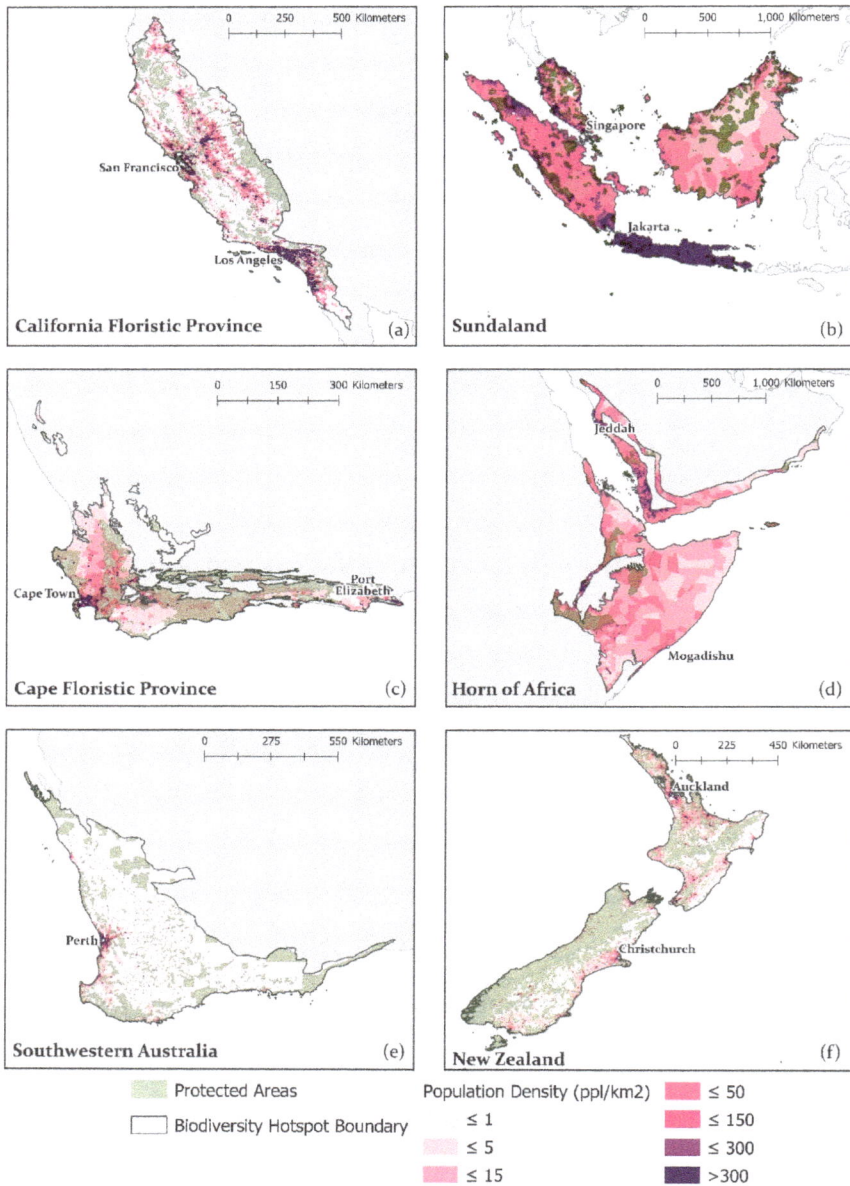

Figure 5. Patterns of protected area and population density in biodiversity hotspots: (**a**) California Floristic Province (15.3% protected area in 2015), (**b**) Sundaland (13.6% protected area in 2015), (**c**) Cape Floristic Province (47.4% protected area in 2015), (**d**) Horn of Africa (8.8% protected area in 2015), (**e**) Southwestern Australia (17.1% protected area in 2015), and (**f**) New Zealand (32.6% protected area in 2015).

Taken together, the various patterns of population density and protected areas offer a picture of the current state of conservation in each hotspot and provide perspective on the challenges that lay ahead for maintaining and/or restoring ecological connectivity. For example, in the California Floristic Province (2) (Figure 5a), the majority of protected areas are located away from areas of high population density (with the exception of a few protected areas around the San Francisco Bay Area), but they are not well connected to each other. In New Zealand (26) (Figure 5f) the protected areas are similarly located away from where people live, but they are much more connected than in California, particularly on the South Island. Conversely, in the Cape Floristic Region (3) (Figure 5c) there is a reasonably well-connected system of protected areas that runs through the city of Cape Town, effectively bisecting the densest part of the city.

Immense challenges lie ahead for biodiversity hotspots where there is little protected area and high population density, such as Sundaland (32) (Figure 5b) and the Horn of Africa (14) (Figure 5d). In these regions there is often a high reliance on natural resources by the local peoples. It is likely that a simple expansion of protected area will be insufficient in all cultural contexts, particularly in areas where illegal logging and poaching remain a major problem. Conservation efforts should, therefore, entail other culturally-appropriate measures to safeguard biodiversity [35,58]. Conservation in areas where people live and work is likely to look different in different cultural contexts.

4.4. Data Limitations

There are numerous acknowledged issues with the World Database of Protected Areas (WDPA) that limit the accuracy of this analysis. One issue is spatial accuracy—the data in the WDPA is compiled by the UNEP in collaboration with a wide range of government and NGO partners. A wide range of sources means that there are a variety of different scales and techniques used to gather the data in the WDPA [39]. As discussed in the methods section, roughly 9% of protected areas in the WDPA are recorded as points rather than polygons. Excluding these areas would, therefore, likely result in a substantial underestimate of protected areas. In order to include these protected areas, buffers were created in accordance with the recorded area for each protected area [39]. While this does not account for the shape and extent of protected areas, it does allow for general spatial/areal analyses, sufficient for this analysis. Finally, there are many protected areas which overlap in the WDPA, an issue that is augmented with the addition of buffers on the point data [39]. As described in the methods section, this issue was resolved by running the dissolve tool on the data in ArcGIS Pro (ESRI: Redlands, California, United States) to eliminate overlap.

4.5. Future Conservation Planning

Factors other than human population density and percent protected area contribute to the status of biodiversity and its conservation and thus must be considered when setting conservation objectives, including (i) threats such as habitat loss, climate change, overexploitation, invasive species [59], and (ii) the effectiveness of protected areas, measured as the strength of protection and size and/or connectivity of the protected areas network. It should also be noted that setting conservation priorities on the basis of biodiversity hotspots defined as those rich in endemic species and threatened by anthropogenic influences is a relatively reactive approach, as they are by definition areas that are already degraded or threatened. Therefore, an approach that focuses on critical biodiversity hotspots should be balanced by a proactive approach that incorporates some of the least-threatened highly diverse regions, in order to protect the best of what's left of the world's biodiversity while human pressures and interests in those areas remain less intense.

Although there are a handful of protected areas around the world within densely populated areas (such as Table Mountain National Park in the Cape Floristic Region (3)), they are not the norm, and the general paradigm is that conservation is an act that happens away from where people live. However, the reality is that hundreds of millions of people live in some of the most biologically diverse places on Earth. A fundamental shift in what we think conservation is supposed to look like is needed. Conservation is not simply something that happens 'over there', on lands away from humans, but rather it is possible and, as these results indicate, necessary, to meet biodiversity conservation targets where many people live and work, as well as in more remote regions. To adequately safe guard biodiversity we need to adopt new approaches to conservation that recognize that we are part of biodiversity and that areas of high human population density do not need to be places of conflict with nature, but rather they should be places of coexistence [45–47]. To achieve this, collaboration and participatory action in conservation planning will be crucial, especially across multiple scales [60,61]. A framework for how this could be carried out on the ground are the landscape conservation cooperatives (LCC) in the United States, which show promise that large scale collaboration across governments, NGOs and others is possible. LCCs aim to connect a variety of actors through a bottom up approach to engagement in systematic conservation planning at regional scales [61] The model could prove to be useful for safeguarding biodiversity on large scales all over the world, including in the hotspots.

5. Conclusions

This study examined two of many factors that affect biodiversity, human population density and percent protected area. Although significant progress has been made through conservation efforts in biodiversity hotspots over the last twenty years, more is needed. By 2015, the Aichi target 11 goal of 17% protection of terrestrial lands and inland waters had been met in 19 of the 36 hotspots. Only two hotspots had achieved the more-ecologically-realistic target of 50%. The majority of hotspots that did not meet the 17% target are in economically disadvantaged nations. Efforts to meet and exceed these targets will require financial support and other resources from the global community. Compounding the challenge of these future efforts will be increasing human presence in the biodiversity hotspots. Hundreds of millions of people currently live in these areas, and ongoing increases in their populations are putting more pressure on already strained ecosystems. To meet conservation targets in the biodiversity hotspots, an approach that recognizes that humans are part of nature and that areas of high population density can become areas where people live in coexistence with the natural world is necessary.

Author Contributions: C.C. designed and conducted the analysis in GIS, performed statistical analyses and drafted the manuscript. K.F.B. contributed to the analysis of results and the writing and editing of the manuscript.

Funding: This work was supported by the Nova Scotia Graduate Scholarship, awarded to C.C.

Acknowledgments: Thank you to the GIS Centre at Dalhousie University for their support of this work. We would also like to thank the Center for International Earth Science Informational Network (CIESN), the United Nations Environmental Programme (UNEP) and the International Union for Conservation of Nature (IUCN) for providing their data open access and free of charge. Without organizations providing easily accessible data, research like this would not be possible. Thank you to the four reviewers who provided valuable constructive feedback on an earlier draft of this paper.

Conflicts of Interest: The authors declare no conflict of interest.

Appendix A

Table A1. Population Density Data for each Biodiversity Hotspot.

Biodiversity Hotspot	Population Density, 1995 (ppl/km²)	Population Density, 2015 (ppl/km²)	Change in Population Density, 1995–2015 (ppl/km²)	Change in Population Density, 1995–2015 (%)	Predicted Population Density, 2020 (ppl/km²)	Predicted Change in Population Density, 2015–2020 (ppl/km²)	Predicted Change in Population Density, 2015–2020 (%)
GLOBAL	38	56	18	47.37	61	5	8.93
GLOBAL (BIODIVERSITY HOTSPOTS ONLY)	76	103	27	35.53	112	9	8.74
Atlantic Forest (1)	77	98	21	27.3	104	6	6.1
California Floristic Province (2)	108	202	94	87.0	213	11	5.4
Cape Floristic Region (3)	48	80	32	66.7	91	11	13.8
Caribbean Islands (4)	150	189	39	26.0	208	19	10.1
Caucasus (5)	67	73	6	9.0	75	2	2.7
Cerrado (6)	11	16	5	45.5	17	1	6.3
Chilean Winter Rainfall and Valdivian Forests (7)	29	42	13	44.8	47	5	11.9
Coastal Forests of Eastern Africa (8)	49	73	24	49.0	85	12	16.4
East Melanesian Islands (9)	12	19	7	58.3	22	3	15.8
Eastern Afromontane (10)	84	143	59	70.2	166	23	16.1
Forests of East Australia (11)	33	41	8	24.2	41	0	0.0
Guinean Forests of West Africa (12)	111	187	76	68.5	216	29	15.5
Himalaya (13)	111	169	58	52.3	197	28	16.6
Horn of Africa (14)	16	29	13	81.3	35	6	20.7
Indo-Burma (15)	128	170	42	32.8	183	13	7.6
Irano-Anatolian (16)	60	69	9	15.0	73	4	5.8
Japan (17)	330	336	6	1.8	337	1	0.3
Madagascar and the Indian Ocean Islands (18)	27	44	17	63.0	52	8	18.2
Madrean Pine–Oak Woodlands (19)	45	45	0	0.0	50	5	11.1
Maputaland-Pondoland-Albany (20)	65	85	20	30.8	90	5	5.9
Mediterranean Basin (21)	100	133	33	33.0	143	10	7.5
Mesoamerica (22)	60	89	29	48.3	99	10	11.2
Mountains of Central Asia (23)	39	55	16	41.0	60	5	9.1
Mountains of Southwest China (24)	39	38	−1	−2.6	40	2	5.3
New Caledonia (25)	11	13	2	18.2	15	2	15.4
New Zealand (26)	13	18	5	38.5	19	1	5.6
North American Coastal Plain (27)	56	87	31	55.4	97	10	11.5
Philippines (28)	242	345	103	42.6	378	33	9.6
Polynesia-Micronesia (29)	63	89	26	41.3	94	5	5.6
Southwest Australia (30)	4	6	2	50.0	6	0	0.0
Succulent Karoo (31)	4	3	−1	−25.0	4	1	33.3
Sundaland (32)	130	172	42	32.3	187	15	8.7
Tropical Andes (33)	34	46	12	35.3	51	5	10.9
Tumbes-Choco-Magdalena (34)	43	63	20	46.5	72	9	14.3
Wallacea (35)	71	100	29	40.8	112	12	12.0
Western Ghats and Sri Lanka (36)	276	335	59	21.4	354	19	5.7

Table A2. Protected Area Data for Biodiversity Hotspots.

Biodiversity Hotspot	Protected Area, 1995 (%)	1995 Protected Area +/− 17%	Protected Area, 2015 (%)	2015 Protected Area +/− 17%	Change in Protected Area, 1995–2015 (%)
GLOBAL	9.8	−7.2	15.4	−1.6	5.6
GLOBAL (HOTSPOTS ONLY)	10.64	−6.36	24.22	+7.2	13.57
Atlantic Forest (1)	16.53	−0.47	41.60	+24.60	25.08
California Floristic Province (2)	13.61	−3.39	15.34	−1.66	1.74
Cape Floristic Region (3)	13.74	−3.26	47.42	+30.42	33.69
Caribbean Islands (4)	5.15	−11.85	31.79	+14.79	26.64
Caucasus (5)	8.30	−8.70	10.09	−6.91	1.78
Cerrado (6)	17.02	+0.02	53.88	+36.88	36.86
Chilean Winter Rainfall and Valdivian Forests (7)	12.36	−4.64	35.97	+18.97	23.61
Coastal Forests of Eastern Africa (8)	11.68	−5.32	21.92	+4.92	10.24
East Melanesian Islands (9)	1.46	−15.54	2.19	−14.81	0.73
Eastern Afromontane (10)	18.77	+1.77	19.63	+2.63	0.86
Forests of East Australia (11)	6.50	−10.50	23.07	+6.07	16.57
Guinean Forests of West Africa (12)	15.31	−1.69	33.85	+16.85	18.54
Himalaya (13)	9.73	−7.27	16.88	−0.12	7.15
Horn of Africa (14)	8.39	−8.61	8.85	−8.15	0.46
Indo-Burma (15)	7.49	−9.51	13.32	−3.68	5.83
Irano-Anatolian (16)	4.27	−12.73	5.01	−11.99	0.74
Japan (17)	27.29	+10.29	29.42	+12.42	2.14
Madagascar and the Indian Ocean Islands (18)	2.78	−14.22	11.69	−5.31	8.92
Madrean Pine-Oak Woodlands (19)	2.75	−14.25	15.12	−1.88	12.37
Maputaland-Pondoland-Albany (20)	7.60	−9.40	10.03	−6.97	2.43
Mediterranean Basin (21)	7.08	−9.92	21.59	+4.59	14.51
Mesoamerica (22)	15.55	−1.45	35.02	+18.02	19.46
Mountains of Central Asia (23)	5.34	−11.66	13.99	−3.01	8.65
Mountains of Southwest China (24)	0.95	−16.05	11.26	−5.74	10.31
New Caledonia (25)	22.41	+5.41	57.74	+40.74	35.33
New Zealand (26)	25.26	+8.26	32.57	+15.57	7.31
North American Coastal Plain (27)	5.17	−11.83	6.89	−10.11	1.71
Philippines (28)	8.86	−8.14	15.44	−1.56	6.58
Polynesia-Micronesia (29)	9.14	−7.86	13.96	−3.04	4.82
Southwest Australia (30)	15.10	−1.90	17.06	+0.06	1.96
Succulent Karoo (31)	3.08	−13.92	31.01	+14.01	27.93
Sundaland (32)	12.79	−4.21	13.58	−3.42	0.79
Tropical Andes (33)	15.46	−1.54	46.72	+29.72	31.25
Tumbes-Choco-Magdalena (34)	10.19	−6.81	23.15	+6.15	12.97
Wallacea (35)	9.88	−7.12	10.90	−6.10	1.02
Western Ghats and Sri Lanka (36)	17.42	+0.42	24.35	+7.35	6.93

Table A3. Correlation Coefficients for Population Density and Protected Area in Biodiversity Hotspots.

	1995 Population Density (ppl/km²)	2015 Population Density (ppl/km²)	2020 Predicted Population Density (ppl/km²)	Population Density Change (1995–2015)	Population Density Change (2015–2020)	1995 Protected Areas (%)	2018 Protected Areas (%)	Protected Areas Change (1995–2018)
1995 Population Density (ppl/km²)	1.00							
2015 Population Density (ppl/km²)	0.97	1.00						
2020 Predicted Population Density (ppl/km²)	0.96	1.00	1.00					
Population Density Change (1995–2015)	0.55	0.73	0.77	1.00				
Population Density Change (2015–2020)	0.51	0.67	0.72	0.88	1.00			
1995 Protected Areas (%)	0.30	0.27	0.26	0.10	0.03	1.00		
2018 Protected Areas (%)	−0.07	−0.09	−0.09	−0.12	−0.09	0.61	1.00	
Protected Areas Change (1995–2018)	−0.26	−0.27	−0.26	−0.20	−0.12	0.19	0.89	1.00

References

1. Dietz, T.; Rosa, E.D.; York, R. Driving the Human Ecological Footprint. *Front. Ecol. Environ.* **2007**, *5*, 13–18. [CrossRef]
2. Kerr, J.T.; Currie, D.J. Effects of Human Activity on Global Extinction Risk Effects of Human Activity on Global Extinction Risk. *Conserv. Biol.* **1995**, *9*, 1528–1538. [CrossRef]
3. McKee, J.K.; Sciulli, P.W.; Fooce, C.D.; Waite, T.A. Forecasting Global Biodiversity Threats Associated with Human Population Growth. *Biol. Conserv.* **2003**, *115*, 161–164. [CrossRef]
4. Liu, J.G.; Daily, G.C.; Ehrlich, P.R.; Luck, G.W. Effects of Household Dynamics on Resource Consumption and Availability. *Nature* **2003**, *421*, 530–533. [CrossRef] [PubMed]
5. Mcdonald, R.I.; Marcotullio, P.J.; Güneralp, B. Urbanization, Biodiversity and Ecosystem Services: Challenges and Opportunities. In *Urbanization, Biodiversity and Ecosystem Services: Challenges and Opportunities*; Elmqvist, T., Fragkias, M., Goodness, J., Guneralp, B., Marcotullio, P.J., McDonald, R.I., Parnell, S., Schewenius, M., Sendstad, M., Seto, K.C., Eds.; Springer: Dordrecht, The Netherlands, 2013; pp. 437–452. ISBN 978-94-007-7088-1.
6. McKinney, M.L. Urbanization, Biodiversity and Conservation. *BioScience* **2002**, *52*, 883–890. [CrossRef]
7. Mikusinski, G.; Angelstam, P. Economic geography, forest distribution and woodpecker diversity in central Europe. *Conserv. Biol.* **1998**, *12*, 200–208. [CrossRef]
8. Brooks, T.M.; Mittermeier, R.A.; da Fonseca, G.A.B.; Gerlach, J.; Hoffmann, M.; Lamoreux, J.F.; Mittermeier, C.G.; Pilgrim, J.D.; Rodrigues, A.S.L. Global Biodiversity Conservation Priorities. *Science* **2006**, *313*, 58–62. [CrossRef] [PubMed]
9. Le Saout, S.; Hoffmann, M.; Shi, Y.; Hughes, A.; Bernard, C.; Brooks, T.M.; Bertzky, B.; Butchart, S.H.M.; Stuart, S.N.; Badman, T.; et al. Protected Areas and Effective Biodiversity Conservation. *Science* **2013**, *342*, 803–805. [CrossRef] [PubMed]
10. Margules, C.; Pressey, R. Systematic Conservation Planning. *Nature* **2000**, *405*, 243–253. [CrossRef] [PubMed]
11. Mittermeier, R.A.; Turner, W.R.; Larsen, F.W.; Brooks, T.; Gascon, C. Chapter 1: Global Biodiversity Conservation: The Critical Role of Hotspots. In *Biodiversity Hotspots: Distribution and Protection of Conservation Priority Areas*; Zachos, F.E., Habel, J.C., Eds.; Springer: Berlin, Germany, 2011; pp. 3–22. ISBN 978-3-642-20992-5.
12. Myers, N. Threatened Biotas: 'Hot Spots' in Tropical Forests. *Environmentalist* **1988**, *8*, 187–208. [CrossRef] [PubMed]
13. Hoeskstra, J.M.; Boucher, T.M.; Ricketts, T.H.; Roberts, C. Confronting a biome crisis: Global disparities of habitat loss and protection. *Ecol. Lett.* **2005**, *8*, 23–29. [CrossRef]
14. Sattersfield, A.J.; Crosby, M.J.; Long, A.J.; Wedge, D.C. *Endemic Bird Areas of the World: Priorities for Biodiversity Conservation*; BirdLife Conservation Series 7; BirdLife International: Cambridge, UK, 1998; ISBN 0946888337.
15. WWF; IUCN. *Centres of Plant Diversity: A Guide and Strategy for Their Conservation*; WWF and IUCN: Gland, Switzerland, 1997; ISBN 2-8317-0199-6.
16. Mittermeier, R.A.; Mittermeierm, C.G.; Robles Gil, P. *Megadiversity: Earth's Biologically Wealthiest Nations*; CEMEX: Mexico City, Mexico, 1997; ISBN 9686397507.
17. Olson, DM.; Dinerstein, E. The global 200: A representation approach to conserving the Earth's most biologically valuable ecoregions. *Conserv. Biol.* **1998**, *12*, 502–515. [CrossRef]
18. Olson, DM.; Dinerstein, E. The global 200: Priority ecoregions for global conservation. *Ann. Missouri Bot.* **2002**, *89*, 199–224. [CrossRef]
19. Mittermeier, R.A.; Mittermeier, C.G.; Brooks, T.M.; Pilgrim, J.D.; Konstant, W.R.; da Fonseca, G.A.B.; Kormos, C. Wilderness and biodiversity conservation. *Proc. Natl. Acad. Sci. USA* **2003**, *100*, 10309–10313. [CrossRef] [PubMed]
20. Bryant, D.; Nielsen, D.; Tangley, L.; Sizer, N.; Miranda, M.; Brown, P.; Johnson, N.; Malk, A.; Miller, K. *The Last Frontier Forests: Ecosystems and Economies on the Edge: What Is the Status of the World's Remaining Large, Natural Forest Ecosystem?* World Resources Institute: Washington, DC, USA, 1997.
21. Sanderson, E.W.; Jaiteh, M.; Levy, M.A.; Redford, K.H.; Wannebo, A.V.; Woolmer, G. The human foorprint and the last of the wild: The human footprint is a global map of human influence on the land surface, which suggests that human beings are stewards of nature, whether we like it not. *Bioscience* **2002**, *52*, 891–904. [CrossRef]

22. Myers, N. The biodiversity challenge: Expanded hot-spots analysis. *Environmentalist* **1990**, *10*, 243–256. [CrossRef] [PubMed]

23. Mittermeier, R.A.; Myers, N.; Mittermeier, C.G.; Robles Gil, P. *Hotspots: Earth's Biologically Richest and Most Endangered Terrestrial Ecoregions*; CEMEX, S.A. Agrupacion Sierra Madre, S.C: Mexico City, Mexico, 1999; p. 431. ISBN 9686397582.

24. Myers, N.; Mittermeier, R.A.; Mittermeier, C.G.; da Fonseca, G.A.B.; Kent, J. Biodiversity hotspots for conservation priorities. *Nature* **2000**, *403*, 853–858. [CrossRef] [PubMed]

25. Cincotta, R.P.; Wisnewski, J.; Engelman, R. Human Population in the Biodiversity Hotspots. *Nature* **2000**, *404*, 990–992. [CrossRef] [PubMed]

26. Mittermeier, R.A.; Gil, P.G.; Hoffman, M.; Pilgrim, J.; Brooks, T.M.; Mittermeier, C.G.; Lamoreux, J.; da Fonseca, G.A.B. Hotspots revisted: Earth's biologically richest and most endangered terrestrial ecoregions. *UNEP*. 2004. Available online: http://hdl.handle.net/20.500.11822/15160 (accessed on 23 August 2018).

27. Williams, K.J.; Ford, A.; Rosauer, D.F.; De Silva, N.; Mittermeier, R.; Bruce, C.; Larsen, F.W.; Margules, C. Forests of east Australia: The 35th biodiversity hotspot. In *Biodiversity Hotspots*; Zachos, F.E., Habel, J.C., Eds.; Springer: Berlin, Germany, 2011; pp. 295–310. ISBN 978-3-642-20991-8.

28. Critical Ecosystem Partnership. Biodiversity Hotspots Defined. 2016. Available online: https://www.cepf.net/our-work/biodiversity-hotspots/hotspots-defined (accessed on 2 September 2018).

29. Convention on Biological Diversity. *Quick Guide to the Aichi Biodiversity Targets: Protected Areas Increased and Improved (Target. 11)*; UNEP: Cambridge, UK, 2012.

30. Baillie, J.; Zhang, Y. Space for Nature. *Science* **2018**, *361*, 1051. [CrossRef] [PubMed]

31. Noss, R.F.; Cooperrider, A.Y. *Saving Nature's Legacy: Protecting and Restoring Biodiversity*; Island Press: Washington, DC, USA, 1994; p. 416. ISBN 1559632488.

32. Noss, R.; Dobson, A.P.; Baldwin, R.; Beier, P.; Davis, C.R.; Dellasalla, D.A.; Francis, J.; Locke, H.; Nowak, K.; Lopez, R.; et al. Bolder thinking for conservation. *Conserv. Biol.* **2012**, *26*, 1–4. [CrossRef] [PubMed]

33. Soule, M.E.; Sanjayan, M.A. Conservation targets: Do they help? *Science* **1998**, *279*, 2060–2061. [CrossRef] [PubMed]

34. Luck, G.W. A Review of the Relationships between Human Population Density and Biodiversity. *Biol. Rev.* **2007**, *82*, 607–645. [CrossRef] [PubMed]

35. Mcdonald, R.I.; Kareiva, P.; Forman, R.T.T. The Implications of Current and Future Urbanization for Global Protected Areas and Biodiversity Conservation. *Biol. Conserv.* **2008**, *141*, 1695–1703. [CrossRef]

36. Venter, O.; Sanderson, E.W.; Magrach, A.; Allan, J.R.; Beher, J.; Jones, K.R.; Possingham, H.P.; Laurance, W.F.; Wood, P.; Fekete, B.M.; et al. Sixteen Years of Change in the Global Terrestrial Human Footprint and Implications for Biodiversity Conservation. *Nat. Commun.* **2016**, *7*, 1–11. [CrossRef] [PubMed]

37. Weinzettel, J.; Vacjar, D.; Medkova, H. Human footprint in biodiversity hotspots. *Front. Ecol. Evol.* **2018**, *16*, 447–452. [CrossRef]

38. (CIESIN), Center for International Earth Science Informational Network, Columbia University. Gridded Population of the World, Version 4 (GPWv4), Revision 10 Data Sets. Palisades, NY, USA. 2017. Available online: http://sedac.ciesin.columbia.edu/data/set/gpw-v4-population-count-adjusted-to-2015-unwpp-country-totals-rev10/data-download (accessed on 20 June 2018).

39. (CIESIN), Center for International Earth Science Informational Network, Columbia University. 4 Documentation for the Gridded Population of the World, Version 4 (GPWv4), Revision 10 Data Sets. Palisades, NY, USA. 2017. Available online: http://sedac.ciesin.columbia.edu/downloads/docs/gpw-v4/gpw-v4-documentation-rev10.pdf (accessed on 20 June 2018).

40. U.N. Environment Programme World Conservation Monitoring Centre. *World Database on Protected Areas*; UNEP, IUCN: Cambridge, UK, 2018; Available online: https://www.protectedplanet.net/ (accessed on 5 July 2018).

41. U.N. Environment Programme World Conservation Monitoring Centre. *World Database on Protected Areas*; UNEP, IUCN: Cambridge, UK, 2015; Available online: http://wcmc.io/WDPA_Manual (accessed on 5 July 2018).

42. Araújo, M.B. The Coincidence of People and Biodiversity in Europe. *Global Ecol. Biogeogr.* **2003**, *12*, 5–12. [CrossRef]

43. Balmford, A.; Moore, J.L.; Brooks, T.; Burgess, N.; Hansen, L.A.; Williams, P.; Rahbek, C. Conservation Conflicts across Africa. *Science* **2001**, *291*, 2616–2619. [CrossRef] [PubMed]

44. Luck, G.W.; Ricketts, T.H.; Daily, G.C.; Imhoff, M. Alleviating Spatial Conflict between People and Biodiversity. *Proc. Natl. Acad. Sci. USA* **2004**, *101*, 182–186. [CrossRef] [PubMed]

45. Wikramanayake, E.; McKnight, M.; Dinerstein, E.; Joshi, A.; Gurung, B.; Smith, D. Designing a conservation landscape for tigers in human-dominated environments. *Conserv. Biol.* **2004**, *18*, 839–844. [CrossRef]

46. Goddard, M.; Dougill, A.J.; Benton, T.G. Scaling up from gardens: Biodiversity conservation in urban environments. *Trends Ecol. Evol.* **2010**, *25*, 90–98. [CrossRef] [PubMed]

47. Miller, J.R.; Hobbs, R.J. Conservation where people live and work. *Conserv. Biol.* **2002**, *16*, 330–337. [CrossRef]

48. World Bank. 2018. Available online: https://data.worldbank.org/indicator/NY.GDP.MKTP.CD?end=2017&start=2017&view=map (accessed on 10 September 2018).

49. Bellard, C.; Leclerc, C.; Courchamp, F. Impact of Sea Level Rise on the 10 Insular Biodiversity Hotspots. *Glob. Ecol. Biogeog.* **2014**, *23*, 203–212. [CrossRef]

50. Stedman-Edwards, P. *Socioeconomic Root Causes of Biodiversity Loss*; An Analytical Approach Paper for Case Studies; Worldwide Fund for Nature: Washington, DC, USA, 1997; 90p, Available online: http://awsassets.panda.org/downloads/analytic.pdf (accessed on 20 September 2018).

51. Wilson, E.O. *Half Earth: Our Planet's Fight for Life*; Liveright: New York, NY, USA, 2016; ISBN 978-1631492525.

52. Locke, H. Nature needs (at least) half: A necessary new agenda for protected areas. In *Protecting the Wild: Parks and Wilderness, the Foundation for Conservation*; Wuerthner, G., Crist, E., Butler, T., Eds.; Island Press: Washington, DC, USA, 2015; pp. 3–15.

53. Barnes, M.D.; Glew, L.; Wyborn, C.; Craigie, I.D. Prevent perverse outcomes from global protected area policy. *Nat. Ecol. Evol.* **2018**, *2*, 759–762. [CrossRef] [PubMed]

54. Turner, M.G.; Gardner, R.H. *Landscape Ecology in Theory and Practice*; Springer: New York, NY, USA, 2001; ISBN 978-1-4939-2793-7.

55. Theobald, D.M.; Reed, S.E.; Fields, K.; Soule, M. Connecting natural landscapes using a landscape permeability model to prioritize conservation activities in the United States. *Conserv. Lett.* **2012**, *5*, 123–133. [CrossRef]

56. Mawdsley, J.R.; O'Malley, R.; Ojima, D.S. A review of climate-change adaptation strategies for wildlife management and biodiversity conservation. *Conserv. Biol.* **2009**, *23*, 1080–1089. [CrossRef] [PubMed]

57. Alberti, M. The effects of urban patterns on ecosystem function. *Int. Reg. Sci.* **2005**, *28*, 168–192. [CrossRef]

58. Ervin, J. Rapid Assessment of Protected Area Management Effectiveness in Four Countries. *BioScience* **2003**, *53*, 833–841. [CrossRef]

59. Brook, B.W.; Sodhi, N.S.; Bradshaw, C.J.A. Synergies among Extinction Drivers under Global Change. *Trends Ecol. Evol.* **2008**, *23*, 453–460. [CrossRef] [PubMed]

60. Leonard, P.B.; Baldwin, R.F.; Hanks, R.D. Landscape-scale conservation design across biotic realms: Sequential integration of aquatic and terrestrial landscapes. *Sci. Rep.* **2017**, *7*, 1–12. [CrossRef] [PubMed]

61. Baldwin, R.F.; Trombulak, S.C.; Leonard, P.B.; Noss, R.F.; Hilty, J.A.; Possingham, H.P.; Scarlett, L.; Anderson, M.G. The future of landscape conservation. *BioScience* **2018**, *68*, 60–63. [CrossRef] [PubMed]

Article

Tropical Protected Areas Under Increasing Threats from Climate Change and Deforestation

Karyn Tabor [1,*], Jennifer Hewson [1], Hsin Tien [1], Mariano González-Roglich [1], David Hole [1] and John W. Williams [2]

[1] Betty & Gordon Moore Center for Science, Conservation International, Arlington, VA 22202, USA; jhewson@conservation.org (J.H.); tracy.hs.tien@gmail.com (H.T.); mgonzalez-roglich@conservation.org (M.G.-R.); dhole@conservation.org (D.H.)

[2] Department of Geography, University of Wisconsin, Madison, WI 53706, USA; jww@geography.wisc.edu

* Correspondence: ktabor@conservation.org; Tel.: +1-703-341-2560

Received: 29 June 2018; Accepted: 25 July 2018; Published: 28 July 2018

Abstract: Identifying protected areas most susceptible to climate change and deforestation represents critical information for determining conservation investments. Development of effective landscape interventions is required to ensure the preservation and protection of these areas essential to ecosystem service provision, provide high biodiversity value, and serve a critical habitat connectivity role. We identified vulnerable protected areas in the humid tropical forest biome using climate metrics for 2050 and future deforestation risk for 2024 modeled from historical deforestation and global drivers of deforestation. Results show distinct continental and regional patterns of combined threats to protected areas. Eleven Mha (2%) of global humid tropical protected area was exposed to the highest combined threats and should be prioritized for investments in landscape interventions focused on adaptation to climate stressors. Global tropical protected area exposed to the lowest deforestation risk but highest climate risks totaled 135 Mha (26%). Thirty-five percent of South America's protected area fell into this risk category and should be prioritized for increasing protected area size and connectivity to facilitate species movement. Global humid tropical protected area exposed to a combination of the lowest deforestation and lowest climate risks totaled 89 Mha (17%), and were disproportionately located in Africa (34%) and Asia (17%), indicating opportunities for low-risk conservation investments for improved connectivity to these potential climate refugia. This type of biome-scale, protected area analysis, combining both climate change and deforestation threats, is critical to informing policies and landscape interventions to maximize investments for environmental conservation and increase ecosystem resilience to climate change.

Keywords: protected areas; climate change; deforestation; tropics; biodiversity; conservation

1. Introduction

Protected areas (PAs) represent a cornerstone strategy for preserving global biodiversity. PAs and connective corridors have, to date, been established over approximately 15% of the global terrestrial surface to provide refuge and to facilitate migrations and diversification of the gene pools among secluded populations [1,2]. Protecting important ecosystems is also essential to preserve ecosystem services and for climate change mitigation and adaptation strategies [3]. Furthermore, improving habitat extent and connectivity, a barrier to species dispersal, is vital to improving the adaptive capacity of species to climate change [4,5].

PAs have proven effective in reducing habitat destruction and deterring some illegal activities including poaching, illegal logging, and cattle grazing [6]. However, the expansion and maintenance of a robust protected area network that meets global conservation goals has been, and continues to be, challenging [1,7,8]. Chronic underfunding, shorter-term investments, lack of political support, and limited local engagement contribute to the current trends of PA deterioration [9]. In addition, there is an urgent need to assess the conservation outcomes of protected areas to design more effective interventions and improve management of protected areas [10].

Climate change presents an additional challenge to the effectiveness of PAs because the global protected area network accounts for the current distribution of species and habitats, not potential future distributions. Static PAs may be ineffective when trying to protect biodiversity during a century likely to be characterized by shifting species ranges, elevational migrations, and possibly extinctions due to climate change [11–14]. Segurado et al. [15] estimated 6–11% of species will be pushed out of reserved areas by mid 21st century. Loarie et al. [16] estimated only 8% of PAs have the climatic heterogeneity required to accommodate species shifts in response to anticipated climate change over the next 80 years.

A combination of restricted migration due to habitat reduction and fragmentation, and with the projected rapid warming or drying are likely to render species more vulnerable than during previous historical episodes of climate change [17,18]. Some projections of climate change impacts on species estimate an average of 8–16% global extinction risk depending on the representative concentration pathway [19]. Furthermore, two-thirds of Earth's species are concentrated in the tropics, predominantly in tropical forests [20] and these forests are under significant threat from agriculture conversion and resource extraction. Only one quarter of the ~1990 extents of these humid tropical forests remain, and they are reduced by an additional 5.5 million hectares each year [21]. While an estimated 23% of tropical forests are under protection, deforestation continues to occur inside tropical PAs [22].

Solutions to protecting species from climate change are dependent on land use planning and management; for example, protected area network expansion, protecting climate refugia, and climate-change corridors [23–26]. Planning for PA expansion or corridors requires knowledge of both climate risks and deforestation risks to strategically invest in effective land use planning for habitat preservation or restoration. Researchers have employed a variety of modeling techniques to inform the potential impacts of anthropogenic change on biodiversity [27]. Assessing biodiversity threats from deforestation is based directly on observed data typically from satellite remote sensing [28–32]. Using observed rates of change with information on the driving forces of land cover change allows for deforestation risk assessments or projections of future forest cover scenarios into the near-future [33]. The biodiversity response to habitat change is well documented and the consequences are severe. The uncertainty related to deforestation risk projections are tied to uncertainties in changing geopolitical and socioeconomic factors that directly affect drivers of deforestation making long-term forest cover projections (greater than 20 years) ineffective. Climate models are optimized for longer-term future projections (50–100 years), however, projecting the impacts of climate change on biodiversity is challenging given the uncertainties of how vulnerable or adaptable individual species are to climate change [34,35]. Thus, many studies avoid specifying climate tolerance of species, indirectly assessing climate change impacts on species by quantifying climate change on habitat or in restricted habitats such as PAs [16,36–38].

In recent years, researchers have applied climate metrics applied to measure biodiversity exposure to climate change risks [39]. For example, climate metrics measuring climate dissimilarity (i.e., disappearing and novel climates) are global risk surfaces indicating areas of extreme climatic regime shifts. Disappearing climates indicate areas where the late 20th century climate will cease to exist and therefore species adapted to these climates will go extinct if unable to adapt to new climatic conditions. These species cannot simply migrate to a new location because there are no future climate regimes that meet their 20th century niche requirements [40]. Novel climates, similarly, are too different from any 20th century climate to support the current species if the species are unable to adapt to the new climate. Novel climates are particularly challenging for ecological forecasting because of the uncertainty of species adaptability to these new climates that did not exist in the 20th century [39–42].

Land use change and climate change are interlinked as human-induced forces disrupting ecological processes and driving global change [43–45]. Yet to date, only a handful of studies have addressed the combined impacts of future climate change and land cover change (e.g., [26,43,46,47] despite the risk of extinction and negative economic and human well-being affects from these stressors [18,48–50]. Evaluating the vulnerability of the protected area system to both climate change and habitat loss is essential to inform the prioritization of conservation investments, adaptation and mitigation strategies, and land use planning initiatives that aim to preserve biodiversity and promote long-term resilience. These interventions are crucial towards achieving a range of international initiatives and agreements including Reducing Emissions from Deforestation and Degradation (REDD+), the Aichi biodiversity targets and the United Nations Sustainable Development Goals (SDGs). In this study, we assessed the exposure of PAs in the humid tropical forest biome to both deforestation risk derived from satellite monitoring and climate exposure metrics to compare how the exposure of PAs varies among continents and regions to inform transboundary landscape management solutions. Understanding the potential impacts of anthropogenic change on PAs is critical to informing where and how to invest in biodiversity conservation.

2. Materials and Methods

2.1. Protected Areas

We used the World Database of Protected Areas (WDPA) [51] and selected PAs within the humid tropical forest biome, based on the tropical/subtropical moist broadleaf forest class defined in the Terrestrial Ecoregions of the World [52]. We defined forest as areas with greater than 10% tree cover in 2000 [30]. This threshold captures low-density dry forests, in addition to denser forests. PAs without any forest cover meeting the definition were omitted from the analysis. We also omitted PAs located outside the spatial extent of the climate data used in the analysis, as described below. This yielded 7672 remaining PAs of the 8161 total terrestrial PAs in the humid tropical forest biome.

2.2. Deforestation Risk

For exposure to anthropogenic land cover change, we used a potential deforestation risk layer generated using historical deforestation (2000–2014) and a suite of potential deforestation drivers [53]. The historical deforestation was based-on 30-m global forest loss data from Hansen et al. [30] using a forest definition of 10% tree cover [54] in 2000. The tree cover dataset, by definition, included both natural and production forests. We used a set of potential drivers of deforestation (Table 1) as explanatory variables to predict deforestation. First, we aggregated all deforestation and driver data to 1km due to the resolution of the majority of the deforestation driver datasets. Next, we tested these variables for predicating deforestation and selected only variables with a significant relationship to historical deforestation. We used a multi-layer perceptron model in TerrSet Land Change Modeler software (https://clarklabs.org/terrset/) to generate a layer of future deforestation risk [33]. The layer represented the modeled percent likelihood of potential deforestation, scaled from 0–1, by 2024.

Table 1. Driver variables used to generate potential deforestation risk.

Variable	Source
Distance to railroads	Vmap0 [55]
Distance to roads	Vmap0 [55]
Distance to trails	Vmap0 [55]
Distance to urban areas	[56]
Elevation	Global Multi-resolution Terrain Elevation Data [57]
Slope	Derived from Elevation
Above Ground Biomass	GeoCarbon [58]
Human Influence Index	[59]

Table 1. *Cont.*

Variable	Source
Crop suitability	GLUES [60]
Irrigation area	FAO [61]
Global Opportunity Cost	[62]
Annual Precipitation	Bioclim [63]
Annual Mean temperature	Bioclim [63]
Protected Areas	[51]
Ecoregion Biomes	[52]

2.3. Climate Exposure

We analyzed the relative exposure to climate change using two climate metrics generated from 10-min, downscaled future climate projections. The two metrics were: (1) disappearing climate risk and (2) novel climate risk [64]. We calculated both metrics for 2050 with a high emission scenario, using the IPCC AR4 A2 scenario, similar to AR5 RCP8.5 [65]. The metrics represented dissimilarity measurements of the squared Euclidean distance between seasonal (June–August and December–February) temperature and precipitation variables in the 20th century climate and mid-20th century climate [40,66]. The disappearing climate metric was a measure of dissimilarity between a pixel's late 20th century climate and its closest matching pixel in the global set of 21st-century climates. The novel climate metric represented the dissimilarity between a pixel's future climate and its closest matching pixel in the global set of late 20th-century climates. We calculated mean disappearing and mean novel climate metrics for each of the 7672 PAs. High disappearing risk for a PA indicated that a PA's current climate was very different from any climate that will exist anywhere globally in the mid-21st century, therefore at risk for disappearing. High novel risk for a PA indicated that the future climate was very different than any climates that existed in the 20th-century and therefore emerged as new climates previously unknown in the 20th century.

2.4. Methods

We rescaled the disappearing and novel climate layers from 0 to 1 using the minimum for each as the lower threshold and the 99th percentile as the upper threshold to remove outliers. We then displayed the three outputs: deforestation risk, disappearing climate risk, and novel climate risk as color display channels (red-green-blue). This allowed visual exploration of the spatial patterns emerging over the global tropics when the three risk layers were combined. We analyzed the combined risk by calculating the distribution of the mean deforestation risk, mean disappearing climate risk, and mean novel climate risk for each PA using the original data values, not scaled from 0–1. The highest risk PAs were considered the 75th percentile and the 25th percentile were considered the lowest risk PAs (Table 2). We then identified PAs with the following combinations of risk: highest deforestation risk and highest climate risk (either highest disappearing or highest novel); highest deforestation risk and lowest climate risk; lowest deforestation risk and highest climate risk; and lowest deforestation risk and lowest climate risk.

Table 2. Thresholds for humid tropical biome protected areas (Pas) with highest and lowest climate and deforestation risks.

Risk Category	Threshold	Area (Mha)	Percent of Total Area of PAs
Highest Deforestation Risk	>0.41	29.0	6%
Lowest Deforestation Risk	<0.07	312.5	61%
Highest Disappearing Climate Risk	>2.00	62.3	12%
Lowest Disappearing Climate Risk	<1.10	133.7	26%
Highest Novel Climate Risk	>2.41	165.3	32%
Lowest Novel Climate Risk	<1.18	74.9	15%

3. Results & Discussion

3.1. Global Patterns

Our results indicated distinct patterns of relative risk for climate change and deforestation when comparing the spatial variation of the combined threats of deforestation and climate risk for global tropical forests (Figure 1). Areas exposed to the highest deforestation risk included the Atlantic Forests in Brazil; the Yucatan peninsula in Mexico; northern Thailand, Laos, and Vietnam; Sumatra and coastal Borneo; East Africa; and western Madagascar. The Brazilian Amazon had a very low deforestation risk most likely driven by the reduction in deforestation in the Amazon in the latter half of the deforestation time-period (2007–2014) [67,68]. Deforestation impacts PAs through reducing habitat for species of concern, degrading ecosystem services, and decreasing the ecosystem resilience to climate change. The highest risk deforestation areas that are also high-movement areas for species dispersal under climate change should be prioritized for protection. Once such area indicated in our results was the Atlantic Forest of Brazil due to the area's past and projected future role for high levels of species movements under climate change conditions [47].

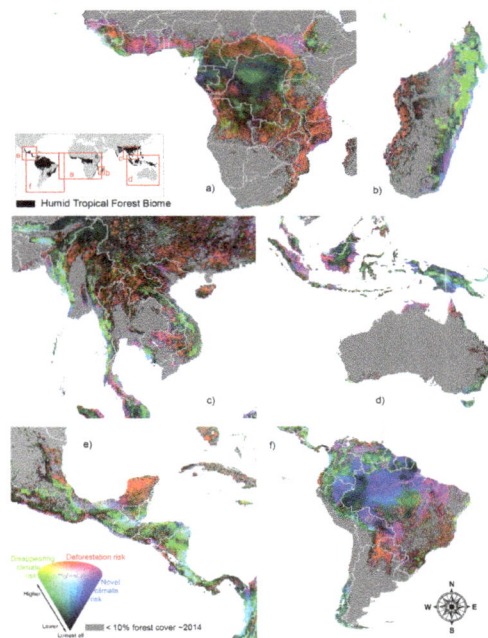

Figure 1. Global tropical forest risk combinations. This map shows the combined deforestation and climate risks for all tropical forests with >10% forest cover in 2014. The data are displayed as deforestation risk (red channel), disappearing climate risk (green channel), and novel climate risk (blue channel), for (**a**) sub-Saharan Africa, (**b**) Madagascar, (**c**) Southeast Asia, (**d**) Asia-Pacific and Australia, (**e**) North America, and (**f**) South America. Combinations of the gradient of values for all three risks revealed spatially distinct patterns. Yellow indicates a combination of high deforestation, high disappearing, and low novel climate risks; purple indicates a combination of high deforestation, high novel climate, and low disappearing climate risks; and cyan is a combination of low deforestation and high combined climate risks. Areas at high risk for both deforestation and climate change are shown in white and areas with both low deforestation and climate change risks appear dark grey. The global locator map depicts the humid tropical forest biome in dark gray.

The climate exposure risks results revealed the highest risk areas for disappearing climates were in higher altitude areas (e.g., Andes, eastern Sumatra, Papua New Guinea, central Mexico, and the eastern forests in Madagascar). Novel climate risks were highest in lower elevation tropical areas (e.g., Amazon basin; the Tabasco region in Mexico; southern Myanmar; the lowlands of Sumatra and New Guinea; Liberia; Gabon; and the south-east coast of Madagascar). These findings were similar to other studies that used climate dissimilarity measures [16,40]. Species in PAs exposed to the disappearing climate risk in these higher elevation areas may be able to move up or down slope to find new habitat if habitat remains intact. This will be challenging in areas with high deforestation risk (e.g., mountain forests in Eastern Madagascar, Colombia Andes, and the highlands of Zambia and Angola; Figure 1). Another challenge is species response to novel climate risk in vast lowland basins like the Amazon, where despite conservation efforts to protect habitat and corridors, species migration may be outpaced by the velocity of change [16]. Areas exposed to the highest risks of disappearing climate, novel climate, and deforestation (e.g., southeast Amazon; southwest Cambodia; and Liberia and Coastal Cameroon) will require intensified landscape interventions and planning for transboundary landscape connectivity. Intensive measures beyond landscape management (e.g., assisted migration, rewilding, and ex-situ conservation) [5] may be required in areas of highest climate risk and deforestation risk.

3.2. Protected Areas

The spatial distribution of humid tropical forest biome PAs in our analysis varied by continent, although almost half of the PAs in the study were in Asia, when considering area of PAs, almost three-quarters of the total area was located in South America (Table 3). Globally, 29 Mha (6%) of PAs were exposed to a deforestation risk greater than 40% (the highest-risk threshold) (Figure 2). The low proportion of PAs exposed to highest deforestation threat was expected because PAs historically have lower deforestation rates than unprotected areas and PAs often are more remote and exposed to less threats [9,69]. Further exploration with a lower threshold revealed that 132 Mha (26%) of PAs was exposed to greater than 10% chance of deforestation. These results are more conservative than the results of Jones et al. [70] indicating that 32.8% of global PAs are exposed to intense human pressure, or "human footprint". The human footprint score is not a measure of deforestation, but instead pressure that may compromise ecosystem intactness and function. Our results suggested that in most regions, PAs can continue to serve in their intended role as reserves for species threatened by deforestation based on observed, historical deforestation rates and not solely proxy measures.

Table 3. Distribution of PAs across the humid tropical forest biome.

Continent	Number of PAs	Total Area (Mha)	Percent of PA's	Percent by Area
Africa	1444	64.6	19%	13%
Asia	2079	70.8	27%	14%
Australia	370	4.4	5%	1%
North America	1290	26.2	17%	5%
South America	2489	347.0	32%	68%
Total	7672	513.0	100%	100%

The continent with the greatest total area of PA exposed to the highest deforestation risk was South America, with a total of 8.8 Mha at risk. North America had the highest percent of area under protection with high deforestation risk (27%) (Figure 2). In addition to the high biodiversity value inherent in the PAs, these areas also represent high biomass value and ecosystem service provision zones, thus investments will have the maximum impact on protecting threatened ecosystems and mitigating climate change. Landscape interventions stemming from international policies addressing climate mitigation such as REDD+ may be a viable option. REDD+ aims to reduce deforestation and degradation by providing countries with a range of incentives and result-based payments in exchange for emissions reductions. Many REDD+ investments focus on reducing forest dependency of local communities by providing alternative livelihoods to reduce deforestation pressures [71]. Conservation investments that focus on improving livelihoods are important because communities vulnerable to climate change rely more on forest products as a safety-net during times of low agricultural production [72]. Other landscape interventions include improving management effectiveness (i.e., "upgrading") and increasing landscape connectivity (i.e., "upsize and interconnect") to enable species movement away from threats [9]. Restoring habitat is another intervention intended to increase ecosystem resiliency to fire propagation [73]. Another key management tool is near real-time monitoring of ecosystem threats utilizing community-based monitoring, remotely-sensed data, or in-situ sensors. These systems enable PA managers with time-sensitive alerts on deforestation threats of fires and illegal extractives to strategically patrol protected areas and enable rapid response to prevent further ecosystem destruction (e.g., [74–77]).

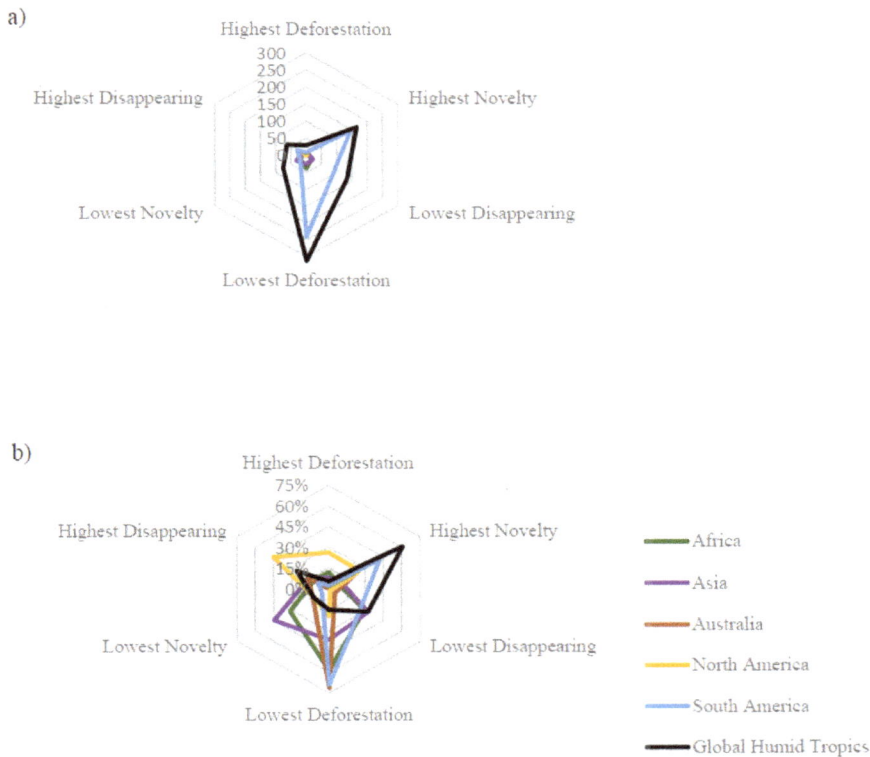

Figure 2. Distribution of PA risks. These two radar graphs show the area of PAs exposed to the highest and lowest deforestation, disappearing climate, and novel climate risks. The figure shows (**a**) the total area of PAs in Mha exposed to the highest and lowest risks by continent and (**b**) the percent area of PAs per continent and humid tropics biome exposed to the highest and lowest risks.

When examining combined climate risks, 226 Mha (44%) were exposed to the highest risk of either highest disappearing or highest novel risk; parsed into 62 Mha (12%) at highest risk of high disappearing climates and 165 M ha (32%) at highest risk of novel climates. PAs at highest risk for disappearing climate indicate areas where species may go extinct [40]. Forty-five percent of North America PA area was exposed to the highest disappearing climate risk whereas 42% of South America PA area was exposed to highest novel climate risk (Figure 2). Garcia et al. [39] and Williams et al. [40] both concluded that novel climates are more likely in the tropics and subtropics and our study indicated that the Amazon basin may be a hotspot for novel climates. This may be because the area is already among the warmest in the world, and also has relatively little seasonality and interannual climatic variability. Hence the 21st-century warming will quickly move the mean temperatures for this region beyond the background range of interannual climates found here or anywhere else in the range of 20th-century climates [40,78].

Novel climates, while posing similar threats to species as disappearing climates, also cause greater uncertainty for how species may reject or adapt to new climates [79]. While these two metrics of climate exposure may indicate different biological conservation responses, landscape management responses are similar as species in both highest-risk disappearing and novel areas may benefit from a reduction of global emissions to mitigate climate change in addition to local forest conservation/climate adaptation planning because intact, healthy ecosystems may increase species' resilience to climate change [80]. Increasing landscape connectivity (upsize and interconnect) will assist species migrations to suitable habitats. In addition to these practices applied to high deforestation risk areas, these PAs could benefit from in-situ monitoring of species response to climate change [81]. Asia had a large proportion of PAs with the lowest risks of disappearing climate (32%) and lowest novel climates (45%), and thus, may be a suitable area for conservation investments increasing connectivity to allow species dispersal to these climate refugia [23].

The intersection of PAs with the highest deforestation and highest combined climate risks revealed 11 Mha (2%) of the global tropical PAs with extreme exposure to the dual threats, and half of this area was in South America (Figure 3a). The continent with the largest proportion of area affected was North America (6%). These PAs should be immediately targeted for conservation investments. Although traditional policies and mechanisms to reduce habitat loss may be effective, the additional climate change risks to these PAs may require newly designed policy mechanisms and practices adapted to unique climate stressors [5].

Sixteen Mha of global humid tropical PA was exposed to highest deforestation and lowest combined climate risks. Forty percent of this area was in South America, but Africa accounted for another 33% of the total area. (Figure 3b). These areas should be targeted for investments in traditional landscape interventions to reduce deforestation and should be considered high priority sites for lower-risk investments to upscale and interconnect ecosystems to promote species movement to these potential climate refugia.

The area of global humid tropical PAs exposed to the lowest deforestation and highest combined climate risks totaled 135 Mha (26%); an overwhelming 35% of this area was located in South America (Figure 3c). Australia also reflected a high proportion of area in PAs with the lowest deforestation and highest combined climate risk (22%). These PAs should be targeted for further ecosystem assessment for climate change sensitivity. With decreased deforestation pressures, these PAs may be prioritized for increasing the size or the connectivity of PAs to allow for migration of species sensitive to climate change [47].

The area of global humid tropical forested PAs exposed to the combination of lowest deforestation and lowest combined climate risk totaled 89 Mha (17%) and was mostly located in South America (60%); however, it was proportionally highest in Africa (34%) and Asia (17%) (Figure 3d). These areas may serve as low risk investments where management may be less costly and the areas may provide climate refuge.

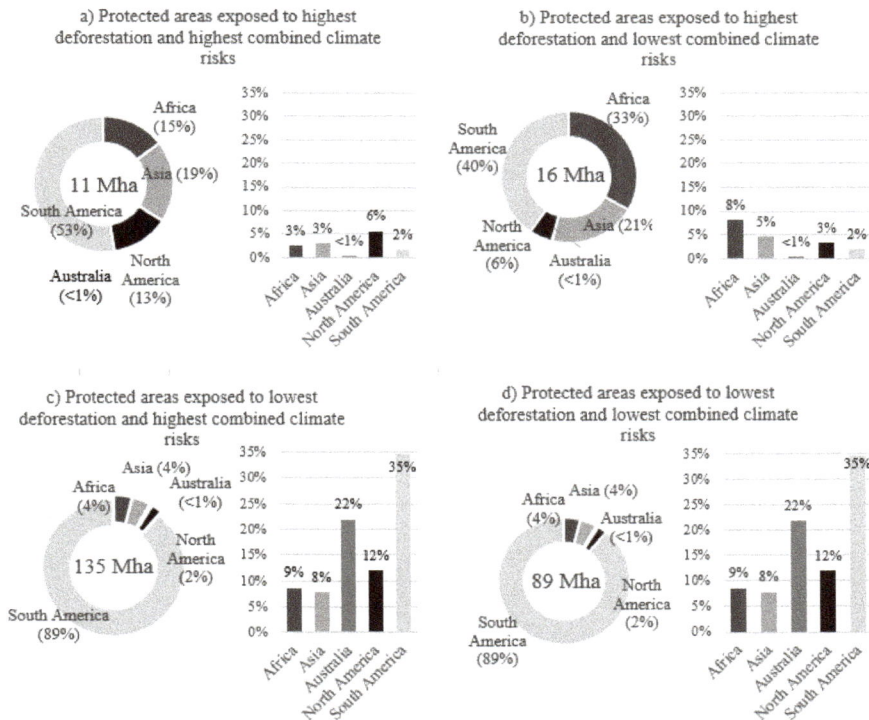

a) Protected areas exposed to highest deforestation and highest combined climate risks

b) Protected areas exposed to highest deforestation and lowest combined climate risks

c) Protected areas exposed to lowest deforestation and highest combined climate risks

d) Protected areas exposed to lowest deforestation and lowest combined climate risks

Figure 3. PAs exposed to combined risks by continent. This quadrant of figures shows the distribution of deforestation and combined climate risks by continent for: (**a**) highest deforestation and highest combined climate risks; (**b**) highest deforestation and lowest combined climate risks; (**c**) lowest deforestation and highest combined climate risks; and (**d**) lowest deforestation and lowest combined climate risks. Combined climate risks included either novel or disappearing risk. The number in the center of the pie charts indicates the total area affected of global humid tropical forest PAs. The number next to each continent name in the pie chart indicates the combined deforestation and climate risk as a percent of the total area of global humid tropical PAs affected. The bar charts show the proportion of area of PAs per continent in each risk category.

3.3. Caveats

In this study, we did not indicate which PAs were at risk to biological diversity loss, rather we compared the relative exposure of risk using climate metrics that represented "threats and opportunities for biodiversity" under climate change [39]. The 25th and 75th percentiles indicated the extremes of highest and lowest risks to PAs in the global set of humid tropical forested PAs. We also did not assess the risk to biological diversity, as an individual species' risk to climate change is determined by its exposure, sensitivity and adaptive capacity based on historical information and some species may be able to persist in a broader range of climate conditions than previously known [79]. We did highlight areas exposed to the greatest threats of deforestation and climate change that require urgent attention and we also indicated areas buffered from these combined threats, which may persist as biological refuges into the future.

4. Conclusions

This study focused on the impacts of climate change and forest change on PAs within the humid tropical forest biome to inform PA investments. The effects of deforestation on biological diversity in the humid tropics and elsewhere are widely documented and these threats can be managed by identifying effective conservation and sustainable land management approaches.

Implementing policies to address climate change is less straightforward given the uncertainties associated with efforts to mitigate future climate change and reduce deforestation loss; projections of how species and ecosystems will adapt (or fail to adapt) to novel climates, and vulnerability of ecosystems to disappearing climates. Therefore, it is essential when addressing climate change impacts on biodiversity to address the immediate threat of habitat loss and improve and preserve ecosystem resilience. For PAs at risk to both threats, a suite of investments should be implemented in the landscape from improved management, upsizing and interconnecting PAs, habitat restoration, socio-economic investments, capacity building, and adopting new technologies for improved monitoring.

The advantage of a biome-wide study is to both shift the focus of protected area management from within a country's boundary to the transboundary threats, and to guide international policies and investments to consider the dual threats of both of climate change and land cover change. The scale of this study can inform global investors where and how to maximize investments for biodiversity conservation, climate mitigation, and preserving climate resilient ecosystems. Finer-scale analyses based on regional and national-scale projections of deforestation are recommended to inform investments for individual PAs. Considering the combined risks of PAs to both climate change and habitat loss can be vital to for national-level spatial priority setting.

Author Contributions: For research articles with several authors, a short paragraph specifying their individual contributions must be provided. The following statements should be used "Conceptualization, K.T., J.H. and D.H.; Methodology, K.T., J.H., M.G.-R., D.H. and J.W.W.; Software, K.T. and J.W.W.; Validation, K.T., H.T., J.W.W. and M.G.-R.; Formal Analysis, K.T., J.W.W. and H.T.; Investigation, K.T. and J.W.W.; Resources, K.T.; Data Curation, K.T. and H.T.; Writing-Original Draft Preparation, K.T.; Writing-Review & Editing, K.T., J.H., H.T., M.G.-R., D.H. and J.W.W.; Visualization, K.T., H.T. and M.G.-R.; Supervision, K.T.; Project Administration, K.T.; Funding Acquisition, K.T.", please turn to the CRediT taxonomy for the term explanation. Authorship must be limited to those who have contributed substantially to the work reported.

Funding: This research was made possible with a Walton Family Foundation grant and a gift from Betty and Gordon Moore.

Acknowledgments: The authors would like to thank Kellee Koenig and Cameron Walkup for their help preparing the manuscript.

Conflicts of Interest: The authors declare no conflict of interest.

References

1. Rodrigues, A.S.L.; Andelman, S.J.; Bakarr, M.I.; Boitani, L.; Brooks, T.M.; Cowling, R.M.; Fishpool, L.D.C.; da Fonseca, G.A.B.; Gaston, K.J.; Hoffmann, M.; et al. Effectiveness of the global protected area network in representing species diversity. *Nature* **2004**, *428*, 9–12. [CrossRef] [PubMed]

2. UNEP-WCMC; IUCN. *Protected Planet Report 2016*; UNEP-WCMC: Cambridge, UK; IUCN: Gland, Switzerland, 2016; Available online: https://wdpa.s3.amazonaws.com/Protected_Planet_Reports/2445GlobalProtectedPlanet2016_WEB.pdf (accessed on July 30 2016).

3. Dudley, N.; Stolton, S.; Belokurov, A.; Krueger, L.; Lopoukhine, N.; MacKinnon, K.; Sandwith, T.; Sekhran, N. *Natural Solutions: Protected Areas Helping People Cope with Climate Change*; IUCNWCPA, TNC, UNDP, WCS, The World Bank and WWF: Gland, Switzerland; Washington, DC, USA; New York, NY, USA, 2010; Available online: https://www.iucn.org/downloads/natural_solutions.pdf (accessed on 31 September 2017).

4. Hurlbert, A.H.; Jetz, W. Species richness, hotspots, and the scale dependence of range maps in ecology and conservation. *Proc. Natl. Acad. Sci. USA* **2007**, *104*, 13384–13389. [CrossRef] [PubMed]

5. Dawson, T.P.; Jackson, S.; House, J.I.; Prentice, I.C.; Mace, G.M. Beyond Predictions: Biodiversity Conservation in a Changing Climate. *Science* **2011**, *332*, 53–58. [CrossRef] [PubMed]

6. Bruner, A.G.; Gullison, R.E.; Rice, R.E.; da Fonseca, G.A. Effectiveness of parks in protecting tropical biodiversity. *Science* **2001**, *291*, 125–128. [CrossRef] [PubMed]

7. Lele, S.; Wilshusen, P.; Brockington, D.; Seidler, R.; Bawa, K. Beyond exclusion: Alternative approaches to biodiversity conservation in the developing tropics. *Curr. Opin. Environ. Sustain.* **2010**, *2*, 94–100. [CrossRef]

8. Laurance, W.F.; Useche, D.C.; Rendeiro, J.; Kalka, M.; Bradshaw, C.J.A.; Sloan, S.P.; Laurance, S.G.; Campbell, M.; Abernethy, K.; Alvarez, P.; et al. Averting biodiversity collapse in tropical forest protected areas. *Nature* **2012**, *489*, 290–293. [CrossRef] [PubMed]

9. Pringle, R.M. Upgrading protected areas to conserve wild biodiversity. *Nature* **2017**, *546*, 91–99. [CrossRef] [PubMed]

10. Geldman, J.; Barnes, M.; Coad, L.; Craigie, I.D.; Hockings, M.; Burgess, N.D. Effectiveness of terrestrial protected areas in reducing habitat loss and population declines. *Biol. Conserv.* **2013**, *161*, 230–238. [CrossRef]

11. Walther, G.-R.; Post, E.; Convey, P.; Menzel, A.; Parmesan, C.; Beebee, T.J.C.; Fromentin, J.-M.; Hoegh-Guldberg, O.; Bairlein, F. Ecological responses to recent climate change. *Nature* **2002**, *416*, 389–395. [CrossRef] [PubMed]

12. Parmesan, C.; Yohe, G. A globally coherent fingerprint of climate change impacts across natural systems. *Nature* **2003**, *421*, 37–42. [CrossRef] [PubMed]

13. Root, T.L.; Price, J.T.; Hall, K.R.; Schneider, S.H. Fingerprints of global warming on wild animals and plants. *Nature* **2003**, *421*, 57–60. [CrossRef] [PubMed]

14. Thomas, C.D.; Cameron, A.; Green, R.E.; Bakkenes, M.; Beaumont, L.J.; Collingham, Y.C.; Erasmus, B.F.N.; de Siqueira, M.F.; Grainger, A.; Hannah, L.; et al. Extinction risk from climate change. *Nature* **2004**, *427*, 145–148. [CrossRef] [PubMed]

15. Segurado, P.; Araújo, M.B. An evaluation of methods for modelling species distributions. *J. Biogeogr.* **2004**, *31*, 1555–1568. [CrossRef]

16. Loarie, S.R.; Duffy, P.B.; Hamilton, H.; Asner, G.P.; Field, C.B.; Ackerly, D.D. The velocity of climate change. *Nature* **2009**, *462*, 1052–1055. [CrossRef] [PubMed]

17. Davies, T.J.; Purvis, A.; Gittleman, J.L. Quaternary climate change and the geographic ranges of mammals. *Am. Nat.* **2009**, *174*, 297–307. [CrossRef] [PubMed]

18. Corvalen, C.; Hales, S.; McMichael, A.J.; Bulter, C. *Ecosystems and Human Well-Being, Health Synthesis*; Wolrd Health Organization: Geneva, Switzerland, 2005.

19. Urban, M.C. Accelerating extinction risk from climate change. *Science* **2015**, *348*, 571–573. [CrossRef] [PubMed]

20. Pimm, S.L.; Raven, P. Extinction by numbers. *Nature* **2000**, *403*, 843–845. [CrossRef] [PubMed]

21. Keenan, R.J.; Reams, G.A.; Achard, F.; De Freitas, J.V.; Grainger, A.; Lindquist, E. Forest Ecology and Management Dynamics of global forest area: Results from the FAO Global Forest Resources Assessment 2015. *For. Ecol. Manag.* **2015**, *352*, 9–20. [CrossRef]

22. Spracklen, B.D.; Kalamandeen, M.; Galbraith, D.; Gloor, E.; Spracklen, D.V. A Global Analysis of Deforestation in Moist Tropical Forest Protected Areas. *PLoS ONE* **2015**, *342*, e0143886. [CrossRef] [PubMed]

23. Keppel, G.; Van Niel, K.P.; Wardell-Johnson, G.W.; Yates, C.J.; Byrne, M.; Mucina, L.; Schut, A.G.T.; Hopper, S.D.; Franklin, S.E. Refugia: Identifying and understanding safe havens for biodiversity under climate change. *Glob. Ecol. Biogeogr.* **2012**, *21*, 393–404. [CrossRef]

24. Alagador, D.; Cerdeira, J.O. Climate change, species range shifts and dispersal corridors: An evaluation of spatial conservation models. *Methods Ecol. Evolut.* **2016**, *7*, 853–866. [CrossRef]

25. Hole, D.G.; Willis, S.; Pain, D.J.; Fishpool, L.D.; Butchart, S.H.M.; Collingham, Y.C.; Rahbek, C.; Huntley, B. Projected impacts of climate change on a continent-wide protected area network. *Ecol. Lett.* **2009**, *12*, 420–431. [CrossRef] [PubMed]

26. Nuñez, T.A.; Lawler, J.J.; Mcrae, B.H.; Pierce, J.D.; Krosby, M.; Kavanagh, D.M.; Singleton, P.H.; Tewksbury, J.J. Connectivity Planning to Address Climate Change. *Conserv. Biol.* **2013**, *27*, 407–416.

27. Pacifici, A.M.; Foden, W.B.; Visconti, P.; Watson, J.E.M.; Butchart, S.H.M.; Kovacs, K.M.; Scheffers, B.R.; Hole, D.G.; Martin, T.G.; Akçakaya, H.R.; et al. Assessing species vulnerability to climate change. *Nat. Clim. Chang.* **2015**, *5*, 215–225. [CrossRef]

28. Hansen, A.J.; DeFries, R. Ecological mechanisms linking protected areas to surrounding lands. *Ecol. Appl.* **2007**, *17*, 974–988. [CrossRef] [PubMed]

29. Harper, G.J.; Steininger, M.K.; Tucker, C.J.; Juhn, D.; Hawkins, F. Fifty years of deforestation and forest fragmentation in Madagascar. *Environ. Conserv.* **2007**, *34*, 325–333. [CrossRef]

30. Hansen, M.C.; Potapov, P.V.; Moore, R.; Hancher, M.; Turubanova, S.A.; Tyukavina, A.; Thau, D.; Stehman, S.V.; Goetz, S.J.; Loveland, T.R.; et al. High-resolution global maps of 21st-century forest cover change. *Science* **2013**, *342*, 850–853. [CrossRef] [PubMed]

31. Burgess, N.D.; Malugu, I.; Sumbi, P.; Kashindye, A.; Kijazi, A.; Tabor, K.; Mbilinyi, B.; Kashaigili, J.; Maxwell Wright, T.; Gereau, R.E.; et al. Two decades of change in state, pressure and conservation responses in the coastal forest biodiversity hotspot of Tanzania. *ORYX* **2017**, *51*, 77–86. [CrossRef]

32. Tabor, K.; Burgess, N.D.; Mbilinyi, B.P.; Kashaigili, J.J.; Steininger, M.K. Forest and woodland cover and change in coastal Tanzania and Kenya. *J. East Afr. Nat. Hist.* **2010**, *99*, 19–45. [CrossRef]

33. Eastman, J.R.; Solórzano, L.A.; van Fossen, M. *Transition Potential Modeling for Land-Cover Change*; GIS, Spatial Analysis, and Modeling; Maguire, J.D., Batty, M., Goodchild, M., Eds.; ESRI Press: Redlands, CA, USA, 2005; pp. 357–385.

34. Hijmans, R.J.; Graham, C.H. The ability of climate envelope models to predict the effect of climate change on species distributions. *Glob. Chang. Biol.* **2006**, *12*, 2272–2281. [CrossRef]

35. Foden, W.B.; Butchart, S.H.M.; Stuart, S.N.; Vié, J.-C.; Akçakaya, H.R.; Angulo, A.; DeVantier, L.M.; Gutsche, A.; Turak, E.; Cao, L.; et al. Identifying the World's Most Climate Change Vulnerable Species: A Systematic Trait-Based Assessment of all Birds, Amphibians and Corals. *PLoS ONE* **2013**, *8*, e65427. [CrossRef] [PubMed]

36. Araujo, M.B.; Cabeza, M.; Thuiller, W.; Hannah, L.; Williams, P.H. Would climate change drive species out of reserves? An assessment of existing reserve-selection methods. *Glob. Chang. Biol.* **2004**, *10*, 1618–1626. [CrossRef]

37. Hannah, L.; Midgley, G.; Hughes, G.; Bomhard, B. The View from the Cape: Extinction Risk, Protected Areas, and Climate Change. *Bioscience* **2005**, *55*, 231–242. [CrossRef]

38. Hewson, J.; Ashkenazi, E.; Andelman, S.; Steininger, M. Projected impacts of climate change on protected areas. *Biodiversity* **2008**, *9*, 100–105. [CrossRef]

39. Garcia, R.A.; Cabeza, M.; Rahbek, C.; Araújo, M.B. Change and Their Implications for Biodiversity Multiple Dimensions of Climate. *Science* **2014**, *344*, 1247579. [CrossRef] [PubMed]

40. Williams, J.W.; Jackson, S.T.; Kutzbach, J.E. Projected distributions of novel and disappearing climates by 2100 AD. *Proc. Natl. Acad. Sci. USA* **2007**, *104*, 5738–5742. [CrossRef] [PubMed]

41. Ordonez, A.; Williams, J.W.; Svenning, J. Mapping climatic mechanisms likely to favour the emergence of novel communities. *Nat. Clim. Chang.* **2016**, *6*, 1104–1109. [CrossRef]

42. Radeloff, V.C.; Williams, J.W.; Bateman, B.L.; Burke, K.D.; Carter, S.K.; Childress, E.S.; Cromwell, K.J.; Gratton, C.; Hasley, A.O.; Kraemer, B.M.; et al. The rise of novelty in ecosystems. *Ecol. Appl.* **2015**, *25*, 2051–2068. [CrossRef] [PubMed]

43. Jetz, W.; Wilcove, D.S.; Dobson, A.P. Projected Impacts of Climate and Land-Use Change on the Global Diversity of Birds. *PLoS Biol.* **2007**, *5*, e157. [CrossRef] [PubMed]

44. Hansen, A.; Neilson, R.; Dale, V.; Flather, C.; Iverson, L.; Currie, D.J.; Shafer, S.; Cook, R.; Bartlein, P.J. Global change in forests: Responses of species, communities, and biomes. *Bioscience* **2001**, *51*, 765–779. [CrossRef]

45. Dale, V.H. The Relationship between Land-Use Change and Climate Change. *Ecol. Appl.* **1997**, *7*, 753–769. [CrossRef]

46. Asner, G.P.; Loarie, S.R.; Heyder, U. Combined effects of climate and land-use change on the future of humid tropical forests. *Conserv. Lett.* **2010**, *3*, 395–403. [CrossRef]

47. Lawler, J.J.; Ruesch, A.S.; Olden, J.D.; McRae, B.H. Projected climate-driven faunal movement routes. *Ecol. Lett.* **2013**, *16*, 1014–1022. [CrossRef] [PubMed]

48. M.E.A. Current State & trends assessment. In *Millennium Ecosystem Assessment*; Island Press: Washington, DC, USA, 2005.

49. Fischlin, A.; Midgley, G.F.; Price, J.T.; Leemans, R.; Gopal, B.; Turley, C.; Rounsevell, M.; Dube, P.; Tarazona, J.; Velichko, A. Ecosystems, their properties, goods, and services. In *Climate Change 2007: Impacts, Adaptation and Vulnerability Contribution of Working Group II to the Fourth Assessment Report of the Intergovernmental Panel on Climate Change*; Parry, M., Canziani, O., Palutikof, J., van der Linden, P., Hanson, C., Eds.; Cambridge University Press: Cambridge, UK, 2007; pp. 211–272.

50. Hoffmann, M.; Hilton-taylor, C.; Angulo, A.; Böhm, M.; Brooks, T.M.; Butchart, S.H.M.; Carpenter, K.E.; Chanson, J.; Collen, B.; Cox, N.A.; et al. The Impact of Conservation on the Status of the World's Vertebrates. *Science* **2010**, *330*, 1503–1509. [CrossRef] [PubMed]

51. IUCN; UNEP-WCMC. The World Database on Protected Areas (WDPA). Available online: www. protectedplanet.net (accessed on 31 July 2017).

52. Olson, D.M.; Dinerstein, E.; Wikramanayake, E.D.; Burgess, N.D.; Powell, G.V.N.; Underwood, E.C.; D'amico, J.O.; Itoua, I.; Strand, H.E.; Morrison, J.C.; et al. Terrestrial Ecoregions of the World: A New Map of Life on Earth. *Bioscience* **2001**, *51*, 933–938. [CrossRef]

53. Hewson, J.; Crema, S.; González-Roglich, M.; Tabor, K.; Harvey, C. Projecting global and continental forest loss: New high-resolution datasets of deforestation risk. *Environ. Res. Lett.* submitted.

54. FAO. *FRA 2015 Terms and Definitions*; Forest Resources Assessment Working Paper 180; FAO: Rome, Italy, 2015.

55. NGA. *Vector Map Level 0 (Digital Chart of the World)*; National Geospatial-Intelligence Agency: Springfield, VA, USA, 2000.

56. Arino, O.; Ramos Perez, J.J.; Kalogirou, V.; Bontemps, S.; Defourny, P.; Van Bogaert, E. Global Land Cover Map for 2009 (GlobCover 2009). PANGAEA: European Space Agency (ESA) & Université catholique de Louvain (UCL): 2012. Available online: https://doi.org/10.1594/PANGAEA.787668 (accessed on 27 July 2018).

57. USGS; NGA. *Global Multi-Resolution Terrain Elevation Data 2010 (GMTED2010)*; U.S. Geological Survey: Reston, VA, USA, 2011.

58. Avitabile, V.; Herold, M.; Heuvelink, G.B.M.; Lewis, S.L.; Phillips, O.L.; Asner, G.P.; Armston, J.; Ashton, P.S.; Banin, L.; Bayol, N.; et al. An integrated pan-tropical biomass map using multiple reference datasets. *Glob. Chang. Biol.* **2016**, *22*, 1406–1420. [CrossRef] [PubMed]

59. WCS; CIESIN. *Last of the Wild Project, Version 2*; Global Hum; NASA SEDAC: Palisades, NY, USA, 2005.

60. Zabel, F.; Putzenlechner, B.; Mauser, W. Global agricultural land resources—A high resolution suitability evaluation and its perspectives until 2100 under climate change conditions. *PLoS ONE* **2014**, *9*, e107522. [CrossRef] [PubMed]

61. Siebert, S.; Henrich, V.; Frenken, K.; Burke, J. *Global Map of Irrigation Areas Version 5*; Rheinische Friedrich-Wilhelms-University: Bonn, Germany; Food and Agriculture Organization of the United Nations: Rome, Italy, 2013.

62. Naidoo, R.T.I. Global-scale mapping of economic benefits from agricultural lands: Implications for conservation priorities. *Biol. Conserv.* **2007**, *140*, 40–49. [CrossRef]

63. Hijmans, R.J.; Cameron, S.E.; Parra, J.L.; Jones, P.G.; Jarvis, A. Very high resolution interpolated climate surfaces for global land areas. *Int. J. Climatol.* **2005**, *1978*, 1965–1978. [CrossRef]

64. Tabor, K.; Williams, J.W. Globally downscaled climate projections for assessing the conservation impacts of climate change. *Ecol Appl.* **2010**, *20*, 554–565. [CrossRef] [PubMed]

65. Moss, R.H.; Edmonds, J.A.; Hibbard, K.A.; Manning, M.R.; Rose, S.K.; Van Vuuren, D.P.; Carter, T.R.; Emori, S.; Kainuma, M.; Kram, T.; et al. The next generation of scenarios for climate change research and assessment. *Nature* **2010**, *463*, 747–756. [CrossRef] [PubMed]

66. Mahony, C.R.; Cannon, A.J.; Wang, T.; Aitken, S.N. A closer look at novel climates: New methods and insights at continental to landscape scales. *Glob. Chang. Biol.* **2017**, *23*, 3934–3955. [CrossRef] [PubMed]

67. Nepstad, D.; McGrath, D.; Stickler, C.; Alencar, A.; Azevedo, A.; Swette, B.; Bezerra, T.; DiGiano, M.; Shimada, J.; da Motta, R.S.; et al. Slowing Amazon deforestation through public policy and interventions in beef and soy supply chains. *Science* **2014**, *344*, 1118–1123. [CrossRef] [PubMed]

68. Austin, K.; González-Roglich, M.; Schaffer-Smith, D.; Schwantes, A.; Swenson, J. Trends in size of tropical deforestation events signal increasing dominance of industrial-scale drivers. *Environ. Res. Lett.* **2017**, *12*, 054009. [CrossRef]

69. Bruner, A.G.; Gullison, R.E.; Balmford, A. Financial Costs and Shortfalls of Managing and Expanding Protected-Area Systems in Developing Countries. *Bioscience* **2004**, *54*, 1119–1126. [CrossRef]

70. Jones, K.R.; Venter, O.; Fuller, R.A.; Allan, J.R.; Maxwell, S.L.; Negret, P.J.; Watson, J.E.M. One-third of global protected land is under intense human pressure. *Science* **2018**, *360*, 788–791. [CrossRef] [PubMed]

71. Angelsen, A.; Brown, S.; Loisel, C.; Peskett, L.; Streck, C.; Zarin, D. Reducing Emissions from Deforestation and Forest Degradation (REDD): An Options Assessment Report. 2009. Available online: http://www.redd-oar.org/links/REDD-OAR_en.pdf (accessed on 27 July 2018).

72. Schaafsma, M.; Morse-Jones, S.; Posen, P.; Swetnam, R.D.; Balmford, A.; Bateman, I.J.; Burgess, N.D.; Chamshama, S.A.O.; Fisher, B.; Freeman, T.; et al. The importance of local forest benefits: Economic valuation of Non-Timber Forest Products in the Eastern Arc Mountains in Tanzania. *Glob. Environ. Chang. Policy Dimens.* **2014**, *24*, 295–305. [CrossRef]

73. Johnstone, J.F.; Allen, C.D.; Franklin, J.F.; Frelich, L.E.; Harvey, B.J.; Higuera, P.E.; Mack, M.C.; Meentemeyer, R.K.; Metz, M.R.; Perry, G.L.W.; et al. Changing disturbance regimes, ecological memory, and forest resilience. *Front. Ecol. Environ.* **2016**, *14*, 369–378. [CrossRef]

74. Pratihast, A.K.; DeVries, B.; Avitabile, V.; de Bruin, S.; Herold, M.; Bergsma, A. Design and Implementation of an Interactive Web-Based Near Real-Time Forest Monitoring System. *PLoS ONE* **2016**, *11*, e0150935. [CrossRef] [PubMed]

75. Davies, D.; Murphy, K.; Michael, K.; Becker-Reshef, I.; Justice, C.; Boller, R.; Braun, S.A.; Schmaltz, J.E.; Wong, M.M.; Pasch, A.N.; et al. The use of NASA LANCE imagery and data for near real-time applications. In *Time-Sensitive Remote Sensing*; Springer: New York, NY, USA, 2015; pp. 165–182.

76. Petrica, L. An evaluation of low-power microphone array sound source localization for deforestation detection. *Appl. Acoust.* **2016**, *1*, 162–169. [CrossRef]

77. Musinsky, J.; Tabor, K.; Cano, C.A.; Ledezma, J.C.; Mendoza, E.; Rasolohery, A.; Sajudin, E.R. Conservation impacts of a near real-time forest monitoring and alert system for the tropics. *Remote Sens. Ecol. Conserv.* **2018**, 1–8. [CrossRef]

78. Mora, C.; Frazier, A.G.; Longman, R.J.; Dacks, R.S.; Walton, M.M.; Tong, E.J.; Sanchez, J.J.; Kaiser, L.R.; Stender, Y.O.; Anderson, J.M.; et al. The projected timing of climate departure from recent variability. *Nature* **2013**, *502*, 183–187. [CrossRef] [PubMed]

79. Wiens, J.A.; Seavy, N.E.; Jongsomjit, D. Protected areas in climate space: What will the future bring? *Biol. Conserv.* **2011**, *144*, 2119–2125. [CrossRef]

80. Asner, G.P.; Tupayachi, R. Accelerated losses of protected forests from gold mining in the Peruvian Amazon. *Environ. Res. Lett.* **2017**, *12*, 094004. [CrossRef]

81. Beaudrot, L.; Ahumada, J.A.; Brien, T.O.; Alvarez-loayza, P.; Boekee, K.; Campos-arceiz, A.; Eichberg, D.; Espinosa, S.; Fegraus, E.; Fletcher, C.; et al. Standardized Assessment of Biodiversity Trends in Tropical Forest Protected Areas: The End is not in Sight. *PLoS Biol.* **2016**, *14*, e1002357. [CrossRef] [PubMed]

land

MDPI

Article

Post-War Land Cover Changes and Fragmentation in Halgurd Sakran National Park (HSNP), Kurdistan Region of Iraq

Rahel Hamad [1,*], Kamal Kolo [2] and Heiko Balzter [3,4]

1   Faculty of Science, Petroleum Geosciences Department, Soran University, Delzyan Campus, Soran 44008, Erbil, Iraq
2   Scientific Research Centre (SRC), Soran University, Delzyan Campus, Soran 44008, Erbil, Iraq; kamal.kolo@soran.edu.iq
3   Centre for Landscape and Climate Research (CLCR), Department of Geography, University of Leicester, University Road, Leicester LE1 7RH, UK; hb91@leicester.ac.uk
4   National Centre for Earth Observation, University of Leicester, University Road, Leicester LE1 7RH, UK
*   Correspondence: rahel.hamad@soran.edu.iq; Tel.: +964-750-413-2550

Received: 8 February 2018; Accepted: 13 March 2018; Published: 19 March 2018

Abstract: Context: The fundamental driving force of land use and land cover (LULC) change is related to spatial and temporal processes caused by human activities such as agricultural expansion and demographic change. Landscape metrics were used to analyze post-war changes in a rural mountain landscape, the protected area of Halgurd-Sakran National Park (HSNP) in north-east Iraq. Therefore, the present work attempts to identify the temporal trends of the most fragmented land cover types between two parts of the national park. **Objectives:** The objectives of this study are to compare two land cover classification algorithms, maximum likelihood classification (MLC) and random forest (RF) in the upper and lower parts of HSCZ, and to examine whether landscape configuration in the park has changed over time by comparing the fragmentation, connectivity and diversity of LULC classes. **Methods:** Two Landsat images were used to analyze LULC fragmentation and loss of habitat connectivity (before and after the Fall of Baghdad in 2003). Seven landscape pattern metrics, percentage of land (PLAND), number of patch (NP), largest patch index (LPI), mean patch size (MPS), euclidian nearest neighborhood distance (ENN_AM), interspersion and juxtaposition (IJI) and cohesion at class level were selected to assess landscape composition and configuration. **Results:** A significant change in LULC classes was noticed in the lower part of the park, especially for pasture, cultivated and forest-lands. The fragmentation trends and their changes were observed in both parts of the park, however, more were observed in the lower part. The inherent causes of these changes are the socio-economic factors created by the 1991–2003 UN post-war economic sanctions. The changes increased during sanctions and decreased afterwards. The fall of Baghdad in 2003, followed by rapid economic boom, marked the greatest cause in land use change, especially in changes-susceptible cultivated areas. **Conclusions:** Shrinkage of forest patches in the lower part of the park increases the distance between them, which contributes to a decline in biological diversity from decreasing habitat area. Lastly, the results confirm the applicability of the combined method of remote sensing and landscape metrics.

Keywords: Halgurd-Sakran National Park; remote sensing; GIS; landscape metrics; fragmentation

1. Introduction

Land is an essential and limited natural resource [1]. Utilizations of landscapes by people often result in a modification of land cover through changing human land use [2–4]. The ecological area or natural environment in which a particular animal or plant species usually lives is described as their habitat [5]. Land use change implies the qualification and quantification change of land cover type "as habitat for organisms and productive land for humans" [6]. Moreover, a decrease in the size of habitat fragments is a significant indicator of habitat fragmentation [7].

Land use/land cover changes are influenced by a variety of factors and they can be categorized into direct (proximate) and indirect (underlying, root) causes. How humans modify land cover represents the direct change, while the interaction of social, economic, political, demographic and culture recognized as indirect or root causes. Moreover, indirect driving forces can be defined as socio-economic drivers, which comprise of socio-political, socio-economic and cultural factors. Socio-economic-political factors drive people to degrade the natural environment [8–10]. The agricultural regions are good indicators for changing in land use and land cover by the conflict and geopolitical forces. Moreover, the year 2003 is the turning point from the poverty to richness and from the dictatorship to the democracy and it also represents the period of the end of sanctions in this study [11].

Landscape fragmentation is a dynamic process and has a large impact on biodiversity as a result of changes of habitats in a landscape through time [12,13]. Landscape fragmentation has several causes such as anthropogenic activities, urban development, natural causes, agricultures, illegal logging, and forest cutting. All these factors alter the landscape structure with its components [14–16]. Biological diversity in a landscape generally declines as fragmentation progresses, when the species extinction rate increases as a result of reducing habitat area and habitat connectivity, thus the species immigration rate will decrease [17].

Landscape connectivity has two components: structural and functional. Structural connectivity is associated with the physical characteristics of the landscape, whereas functional connectivity is related to the movement of the organisms in space and time within a landscape. Studying temporal function connectivity infers the persistence of organisms in time, in the same place, while spatial function connectivity relates to the spatial pattern [18,19]. Habitat corridors are important for species movement and protection. Their management and planning are significant for protecting green belts that can have a major recreation function, especially green corridors along roads [20]. Reductions in habitat areas and connectivity disturb the biodiversity (flora and fauna) [17]. The purpose of the habitat corridor connectivity to allow organisms in a protected area of land is to migrate from one land area to another. This type of conservation is considered when there is a loss of connectivity in habitat land cover types in a landscape usually follow an uneven distribution and are spatially heterogeneous.

The compositional heterogeneity of spatial units in the landscape refers to their number, type, and abundance, whereas the mosaic of spatial arrangements of those units is defined as configurational heterogeneity [20,21].

To gather information on the landscape fragmentation in an area of interest, remotely sensed (RS) images are a versatile tool [22]. Through analyzing landscape patterns, the interaction between anthropogenic factors and the environment can be better understood. All kinds of social and environmental factors change the Earth's surface in different spatial and temporal scales by influencing main ecological functions [23]. Thus, there is a relationship between LULC and the fragmentation of landscapes [24].

Classification is widely used to derive thematic information from (RS) [25]. Remote sensing has been widely used for mapping land cover at a variety of spatio-temporal scales [26]. The categorization of pixels in an image into land cover classes is the general objective of a classification of multispectral data based on the similarity of the spectral signature of a pixel to a set of training signatures (supervised) or based on the data themselves in an unsupervised classification [27]. Traditional image classification techniques include pixel-based (e.g., supervised Maximum Likelihood Classification (MLC) and unsupervised e.g., K-means or ISODATA), sub-pixel-based (e.g., Fuzzy, neural networks regression modeling, etc.), and object-based techniques (e.g., segmentation) [28]. Unsupervised classification does not require training data, unlike supervised classification [27]. In object-based classification, an object starts with the combination of neighboring pixels into similar areas or depends on groups of pixels that represent the shapes and sizes, while in pixel-based classification method each pixel has a class that bases exclusively on the digital number of the pixel itself [29–31]. In contrast to traditional classification methods newer non-parametric algorithms such as random forest (RF) have been developed, which is least sensitive to the parameters and applies multiple decision trees to a set of training data [32]. The RF algorithm [33] is an ensemble classifier that consists of many decision trees that does not overfit. RF can be used for land cover mapping [34] and it often gives better land cover classification accuracies [35,36]. Moreover, RF with its multiple advantages such as handling large numbers of input variables and calculating an error matrix, giving estimates of the importance of the features in the classification fast and robust [32].

Significant human interference with the Earth's surface affects the observed transformations in LULC [37]. In the last few decades, in many areas around the world, the rural land use have changed rapidly [38]. Industrialisation, urbanization, population growth, and economic reforms are major driving forces contributing to land-use change [39]. A rapid socio-economic change can induce LULC change such as farmland abandonment [40]. The complex interrelationship between socio-economic factors and LULC can be further compounded when global climate change is added as a factor [41].

Wars and political conflicts could have reversible or irreversible socio-economic damaging effects on agriculture, land use and land cover. The deliberate desiccation of marshlands in southern Iraq during the 1980–1988 Iraq-Iran war resulted in disastrous and changes to the land cover and use, ecosystem services and anthropogenic activities [42]. The United Nations Security Council enforced a comprehensive set of sanctions on Iraq following the latter's military invasion of Kuwait in August 1990. Consequently, the period 1990–2003 in Iraq generally and in Kurdistan region demonstrated a period of economic war in the form of comprehensive sanctions [11], which had, in addition to the war direct consequences, devastating long-lasting results on population and economy.

The ability to acquire repeated satellite images is useful for monitoring and assessment of natural resources change occurring over time in a specific place. Thus, the use of RS and GIS is a task to address land use and landscape challenges in a wide range of resource management problems. Remote sensing data offers a valuable multi-temporal data on processes and phenomena that act over larger areas. For example, it poses a major data source specifically, for assessing protected areas [43], monitoring biodiversity [44], promoting sustainable urbanization in Greater Dhaka, Bangladesh [45], land use and land cover change detection in the western Nile data of Egypt [46] and Kodaikanal taluk, Tamil nadu [47], impact of land use change on risk erosion and sedimentation in a mixed land use watershed in the Ozark Highlands of the USA [48], applying of RS/GIS to forest fire risk mapping in the Mediterranean coast of Spain [49], using satellite images of studying forest resources in three decades of research development by [50], and monitoring forest cover change and forest degradation using remote sensing and landscape metrics of Nyungwe-Kibira park in Rwanda and Burundi [51].

Halgurd-Sakran National Park (HSNP) belongs to Choman district and consists of three zones; core zone, outer zone and additional outer zone. Our study focused on the core zone (CZ) (Figure 1) [36]. Due to its remote location, mountainous terrain, scant and limited transportation to the districts and sub-districts, there was negligible economic development in HSCZ before the Fall of Baghdad in April 2003. However, because of the attractive natural landscapes and the accelerated general economic growth in Kurdistan after 2003, the HSNP has been selected on the top list of tourism destinations for travelers in Kurdistan region-Erbil governorate since 2008 [36].

The aim of this research is to identify the temporal trends of the most fragmented land cover types between the upper and lower parts of the national park. The comparison of two land cover classification algorithms, MLC and RF in the upper and lower parts of HSCZ was the first specific objective in this study. Whereas, the second specific objective was to examine whether landscape configuration in the park has changed over time, by comparing the fragmentation, connectivity and diversity of LULC classes.

2. Materials and Methods

2.1. Study Area

HSNP is located in the northeast of Erbil-Iraq and shares a boundary with Iran along the Zagros Mountain Range (Figure 1). The climate of the Kurdistan region is very hot and dry in summer and cold and wet in winter. The Kurdish Regional Government (KRG) officially designated this area of interest a protected area since 2008. Significant socio-economic improvements such as urbanization took place after the Fall of Baghdad in 2003 in the area of interest [52,53]. Table 1 presents the population for villages around Choman district, which has grown smoothly over the last six years [54].

Table 1. Rural population for villages around Choman district [38].

Rural Rate	2010	2011	2012	2013	2014	2015
Choman Villages	12,008	12,348	12,689	13,037	13,389	13,746

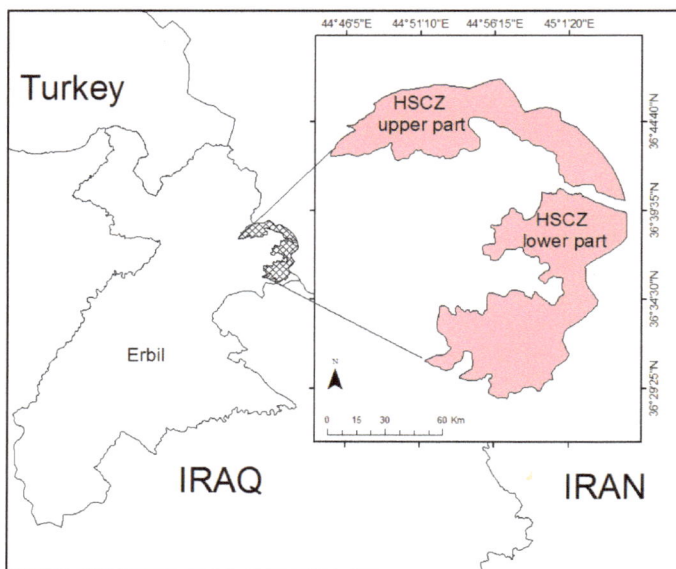

Figure 1. Location map of the study area (HSCZ) in the Kurdistan Region of Iraq.

2.2. Remote Sensing Data

Two satellite images from Landsat 7 Enhanced Thematic Mapper plus (ETM+) for 1998 and Landsat 8 Landsat Data Continuity Mission (LDCM) for 2015 were used in this study (Table 2). Six multispectral bands, excluding the thermal band, were used for Landsat 7 and eight multispectral bands for Landsat 8. The two images were projected into Universal Transverse Mercator (UTM) zone 38 N, and then assembled in a Geographical Information System (GIS). Landsat images were processed and subset to include the entire area of HSCZ. The downloaded images for the study area were at processing level L1T, which means it is geometrically corrected and has a sufficient level. Then, the calibrated radiance values were atmospherically corrected using FLAASH in ENVI 5.3. Landsat images have been used and subset it in ENVI then exported to ArcMap 10.3. Images from 1998 and 2015 were categorized into four land cover classes; namely, bare surface, pasture, cultivated area, and forest. 160 training sites, 40 points for each land cover class (bare surface, pasture, cultivated land, and forest), were chosen for the upper and lower parts to be representative. The selected training sites were principally based on high-resolution imagery from Google Earth, ESRI ArcMap base map, and expert knowledge from the Agricultural Department of the District of Choman. Ancillary data such as slope, aspect, elevation, and NDVI were used in image classification to increase the accuracy and discrimination in land use/land cover mapping [55].

Table 2. Satellite (sub-scene) images used in this study to quantify landscape patterns and land cover change.

Satellite Sensor	Path/Row	Acquisition Date	Resolution	Band Nos.
Landsat 7 ETM+	169/035	13 September 1998	30 m	1, 2, 3, 4, 5, 7
Landsat 8 LDCM	169/035	24 August 2015	30 m	1, 2, 3, 4, 5, 7, 8, 9

Maximum Likelihood Classification is one of the most commonly used algorithms [56], which has been broadly used for the LULC classification worldwide [57]. Pixel-based MLC algorithm relies on a statistical model [56]. The MLC algorithm is categorizing objects into classes with the highest probability that a pixel belongs to a particular class [29–31]. Furthermore, it assumes that the image data for each class is normally distributed according to their spatial and spectral characteristics [56]. During the classification procedure in ArcMap 10.3 the vector data integrates with the satellite image for applying pixel-based method.

On the other hand, the RF classification ensemble method is building a set of classifiers and develops lots of decision trees based on random selection of data and random selection of variables [58], then after taking a vote of their accurate and stable predictions [56]. Random forest splits each node using the best among a subset of predictors randomly chosen at that node. Thus, this procedure produces a wide variety, which usually lead to a better model [58]. Furthermore, the number of variables (M) and the number of trees (T) are two parameters in the forest. Increasing the number of variables lead to increasing the correction between the trees and the classification accuracy of an individual tree in the forest [59]. "As *m* increases correlation and individual tree accuracy also increase and some optimal *m* will give the lowest error rate". In addition, the variable importance and internal structure are also other two additional measures in the random forest. The importance of the predictor features measure by variable importance, which is done by out of bagging (OOB) [60]. In order to estimate the importance of a certain feature, the out of bagging samples are evaluated by computing the accuracy throughout the trees and by counting the votes of each variable in the classification [59,61].

Moreover, building random forests model starts with using roughly two-thirds of the training samples and the rest (one-third) of training data (the out-of-bag samples) is used to estimate the error of the predictions. Horning (2010) showed the overcoming of the over-fitting problem for the training data by using "pruning" in decision trees. This process offers the lowest error by removing terminal nodes. In the decision trees the growing of tree is too large with the terminal nodes. Reducing these nodes will simplify the tree. We have implemented RF using a software package in R language and analyzed Landsat images.

The classification was carried out by making a comparison between MLC and RF. Each class has been prepared for spatial analysis metrics in FRAGSTATS, and then independently validated for their overall accuracy and kappa index (κ). Each class was prepared for spatial metrics analysis in FRAGSTATS 4.2.

2.3. Analyzing Changes for Fragmentation, Connectivity and Diversity

Landscape metrics following McGarigal and Marks (1995) [7], and McGarigal and Cushman (2002) [62] were grouped in four types: fragmentation, isolation, heterogeneity, and connectivity.

FRAGSTATS 4.2 was used to extract the landscape metrics from each land use map of 1998 and 2015. Seven landscape metrics at class level were calculated for analyzing fragmentation, connectivity, and diversity (heterogeneity), (Table 3) [62].

Table 3. Metrics selected at class level for the quantification of landscape patterns in HSCZ.

Metrics *	Units	Description
Percentage of land (PLAND)	%	Amount of the landscape occupied by certain LULC class [7]
Number of patches (NP)	n	Number of patches per class [7]
Largest patch index (LPI)	%	Percentage of landscape accounted for by largest patch [20]
Mean patch size (MPS)	ha	Mean area of patches of the same LULC class [62]
Euclidian Nearest Neighborhood	m	The measure of patch context to quantify patch isolation [62]
		Distance (ENN_AM)
Interspersion and juxtaposition (IJI)	%	Degree of intermixing of class patch types [7]
COHESION (Cohesion)	%	The physical connectedness of the corresponding class patch [62]

* Definitions and equations for calculation of the metrics are provided by McGarigal and Marks (1995), and McGarigal and Cushman (2002).

Metrics that were used for fragmentation were PLAND, NP, LPI, and MPS, while the ENN_AM was used to interpret the isolation of land cover patches. Concerning heterogeneity, the interspersion and juxtaposition was used, and cohesion was applied for connectivity. For this research, mean patch size and number of patches were used to summarize the fragmentation process of LULC categories at class level.

3. Results

3.1. Accuracy Assessment and the Performance Comparison of Two Algorithms

The comparisons of accuracy assessments between two classification algorithms are shown in Table 4. The accuracy assessment of ML classified images shows an overall accuracy for the 1998 image in the upper part of 66.2% with κ = 0.55; and overall accuracy for 2015 of 81% with κ = 0.74. The results for the RF classification in the upper part gave an overall accuracy for the 1998 image of 98% and κ = 0.97 and for 2015 of 98% and κ = 0.97.

Table 4. Comparison between the accuracies of the two classification algorithms assessed by independent validation data in the upper and lower parts for 1998 and 2015 in HSCZ.

Land-Cover Class	MLC		RF	
	1998-Upper Part	**2015-Upper Part**	**1998-Upper Part**	**2015-Upper Part**
Overall accuracy (%)	66.20	81	98	98
κ	0.55	0.74	0.97	0.97
	1998-Lower Part	**2015-Lower Part**	**1998-Lower Part**	**2015-Lower Part**
Overall accuracy (%)	57	84	99	99
κ	0.67	0.79	0.98	0.99

Concerning the lower part of HSCZ the MLC accuracy assessment for image 1998, as presented in Table 5, has 57% overall accuracy and κ = 0.67. The RF classification for 1998 showed 99% overall accuracy and κ = 0.98 and for 2015 the overall accuracy was 99% with κ = 0.99. The classification results for the HSCZ revealed that the RF algorithm was more successful and gives higher classification accuracies than MLC.

3.2. Land Use/Land Cover Changes Based on Random Forest Classification

3.2.1. Land Cover Class Distribution in the Upper and Lower Parts of Core Zone

Figure 2 represents the study area that was categorized into four different classes (bare surface, pasture, cultivated land, and forest) using random forest classification [36]. In general, the lower part between the two time periods showed greater changes between LULC classes. Figure 3 displays the changes that occurred among land use land cover classes from 1998 to 2015 using the Land Change Modeller in TerrSet.

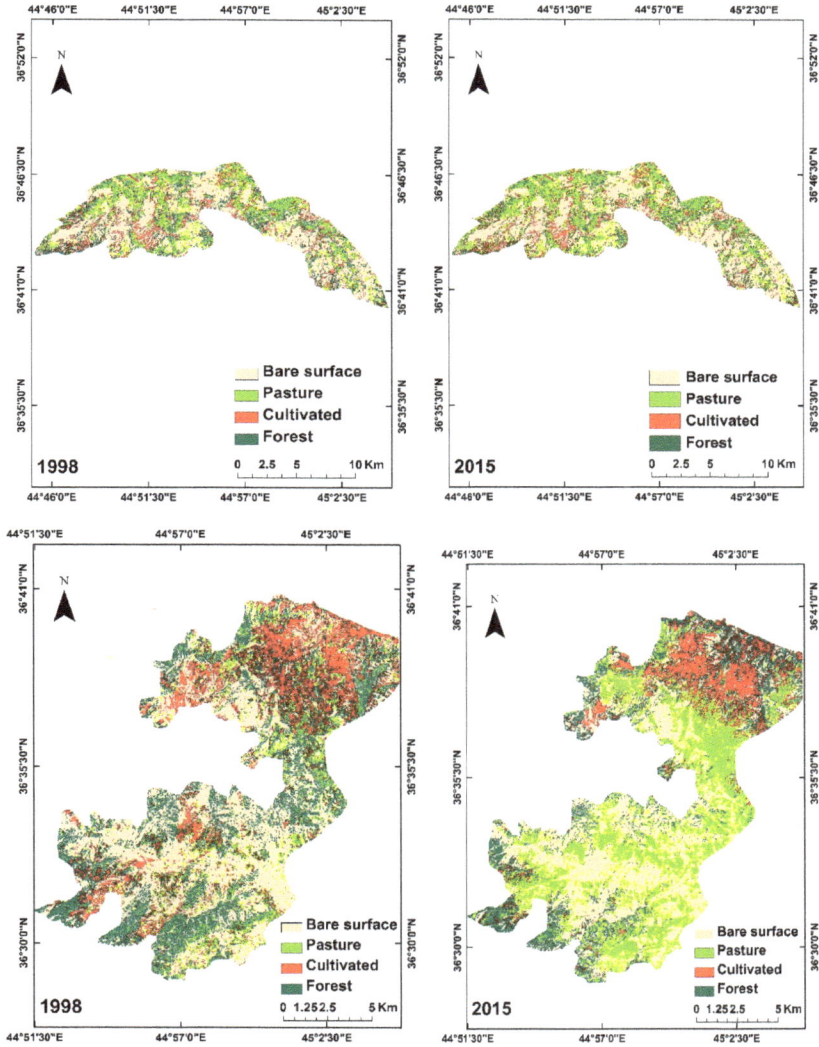

Figure 2. The classified land use and land cover (LULC) maps for upper and lower parts of HSCZ for years 1998 and 2015 from Random Forest classifier with Landsat data, which compare the changes in two parts over time.

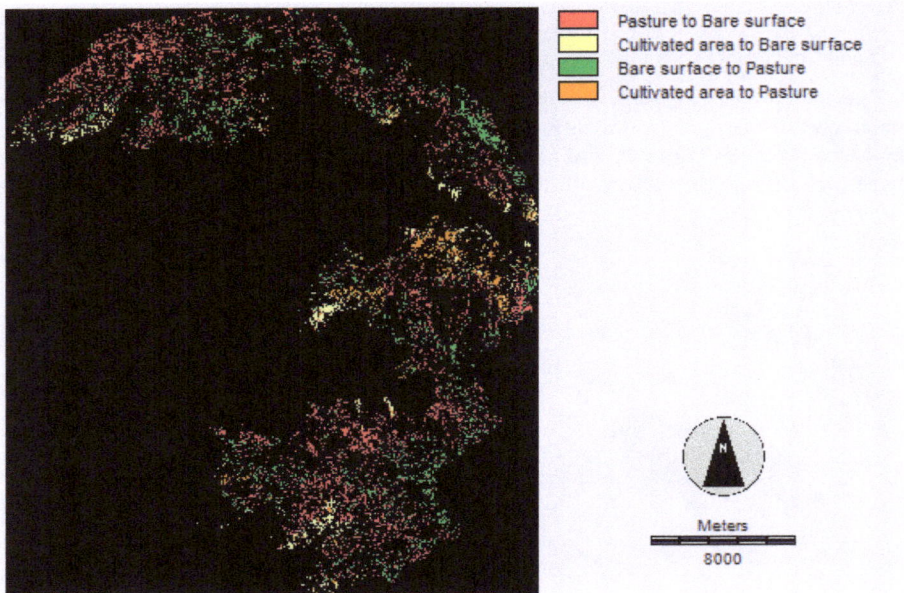

Figure 3. LULC change by category in two time spans (1998–2015) in the upper and lower parts of HSCZ.

3.2.2. Land Use/Land Cover Dynamics

From 1998 to 2015, increase in pasture area was relatively significant as compared to the other LULC classes within the 17-year period of study in the lower part of the park, and this resulted in the increase of its percentage of land (PLAND) (Table 5).

Table 5. Metrics comparison and changes in PLAND at land cover class level for 1998 and 2015 in the upper part (up) and lower part (lp) of HSCZ.

LULC Class	Up-1998	Up-2015	Lp-1998	Lp-2015
Bare surface	51.84	51.49	44.89	43.80
Pasture	22.43	23.47	12.21	30.76
Cultivated area	10.19	8.80	15.95	9.80
Forest	15.52	16.14	26.93	15.76

In the upper part of the park, the barren land demonstrates a lesser decrease in IJI in 2015, which reflects the tendency towards uneven distribution. However, the patch numbers increased significantly for cultivated area with a slight increase for the bare surface, while pasture and forest lands decreased slightly over time Figure 4. The tendency towards uneven distribution (IJI) of barren land can be observed in Figure 5.

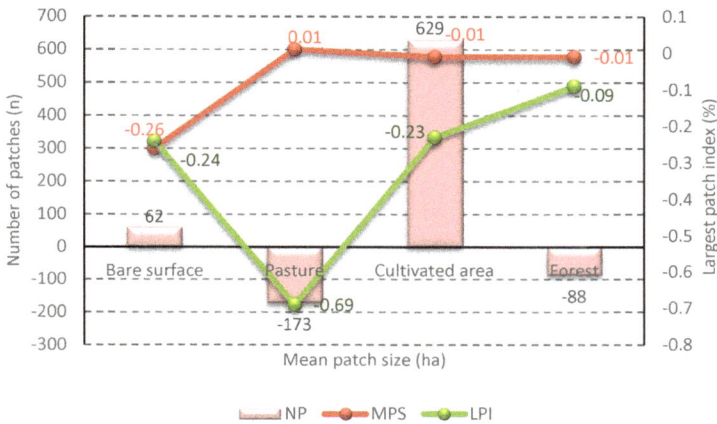

Figure 4. Gains and losses of patch number (NP), mean patch size (MPS) and largest patch index (LPI) of LULC by category in different time periods (1998–2015) in the upper part of the HSCZ.

Likewise, Figures 4 and 5 indicate trend towards more fragmentation, a small tendency to greater isolation and the heterogeneity and more physical connection of the pasture class in 2015. Thus, this analysis shows a slight trend towards increasing fragmentation of pasture land. Whereas, the cultivated area was more isolated, more evenly distributed and more physically connected in 2015 compared to 1998. This change can be related to the reduction of farming in the study area as a result of socio-economic growth. Therefore, the patches were not affected by fragmentation in the upper part of the park. Concerning the forest land, a slight decrease of the ENN_AM, very small changes in LPI and MPS were observed. Thus, the degree of isolation and spatial distribution of forest patches were reduced with less uniform distribution of forest patches from 1998 to 2015. Regardless of the slight increase of NP, LPI and MPS, with a slight decrease of the ENN_AM and cohesion, a fragmentation process of forest land was ongoing in the time period in the upper part of HSCZ.

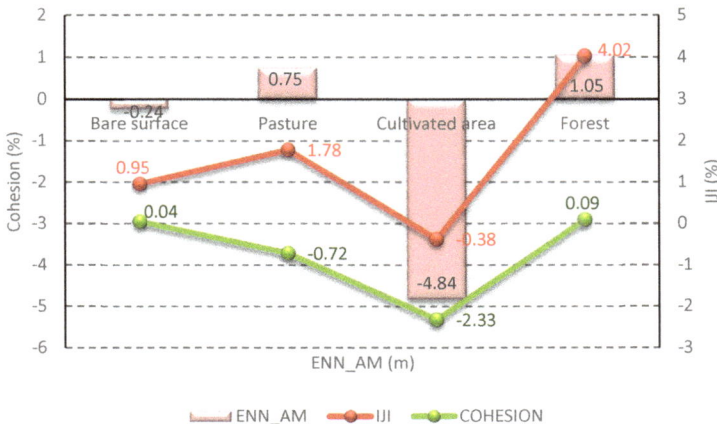

Figure 5. Gains and losses of ENN_AM, IJI and Cohesion of LULC by category in different time periods (1998–2015) in the upper part of the HSCZ.

In the lower part of the park, there was a tendency for barren land to fragment more over this time period with the trend to isolation over time. Figure 6 represents that patch types that were uniformly adjacent to each other and were more clumped in distribution, while the reduction of IJI in 2015 indicates a more uneven distribution of bare cover patch sizes.

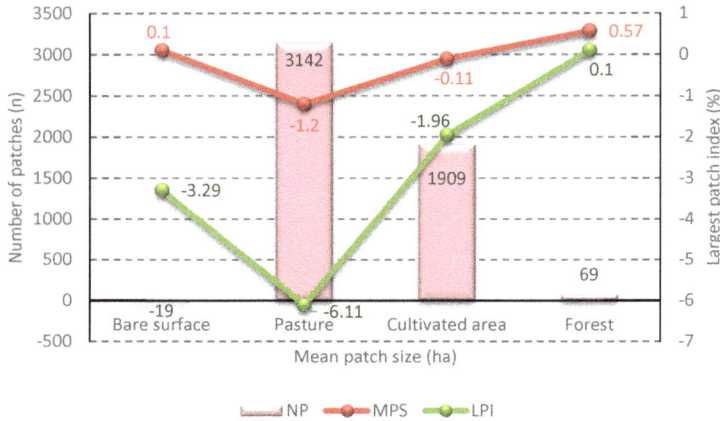

Figure 6. Gains and losses of NP, MPS and LPI of LULC by category in different time periods (1998–2015) in the lower part of the HSCZ.

In general, in the lower part of the park the fragmentation was continuing for barren land by the increase of NP which was also confirmed by a decrease of MPS. Thus, the upper and lower parts are behaving differently concerning the barren land; this could be that the upper part contains fewer villages, higher mountain ranges, and inaccessibility to natural resources than the lower part of the park. A huge reduction was observed for pasture patch number (NP) with a slight increase of LPI and MPS over time (Figure 6). These changes indicate a trend of pasture land becoming more clumped and gradually reduced the distance of pasture patches to the nearest neighbor of the same class. Additionally, the pasture patches were more adjacent to each other in 1998 but were less clumped in distribution in 2015. Over time, a large increase in the cohesion indicating that the pasture land was less cohesive and less physically connected in 1998 in the lower part compared to the same class in 2015 (Figure 6). The decreasing and increasing of patch number (NP) and mean patch size (MPS) expresses the stability and physical connectivity of the pasture class in the lower part of the park over time.

On the other hand, the reduction in cultivated area is accompanied by a significant reduction of NP with an increase for both LPI and MPS (Figure 6). The mean distance ENN_ MN for patches of cultivated land increased similar to the upper part of the park. Patches became more isolated, the distances between patches increased and the numbers of cultivated land patches decreased. A slight increase for IJI value shows a trend towards more a uniform distribution of cultivated land patches. The patches were more connected in 2015 (cohesion) (Figure 7). This reduction of cultivated land in the lower part of the park reflects the socio-economic growth in the region especially after the Fall of Baghdad, April 2003. Concerning, the forest land the NP, LPI, and MPS of patches declined slightly (Figure 6). The increase in the mean distance (ENN_AM) between forest patches reflects the increasing isolation between forest patches. The decrease of IJI indicates changes towards a more uneven distribution among forest patches. Physical connection and aggregation for forest patches were reduced and cohesion slightly decreased, (Figure 7). Overall the lower part of the park was not affected by forest fragmentation, despite a noticeable overall forest loss.

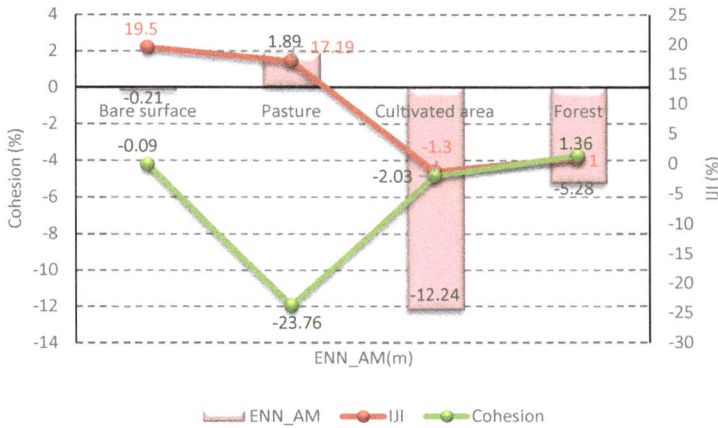

Figure 7. Gains and losses of ENN_AM, IJI and Cohesion of LULC by category in different time periods (199–2015) in the lower part of the HSCZ.

4. Discussion

4.1. Landscape Fragmentation at Class Level

The classification technique used to derive the LULC maps for 1998 and 2015 was random forest classification. One advantage of RF is that it is one of the most accurate learning algorithms available [32,33]. The overall LULC classification accuracy level in the upper and lower parts for two dates were very high for investigating the study area. On the other hand, the proportion of each LULC classes holds a significant information concerning the composition and configuration of landscape mosaic in the upper and lower parts of the park. Four main criteria: fragmentation, isolation, heterogeneity, and connectivity have been selected for measuring the landscape spatial pattern in HSCZ. Then, seven of the most appropriate landscape indices for representing each criterion were selected.

4.1.1. Bare Surface

This shrank of the bare surface size in both parts is accompanying with increasing the IJI. Noticeably, the distribution of adjacencies among bare surface patches trend to become more increasingly uneven in the lower part than the upper part of the park, as the IJI value higher in the upper part than IJI value in the lower part in 2015. This means that the disaggregation or the heterogeneity of the bare surface patches are higher in the lower part than the upper part across the landscape of the park. This variance is related to the NP in both parts. The NP decreased in the upper part in 2015, while the number of bare surface patches in the lower part of the park increased at the same time period. This suggesting that the fragmentation of the bare surface patches in the lower part are higher or a tendency of patches are directed to be fragmented more in the lower part, as a consequence could be of fire, excessive grazing or even of climate conditions. Thus, the fragmentation strengthened in this part of the park. Furthermore, higher ENN_AM values in the lower part indicate more isolated landscape patterns. Thus, the isolation between patches was more isolated in the lower part than the upper part. The scores of the cohesion are the same in both parts indicating the physically connected landscape spatial patterns in both parts are constant in 2015.

4.1.2. Pasture

The PLAND of the pasture in the lower part significantly increase with a slight decrease in the upper part in 2015. This can be related to economic boom which most of population changed their professional work and left the agricultural lands, as lower part contains more villages and population. The connectivity in the lower part improved better for pasture patches than the upper part, this witnessed by the cohesion values. This increase of cohesion value drives the landscape to more ecosystem sustainability. Moreover, the rate of decline is fairly sharp concerning the NP in the lower part, which made the patches of pasture distributed more regularly or clumped in this part of the park in 2015. Whereas, the IJI value in the upper part indicates more irregularly distributed as displayed a slight decrease values between 1998 and 2015. Thus, the pasture patches trend to the fragmentation in the upper part of the park in 2015 and at the same time they trend to accumulation in the lower part. The ENN_MN among patches decreased while the Cohesion increased in both parts of the park in 2015, suggesting that the connectivity of pasture patches had increased and pasture patches are less isolated. Many researchers observed how the increase of pasture lands provide more forage to animals, reduce erosion, filter runoff, absorb rainfall and improve production [63,64].

4.1.3. Cultivated Area

The period between 1998 and 2015 represents the economic boom period in the study area especially, after the Fall of Baghdad. The modification of the landscape metrics were unique of cultivated area in both parts in 2015 by declining their areas in sizes PLAND and number of patches and similarly, by increasing their LPI, MPS, ENN_AM, IJI and Cohesion. Furthermore, the observed decrease in NP of cultivated patches in the upper and lower parts of the park indicates a reduction in human activities on the landscape such as agricultural activities as a result of decrease in poverty between 1998 and 2015. This positive modification led to more sustainable, stable, aggregated and better connected patches across the landscape in 2015. One thing, which was concerned in this area is increasing the isolation ENN_AM among the cultivated patches in both parts. This could be stated to the result of similar isolated patches joined to one another. Similar patches of cultivated lands aggregated as a result of increasing MPS and LPI with better connection.

4.1.4. Forest Class

The decrease in forest patch size in the lower part is accompanying with decreasing the NP, LPI, MPS, IJI and cohesion, but not the ENN_AM. This suggesting that the forest patches in the lower part in 2015 were less fragmented, fewer connected, the distribution of adjacencies among patches becomes uneven and patches were more isolated in 2015. Whereas, the forest patches in the upper part in 2015 indicate to larger patch size, greater fragmentation, lesser isolated, uneven distribution among patches and less connected across the landscape. The class level metrics analysis indicated that the decrease in the number of patches in the lower part of the park in 2015 was not related to the reduction of the LPI and MPS, but to similar isolated patches joined to one another.

This suggested that the modification and fluctuation of forest patches in the lower part of the park is higher than the upper part, this could be attributed to an increase of human activities, population density and illegal logging. Fragmentation of forest patches is of greater concern because it creates a natural imbalance in shape, size and distribution [65] and although it influences the dynamics of species and material in the landscape [20]. Moreover, the negative changes in forest cover provide some evidence of disturbance of ecological sustainability of the natural resources. Ecologists have found that small remnant areas cannot succeed in preserving species diversity [66].

A brief summary, changes in LULC classes were related to modification of spatial patterns, as confirmed by class level metrics. As a whole, class level metrics revealed absence of fragmentation for all classes excluding bare surface between 1998 and 2015 in the lower part of the park. This indicates a reduction in human activities such as agricultural activities and could be the result of previously isolated patches becoming connected.

Concerning the upper part, however, class level metrics indicated the presence of smaller and more isolated patches resulting from ongoing fragmentation. Many studies indicate that the increase in the number of patches are related to the reduction of the LPI [14,67,68]. Thus, the class level metrics analysis indicated that the increase in the number of patches was related to the reduction of the largest patch index (LPI) only for pasture and forest patches but, not for bare surface and cultivated area.

4.2. Drivers and Consequences of Changes on the Landscape

Generally speaking, more changes in landscape patterns and composition were observed and the landscape became more fragmented in the lower part of the park as a result of rapid increase in socio-economic growth between 1998 and 2015. Thus, the Fall of Baghdad shaped and changed the land in the HSCZ. Moreover, the increase in landscape fragmentation is tightly related to the increase in the human population therefore, rural population density would be one of the major factors related to the landscape change and fragmentation in the lower part of the park. Moreover, indirect or underlying causes were identified as social, economic, political and demographic factors, which promoted the economic growth. On the other hand, direct changes were recognized as agriculture and forestry changes.

Before 2003 the agricultural intensification caused landscape degradation, while after 2003 the new political conditions and economic development have led to substantial improvement in the natural environment at least in the upper part of the park. To conclude, HSCZ reflects major geopolitical events after the Fall of Baghdad.

The results indicate that the cultivated area increased during the United Nation sanctions (1998), when all food imports were banned and then decreased with the lifting of the sanctions in 2003 when food imports were resumed. Furthermore, LULC classes are associated with isolated patches in the lower part of the park. All classes were declined except pastureland from 1998 to 2015. Moreover, MPS variables were also declined for bare surface and forest-patches but increased for pasture and cultivated-patches. Concerning the upper part, Euclidian Nearest Neighborhood Distance variables indicated that they're less isolated patches for pasture and forest in 2015, while there is an increase in distance for bare surface and cultivated-lands. However, MPS variables for all LULC classes in the upper part were increased except pastureland in 2015. On the other hand, The villages are influenced seriously by the agricultural economy and economic development for the period of last seventeen years in HSCZ [69,70]. The registered population (Table 1) of the villages were 12,008 people in 2010 and 13,746 in 2015 according to census data from Kurdistan Regional Statistics Office KRSO [54].

4.3. Random Forest Classification

One of the objectives of this study was to compare the performance of MLC and RF to map land cover. The comparison of MLC and RF classifications of Landsat imagery indicate that the RF statistically produced better classification results than MLC. The random forest algorithm could map LULC better compared to MLC algorithm. Adding a number of layers in the RF algorithm tends to increase the accuracy of the classification, such that the RF classifier provided relatively high accuracies of up to 98% for the upper part in 1998 and 2015, and up to 99% in the lower part for two time periods. The RF was a reliable classifier algorithm for the current study and confirmed by obtaining the high overall accuracy of the classification maps.

Furthermore, because the identification of the best measures of fragmentation and habitat loss is impossible [55], we conclude that the process of selecting the most appropriate landscape metrics is critical and several criteria challenges the selecting of landscape metrics such as; (1) the selectivity of metrics must meet the specific objective and question of the main subject of the research; (2) the indicators should be relatively independent or orthogonal of one another; (3) they must be spatio-temporarily sensitive to landscape change; and (4) metrics performance must be known. These aforementioned criteria follow the earlier findings [56–58].

5. Conclusions

The obtained results of accuracy assessment indicate the success of random forest classification in land cover mapping from Landsat images in the current study and bridges landscape ecology and remote sensing to understand the effects of the socio-economic-political factors in the park.

The metrics successfully depicted the changes in the landscape and were considered useful tools for preserving and sustaining landscape diversity. Furthermore, metrics provide numerical data on the landscape and with their calculating, human impacts can be detected on the landscapes over time. Therefore, Halgurd-Sakran National Park core zone was an appropriate area for challenging these landscape metrics for discovering landscape changes. Hence, the landscape structure was quantified, analyzed and mapped to; describe changes between two different times, compare different landscapes, and to connect the landscape pattern to the ecological processes. Of all the metrics computed, NP, MPS, ENN_AM, and COHESION were found to be most suitable for carrying out the landscape change analysis as these metrics describe both the composition and the configuration of landscape.

Halgurd-Sakran National Park core zone has experienced significant geopolitical change related to the Fall of Baghdad in April 2003. Moreover, most of the drivers influencing the LULC change of the park are influenced by political, societal and economic factors. Thus, economic and politic are the main influencing forces in HSCZ. Since this study provides the first preliminary understanding into the concern of driving force impacts in HSCZ, we diagnose the need for future research in order to make available more precise and better-informed and more sustainable landscape decisions.

To conclude, monitoring and comparison of the natural resources in the upper and lower parts of the Halgurd-Sakran National Park core zone is necessary, which provides park managers, designers, planners and other partners, and the public with scientifically dependable data and information on the current state of the environmental selected park resources.

Acknowledgments: Financial support for this research was provided by Soran University. This work forms a part of a study and is supported by the Scientific Research Centre (SRC), Soran University and the Centre for Landscape and Climate Research (CLCR), University of Leicester, Department of Geography, University of Leicester. H. Balzter was supported by the Royal Society Wolfson Research Merit Award, 2011/R3 and the NERC National Centre for Earth Observation in the UK. We would like to thank editors of Remote Sensing journal and the two anonymous reviewers for their comments and suggestions on the manuscript.

Author Contributions: All authors contributed to the conception of the study. R.H. collected the input data, carried out, conceived, and designed the methodology, analyzed the data, and wrote the paper. H.B. and K.K. supervised the research and contributed to the manuscript.

Conflicts of Interest: The authors declare no conflict of interest.

References

1. Rahman, A.; Kumar, S.; Fazal, S.; Siddiqui, M.A. Assessment of land use/land cover change in the North-West District of Delhi using remote sensing and GIS techniques. *J. Indian Soc. Remote Sens.* **2012**, *40*, 689–697. [CrossRef]

2. Natya, S.; Rehna, V. Land Cover Classification Schemes Using Remote Sensing Images: A Recent Survey. *Br. J. Appl. Sci. Technol.* **2016**, *13*, 1–11. [CrossRef]

3. Usman, M.; Liedl, R.; Shahid, M.A.; Abbas, A. Land use/land cover classification and its change detection using multi-temporal MODIS NDVI data. *J. Geogr. Sci.* **2015**, *25*, 1479–1506. [CrossRef]

4. Aspinall, R.J.; Hill, M.J. *Land Use Change: Science, Policy and Management*; CRC Press: Boca Raton, FL, USA, 2007.

5. Van Dyke, F. The conservation of habitat and landscape. In *Conservation Biology: Foundations, Concepts, Applications*; Springer: Dordrecht, The Netherlands, 2008; pp. 279–311.

6. Didham, R.K. Ecological consequences of habitat fragmentation. In *eLS*; John Wiley & Sons, Ltd.: Hoboken, NJ, USA, 2010.

7.   McGarigal, K.; Marks, B.J. *Fragstats: Spatial Pattern Analysis Program for Quantifying Landscape Structure*; General Technical Report PNW-GTR-351; U.S. Department of Agriculture, Forest Service, Pacific Northwest Research Station: Portland, OR, USA, 1995.

8.   Tizora, P.; Le Roux, A.; Mans, G.; Cooper, A. Land Use and Land Cover Change in the Western Cape Province: Quantification of Changes & Understanding of Driving Factors. 2016. Available online: http://hdl.handle.net/10204/8995 (accessed on 1st February).

9.   Krkoška Lorencová, E.; Harmáčková, Z.V.; Landová, L.; Pártl, A.; Vačkář, D. Assessing impact of land use and climate change on regulating ecosystem services in the Czech Republic. *Ecosyst. Health Sustain.* **2016**, *2*, e01210. [CrossRef]

10.   Berakhi, R.O. Implication of Human Activities on Land Use Land Cover Dynamics in Kagera Catchment, East Africa. Master's Degree, Southern Illinois University Carbondale, Carbondale, IL, USA, December 2013.

11.   Gibson, G.R. War and Agriculture: Three Decades of Agricultural Land Use and Land Cover Change in Iraq. Doctoral Dissertation, University Libraries, Virginia Polytechnic Institute and State University, Blacksburg, VA, USA, 9 May 2012.

12.   Vogt, P. Quantifying landscape fragmentation. *Simp. Brasil. Sensoriamento Remoto* **2015**, *17*, 1239–1246.

13.   Midha, N.; Mathur, P. Assessment of forest fragmentation in the conservation priority Dudhwa landscape, India using FRAGSTATS computed class level metrics. *J. Indian Soc. Remote Sens.* **2010**, *38*, 487–500. [CrossRef]

14.   Vorovencii, I. Quantifying landscape pattern and assessing the land cover changes in Piatra Craiului National Park and Bucegi Natural Park, Romania, using satellite imagery and landscape metrics. *Environ. Monit. Assess.* **2015**, *187*, 692. [CrossRef] [PubMed]

15.   Girvetz, E.H.; Thorne, J.H.; Berry, A.M.; Jaeger, J.A. Integration of landscape fragmentation analysis into regional planning: A statewide multi-scale case study from California, USA. *Landsc. Urban Plan.* **2008**, *86*, 205–218. [CrossRef]

16.   Coppedge, B.R.; Engle, D.M.; Masters, R.E.; Gregory, M.S. Avian response to landscape change in fragmented southern Great Plains grasslands. *Ecol. Appl.* **2001**, *11*, 47–59. [CrossRef]

17.   Vannette, R.L.; Leopold, D.R.; Fukami, T. Forest area and connectivity influence root-associated fungal communities in a fragmented landscape. *Ecology* **2016**, *97*, 2374–2383. [CrossRef] [PubMed]

18.   Auffret, A.G.; Plue, J.; Cousins, S.A. The spatial and temporal components of functional connectivity in fragmented landscapes. *Ambio* **2015**, *44*, 51–59. [CrossRef] [PubMed]

19.   Mühlner, S.; Kormann, U.; Schmidt-Entling, M.; Herzog, F.; Bailey, D. Structural versus functional habitat connectivity measures to explain bird diversity in fragmented orchards. *J. Landsc. Ecol.* **2010**, *3*, 52–64. [CrossRef]

20.   Forman, R.T. *Land Mosaics: The Ecology of Landscapes and Regions (1995)*; Cambridge University Press: Cambridge, UK, 1995.

21.   Lovett, G.M.; Jones, C.G.; Turner, M.G.; Weathers, K.C. Conceptual Frameworks: Plan for a Half-Built House. In *Ecosystem Function in Heterogeneous Landscapes*; Springer: New York, NY, USA, 2005; pp. 463–470.

22.   Southworth, J.; Nagendra, H.; Tucker, C. Fragmentation of a landscape: Incorporating landscape metrics into satellite analyses of land-cover change. *Landsc. Res.* **2002**, *27*, 253–269. [CrossRef]

23.   Parsa, V.A.; Yavari, A.; Nejadi, A. Spatio-temporal analysis of land use/land cover pattern changes in Arasbaran Biosphere Reserve: Iran. *Model. Earth Syst. Environ.* **2016**, *2*, 178. [CrossRef]

24.   Linh, N.; Erasmi, S.; Kappas, M. Quantifying land use/cover change and landscape fragmentation in Danang City, Vietnam: 1979–2009. *Int. Arch. Photogramm. Remote Sens. Spat. Inf. Sci.* **2012**, *39-B8*, 501–506. [CrossRef]

25.   Lu, D.; Hetrick, S.; Moran, E. Impervious surface mapping with Quickbird imagery. *Int. J. Remote Sens.* **2011**, *32*, 2519–2533. [CrossRef] [PubMed]

26.   Foody, G.M. Status of land cover classification accuracy assessment. *Remote Sens. Environ.* **2002**, *80*, 185–201. [CrossRef]

27.   Lillesand, T.; Kiefer, R.W.; Chipman, J. *Remote Sensing and Image Interpretation*; John Wiley & Sons: Hoboken, NJ, USA, 2014.

28.   Li, M.; Zang, S.; Zhang, B.; Li, S.; Wu, C. A review of remote sensing image classification techniques: The role of spatio-contextual information. *Eur. J. Remote Sens.* **2014**, *47*, 389–411. [CrossRef]

29. Syed, S.; Dare, P.; Jones, S. Automatic classification of land cover features with high resolution imagery and lidar data: An object-oriented approach. In Proceedings of the National Biennial Conference of the Spatial Sciences Institute, Melbourne, Australia, 12–16 September 2005.

30. Burai, P.; Deák, B.; Valkó, O.; Tomor, T. Classification of herbaceous vegetation using airborne hyperspectral imagery. *Remote Sens.* **2015**, *7*, 2046–2066. [CrossRef]

31. Qian, J.; Zhou, Q.; Hou, Q. Comparison of pixel-based and object-oriented classification methods for extracting built-up areas in arid zone. In Proceedings of the ISPRS Workshop on Updating Geo-spatial Databases with Imagery & The 5th ISPRS Workshop on Dynamic and Multi-dimensional GIS, Urumchi, China, 28–29 August 2007; pp. 163–171.

32. Horning, N. Random Forests: An algorithm for image classification and generation of continuous fields data sets. In Proceedings of the International Conference on Geoinformatics for Spatial Infrastructure Development in Earth and Allied Sciences, Hanoi, Vietnam, 9–11 December 2010.

33. Breiman, L. Random forests. *Mach. Learn.* **2001**, *45*, 5–32. [CrossRef]

34. Liaw, A.; Wiener, M. Classification and regression by randomForest. *R News* **2002**, *2*, 18–22.

35. Balzter, H.; Cole, B.; Thiel, C.; Schmullius, C. Mapping CORINE land cover from Sentinel-1A SAR and SRTM digital elevation model data using Random Forests. *Remote Sens.* **2015**, *7*, 14876–14898. [CrossRef]

36. Hamad, R.; Balzter, H.; Kolo, K. Multi-Criteria Assessment of Land Cover Dynamic Changes in Halgurd Sakran National Park (HSNP), Kurdistan Region of Iraq, Using Remote Sensing and GIS. *Land* **2017**, *6*, 18. [CrossRef]

37. Williams, K.J.; Schirmer, J. Understanding the relationship between social change and its impacts: The experience of rural land use change in south-eastern Australia. *J. Rural Stud.* **2012**, *28*, 538–548. [CrossRef]

38. Schirmer, J. Socio-Economic Impacts of Land Use Change to Plantation Forestry: A Review of Current Knowledge and Case Studies of Australian Experience. In Proceedings of the 2006 IUFRO Forest Plantations Meeting, Charleston, SC, USA, 10–13 October 2006.

39. Long, H.; Tang, G.; Li, X.; Heilig, G.K. Socio-economic driving forces of land-use change in Kunshan, the Yangtze River Delta economic area of China. *J. Environ. Manag.* **2007**, *83*, 351–364. [CrossRef] [PubMed]

40. Figueiredo, J.; Pereira, H.M. Regime shifts in a socio-ecological model of farmland abandonment. *Landsc. Ecol.* **2011**, *26*, 737–749. [CrossRef]

41. Brown, D.; Polsky, C.; Bolstad, P.V.; Brody, S.D.; Hulse, D.; Kroh, R.; Loveland, T.; Thomson, A.M. *Land Use and Land Cover Change*; Pacific Northwest National Laboratory (PNNL): Richland, WA, USA, 2014.

42. Saleh, S.A. Temporal Change Detection of AL-Hammar Marsh—IRAQ Using Remote Sensing Techniques. *Glob. J. Hum. Soc. Sci. Res.* **2012**, *12*, 7–14.

43. Nagendra, H.; Lucas, R.; Honrado, J.P.; Jongman, R.H.; Tarantino, C.; Adamo, M.; Mairota, P. Remote sensing for conservation monitoring: Assessing protected areas, habitat extent, habitat condition, species diversity, and threats. *Ecol. Indic.* **2013**, *33*, 45–59. [CrossRef]

44. Stoms, D.M.; Estes, J. A remote sensing research agenda for mapping and monitoring biodiversity. *Int. J. Remote Sens.* **1993**, *14*, 1839–1860. [CrossRef]

45. Dewan, A.M.; Yamaguchi, Y. Land use and land cover change in Greater Dhaka, Bangladesh: Using remote sensing to promote sustainable urbanization. *Appl. Geogr.* **2009**, *29*, 390–401. [CrossRef]

46. El-Kawy, O.A.; Rød, J.; Ismail, H.; Suliman, A. Land use and land cover change detection in the western Nile delta of Egypt using remote sensing data. *Appl. Geogr.* **2011**, *31*, 483–494. [CrossRef]

47. Prakasam, C. Land use and land cover change detection through remote sensing approach: A case study of Kodaikanal taluk, Tamil nadu. *Int. J. Geomat. Geosci.* **2010**, *1*, 150.

48. Leh, M.; Bajwa, S.; Chaubey, I. Impact of land use change on erosion risk: An integrated remote sensing, geographic information system and modeling methodology. *Land Degrad. Dev.* **2013**, *24*, 409–421. [CrossRef]

49. Chuvieco, E.; Congalton, R.G. Application of remote sensing and geographic information systems to forest fire hazard mapping. *Remote Sens. Environ.* **1989**, *29*, 147–159. [CrossRef]

50. Boyd, D.; Danson, F. Satellite remote sensing of forest resources: Three decades of research development. *Prog. Phys. Geogr.* **2005**, *29*, 1–26. [CrossRef]

51. Kayiranga, A.; Kurban, A.; Ndayisaba, F.; Nahayo, L.; Karamage, F.; Ablekim, A.; Li, H.; Ilniyaz, O. Monitoring forest cover change and fragmentation using remote sensing and landscape metrics in Nyungwe-Kibira park. *J. Geosci. Environ. Prot.* **2016**, *4*, 13. [CrossRef]

52. Robinson, L. *Masters of Chaos: The Secret History of the Special Forces*; PublicAffairs: New York, NY, USA, 2005.

53. Eklund, L.; Persson, A.; Pilesjö, P. Cropland changes in times of conflict, reconstruction, and economic development in Iraqi Kurdistan. *Ambio* **2016**, *45*, 78–88. [CrossRef] [PubMed]
54. KRSO. Ministry of Planning 2016. Kurdistan Regional Statistics Office-KRSO. Available online: http//www.mop.gov.krd (accessed on 25 January 2017).
55. Eiumnoh, A.; Shrestha, R.P. Application of DEM data to Landsat image classification: Evaluation in a tropical wet-dry landscape of Thailand. *Photogramm. Eng. Remote Sens.* **2000**, *66*, 297–304.
56. Ok, A.O.; Akar, O.; Gungor, O. Evaluation of random forest method for agricultural crop classification. *Eur. J. Remote Sens.* **2012**, *45*, 421–432. [CrossRef]
57. Thakkar, A.K.; Desai, V.R.; Patel, A.; Potdar, M.B. Post-classification corrections in improving the classification of Land Use/Land Cover of arid region using RS and GIS: The case of Arjuni watershed, Gujarat, India. *Egypt. J. Remote Sens. Space Sci.* **2017**, *20*, 79–89. [CrossRef]
58. Donges, N. SAP Machine Learning Foundation Working Student. Available online: https://towardsdatascience.com/@n.donges (accessed on 25 Feb 2018).
59. Guan, H.; Yu, J.; Li, J.; Luo, L. Random forests-based feature selection for land-use classification using lidar data and orthoimagery. *Int. Arch. Photogramm. Remote Sens. Spat. Inf. Sci.* **2012**, *39-B7*, 203–208. [CrossRef]
60. Kulkarni, A.D.; Lowe, B. Random forest algorithm for land cover classification. *Pattern Recognit. Lett.* **2016**, *27*, 294–300.
61. Reynolds, J.; Wesson, K.; Desbiez, A.L.; Ochoa-Quintero, J.M.; Leimgruber, P. Using remote sensing and Random Forest to assess the conservation status of critical Cerrado Habitats in Mato Grosso do Sul, Brazil. *Land* **2016**, *5*, 12. [CrossRef]
62. McGarigal, K.; Cushman, S.A.; Neel, M.C.; Ene, E. FRAGSTATS: Spatial Pattern Analysis Program for Categorical Maps. 2002. Available online: http://www.umass.edu/landeco/research/fragstats/fragstats.html (accessed on 25 January 2018).
63. Nkonya, E.; Mirzabaev, A.; Von Braun, J. *Economics of Land Degradation and Improvement: A Global Assessment for Sustainable Development*; Springer: Cham, Switzerland, 2016.
64. Speir, R.A. Managing Runoff and Erosion on Croplands and Pastures. 2009. Available online: http://aware.uga.edu/wp-content/2009/07/Erosion-and-Runoff-Update.pdf (accessed on 25 January 2018).
65. Munguía-Rosas, M.A.; Montiel, S. Patch size and isolation predict plant species density in a naturally fragmented forest. *PLoS ONE* **2014**, *9*, e111742. [CrossRef] [PubMed]
66. Harrison, S.; Bruna, E. Habitat fragmentation and large-scale conservation: What do we know for sure? *Ecography* **1999**, *22*, 225–232. [CrossRef]
67. Kadıoğulları, A.I.; Başkent, E.Z. Spatial and temporal dynamics of land use pattern in Eastern Turkey: A case study in Gümüşhane. *Environ. Monit. Assess.* **2008**, *138*, 289–303. [CrossRef] [PubMed]
68. Kamusoko, C.; Aniya, M. Land use/cover change and landscape fragmentation analysis in the Bindura District, Zimbabwe. *Land Degrad. Dev.* **2007**, *18*, 221–233. [CrossRef]
69. Eklund, L.; Seaquist, J. Meteorological, agricultural and socioeconomic drought in the Duhok Governorate, Iraqi Kurdistan. *Nat. Hazards* **2015**, *76*, 421–441. [CrossRef]
70. Lortz, M.G. Willing to Face Death: A history of Kurdish Military Forces-the Peshmerga-from the Ottoman Empire to Present-Day Iraq. Doctoral Dissertation, Florida State University, Tallahassee, FL, USA, 2005.

land

MDPI

Communication

Proposed Release of Wilderness Study Areas in Montana (USA) Would Demote the Conservation Status of Nationally-Valuable Wildlands

R. Travis Belote

The Wilderness Society, Bozeman, MT 59715, USA; travisbelote@gmail.com; Tel.: +1-406-581-3808

Received: 23 April 2018; Accepted: 30 May 2018; Published: 1 June 2018

Abstract: Wildlands are increasingly lost to human development. Conservation scientists repeatedly call for protecting the remaining wildlands and expanding the land area protected in reserves. Despite these calls, conservation reserves can be eliminated through legislation that demotes their conservation status. For example, legislation introduced to the Congress of the United States recently would demote 29 Wilderness Study Areas (WSAs) from the protections afforded by their existing status. The proposed legislation suggests that the 29 areas are not suitable for a promotion and future inclusion in the National Wilderness Preservation System based on decades-old local evaluations. Local evaluations, notwithstanding, it may be important to consider the value of lands from a national perspective. Without a national perspective, local evaluations alone may lead to overlooking the national significance of lands. With this in mind, I used five qualities of wildland value (wildness, intactness of night sky, lack of human-generated noises, intactness of mammals, and intactness of mammal carnivores of conservation concern) to compare the 29 WSAs to all national parks and wilderness areas located within the contiguous United States. The pool of 29 WSAs was similar to the pool of national parks and wilderness areas with respect to the five qualities assessed, and some of the WSAs were characterized by higher values than most of national parks and wilderness areas. This analysis demonstrates the national significance of the WSAs targeted for demotion of their existing conservation status. Such an approach could be used in future land management legislation and planning to ensure that a national perspective on conservation value is brought to bear on decisions facing federally-managed lands.

Keywords: Wilderness Study Areas; light pollution; noise pollution; wildness; protected areas

---

1. Introduction

Humans are impacting Earth's remaining wildlands at an increasingly rapid rate, and researchers have measured this loss at global and national scales [1,2]. Given the loss of wildlands, conservation scientists continue to make calls to protect what is left of Earth's wild places [3–5]. Wildlands are defined by their lack of human modification and serve to maintain ecological processes, populations of species, and biological diversity [5–7]. Lack of human modification in the way of limited road density, light and noise pollution, and intact biological communities represent important conservation priorities [1].

Protected areas, including national parks and wilderness areas, are an effective means of maintaining wildlands and the values therein [8,9]. Teams of scientists have recommended additional lands be included in protected areas [10], recognizing that existing systems of conservation reserves may be insufficient to sustain species and ecosystems into the future [11]. Despite these calls for additional protected areas, policy makers at times recommend that conservation protections be removed from lands (e.g., Bears Ears National Monument and the Arctic National Wildlife Refuge).

In December 2017, Senator Steve Daines introduced the Protect Public Use of Public Lands Act to the U.S. Congress, which would "release" five Wilderness Study Areas (WSAs) located on U.S. Forest Service lands in Montana. Later in March of 2018, Congressman Greg Gianforte introduced two bills to release 24 additional WSAs, mostly located on Bureau of Land Management (BLM) lands in Montana. The release of these 29 WSAs (Figure 1) would functionally eliminate the management direction that maintains the wild character of these places. WSAs are classified by the Gap Analysis Program (GAP) as GAP status 2 because of the management directives that maintain biological diversity while restricting commercial extractive uses (timber harvesting or mineral and energy extraction) and motorized recreation. GAP status classification ranges from 1 to 4 and is assigned to lands or management areas based on their policies and guidelines (Table 1). The Forest Service WSAs targeted in the proposed bills were originally designated by Congress via the Montana Wilderness Study Act of 1977 to "maintain their presently existing wilderness character and potential for inclusion in the National Wilderness Preservation System". The BLM WSAs were designated as such through administrative planning processes based on mandates in the Federal Land Policy Management Act of 1976 [12] to preserve the wilderness character of lands.

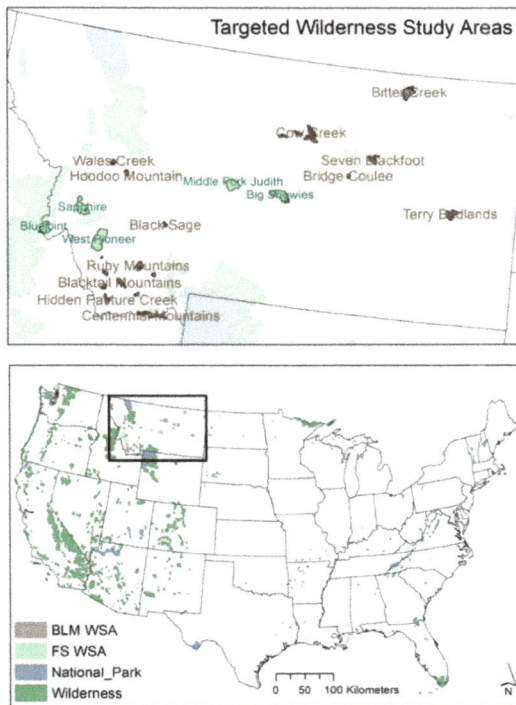

Figure 1. Map of targeted Wilderness Study Areas (WSAs) in Montana (**top**) and all national parks and wilderness areas in the lower 48 states used in this analysis (**bottom**). Not all WSAs are labeled; see Table 2 for full list of WSAs included in proposed legislation aimed at demoting their conservation status.

Release of WSAs would demote the protected area status from GAP 2 to GAP 3 and represents an example of the impermanence of certain highly protected conservation reserves. Conservation reserves can be designated, but such legal protections can also be removed. In the introduced bills proposing release of the 29 WSAs, Senator Daines and Representative Gianforte argue that these areas are not suitable for permanent protection via legislative wilderness designation, and therefore would be more appropriately managed for motorized recreation use and extractive resource use. It was argued that suitability was assessed using local evaluations that occurred during the 1980s. However, before demoting these or any such areas from their current status as WSAs based solely on local assessments, it may be critical to conduct updated assessments and include analyses that quantify their value at national scales [13,14]. Without such an assessment, the value of areas slated for demotion may not be fully appreciated.

Table 1. Gap Analysis Program (GAP) status classification and overview of conservation protection. WSAs set for release in proposed legislation would functionally demote those areas from GAP 2 to GAP 3. BLM: Bureau of Land Management.

GAP Status	Definition	Examples
GAP status 1	An area having permanent protection from conversion of natural land cover and a mandated management plan in operation to maintain a natural state within which disturbance events (of natural type, frequency, intensity, and legacy) are allowed to proceed without interference or are mimicked through management.	National parks, Wilderness Areas in the National Wilderness Preservation System
GAP status 2	An area having permanent protection from conversion of natural land cover and a mandated management plan in operation to maintain a primarily natural state, but which may receive uses or management practices that degrade the quality of existing natural communities, including suppression of natural disturbance.	National monuments, Wilderness Study Areas
GAP status 3	Area having permanent protection from conversion of natural land cover for the majority of area. Subject to extractive uses of either broad, low-intensity type (e.g., logging) or localized intense type (e.g., mining). Confers protection to federally listed endangered and threatened species throughout the area.	National Forests open for timber harvesting and BLM land open for energy development (i.e., outside of wilderness)
GAP status 4	No known public/private institutional mandates/legally recognized easements.	Department of Defense lands

Here, I evaluate the wildland qualities of the WSAs targeted for release and compare them to national parks and designated wilderness areas located in the contiguous United States. National parks and wilderness areas represent highly protected iconic lands in the U.S. that sustain "vignettes of primitive America" and "the biotic associations ... maintained" [15]. National parks and wilderness areas also represent core conservation reserves [16], valuable for protecting wildland values and biodiversity. I compared the relative wildland quality of the targeted WSAs to national parks and wilderness areas to assess their value at a national scale.

To assess the value of the targeted WSAs, I used five maps of data representing wildland qualities and wilderness character: (1) wildness as estimated by human modification [17]; (2) intactness of night skies (opposite of light pollution) [18]; (3) quietness of landscapes (opposite of noise pollution) [19]; (4) intactness of mammal assemblage; and (5) intactness of carnivores of conservation concern [20]. These five metrics represent available mapped data that serve as indicators for the qualities of wildlands associated with ecological integrity available at the extent of the contiguous United States. Wildness, night skies, and quietness have previously been used to assess wilderness character [21,22]. Sustaining lands with high degrees of wildness, dark night skies, lack of human-generated noise pollution, and intact mammal and carnivore communities represent important national or international goals in wildland protection [5,6,23].

Table 2. Wilderness Study Areas targeted for release via recently proposed legislation to the U.S. Senate and House of Representatives. FS: U.S. Forest Service; BLM: Bureau of Land Management.

Targeted Wilderness Study Area	Agency	Hectares
Big Snowies	FS	88,693
Blue Joint	FS	63,407
Middle Fork Judith	FS	80,856
Sapphire	FS	94,740
West Pioneer	FS	153,690
Antelope Creek	BLM	12,912
Axolotl Lakes	BLM	7824
Bell/Limekiln Canyons	BLM	9377
Billy Creek	BLM	3411
Bitter Creek	BLM	60,851
Black Sage	BLM	5963
Blacktail Mountains	BLM	17,530
Bridge Coulee	BLM	6022
Centennial Mountains	BLM	47,870
Cow Creek	BLM	33,658
Dog Creek South	BLM	5140
East Fork Blacktail Deer Creek	BLM	6862
Ervin Ridge	BLM	10,361
Farlin Creek	BLM	1186
Henneberry Ridge	BLM	9581
Hidden Pasture Creek	BLM	15,584
Hoodoo Mountain	BLM	10,919
Ruby Mountains	BLM	26,923
Seven Blackfoot	BLM	20,155
Stafford	BLM	4923
Terry Badlands	BLM	42,742
Twin Coulee	BLM	6839
Wales Creek	BLM	11,457
Woodhawk	BLM	8029

2. Materials and Methods

Wildness was estimated using the map of human modification (Figure 2 [17]). Human modification data are based on land cover, human population density, roads, and other mapped data on ecological condition [17]. These data are highly correlated with an earlier map depicting wildness [24], and have been used as a surrogate for wildness in other work [11]. Data are scaled from 0 (no measured human modification) to 1 (high degree of human modification), but I reverse ordered these so that higher values represent wilder values.

Light pollution is measured during the night from the Visible Infrared Imaging Radiometer Suite (VIIRS), a sensor on board the Suomi National Polar-orbiting Partnership satellite (Figure 2, [25]). This mapped dataset serves as a measure of the intactness of the night sky. From the VIIRS light pollution data, higher values represent more intense light pollution and thus lower wildland quality and greater ecological impacts. Therefore, I reverse ordered the data to represent intactness of the night sky. Because the data are highly skewed, I log-transformed the data. For the final metrics, higher values represent darker, more intact night skies. Similarly, mapped data of human-generated noise pollution is based on field observations and a spatial model using landscape features that influence sound propagation [26,27]. Greater intensity of human noises (higher predicted dBA) is associated with reduced wildland quality and greater ecological impacts (Figure 2). Similar to light pollution data, I reverse ordered these data to represent lack of noise pollution or quietness of the landscape in mapped pixels.

Intactness of wildlife community was estimated by overlaying current and historical distributions of species [20]. I used this overlay to calculate the proportion of mammal species currently present from species historically present. Values closer to 1 are associated with high intactness of wildlife community (i.e., a value of 1 represents lands where all species still occur from those that historically occurred). I also calculated this intactness ratio using 10 mammal carnivores of conservation concern including red wolves (*Canis rufus*), grey wolves (*Canis lupus*), mountain lions (*Puma concolor*), lynx (*Lynx canadensis*), black bear (*Ursus americanus*), grizzly bear (*Ursus arctos*), fisher (*Pekania pennanti*), wolverines (*Gulo gulo*), black footed ferret (*Mustela nigripes*), and swift fox (*Vulpes velox*). These species were chosen based on whether the species or a population of the species was listed as threatened or endangered under the Endangered Species Act, or if the species was listed as sensitive by the U.S. Forest Service. See bottom maps of Figure 2.

After obtaining or calculating these mapped metrics, I overlaid locations of four groups of land management zones: national parks (*N* = 47), wilderness areas (*N* = 709), the targeted WSAs keeping the forest service (*N* = 5), and BLM units (*N* = 24) separated (Figure 1). In cases where wilderness areas and national parks overlapped, wilderness areas took priority. I then calculated the average value of each of the five metrics of wildland quality for all management units using the raster package of R. I calculated summary statistics for each group (i.e., national park, wilderness area, BLM WSA, and U.S. Forest Service (FS) WSA) and plotted the distribution of data using boxplots. This allowed me to compare the median, means, and distribution of each of the five mapped metrics among the four groups of land management zones to evaluate how the targeted WSAs compared to existing highly protected and valuable conservation areas. Finally, I created dot charts of ranks of the targeted WSAs for each of the five metrics and added plots of the median values among national parks and wilderness areas. This simple plotting technique allowed me to easily assess the value of individual targeted WSAs and compare the WSA units to the central tendency of wildland values among national parks and wilderness areas combined.

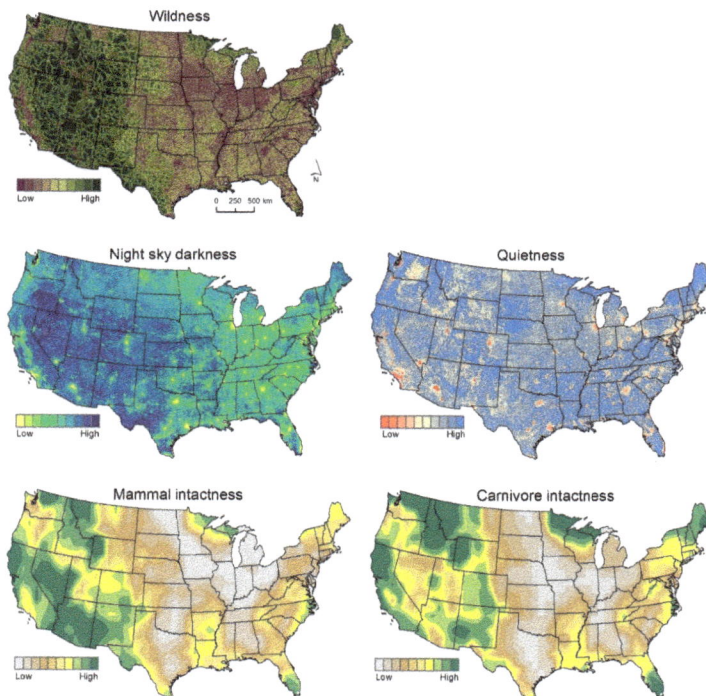

Figure 2. Map of values used to assess wildland quality of Wilderness Study Areas targeted for release.

3. Results

Average and median wildness tended to be higher in the targeted FS and BLM WSAs than in either national parks or wilderness areas (Figure 3). All five FS WSAs and 18 of 24 of the BLM WSAs were wilder than half of national parks and wilderness areas combined (Figure S1). In fact, one FS WSA (Big Snowies) and four BLM WSAs (East Fork Blacktail Deer Creek, Twin Coulee, Centennial Mountains, and Blacktail Mountains) were wilder than 90% of all national parks and wilderness areas combined (Figure S1).

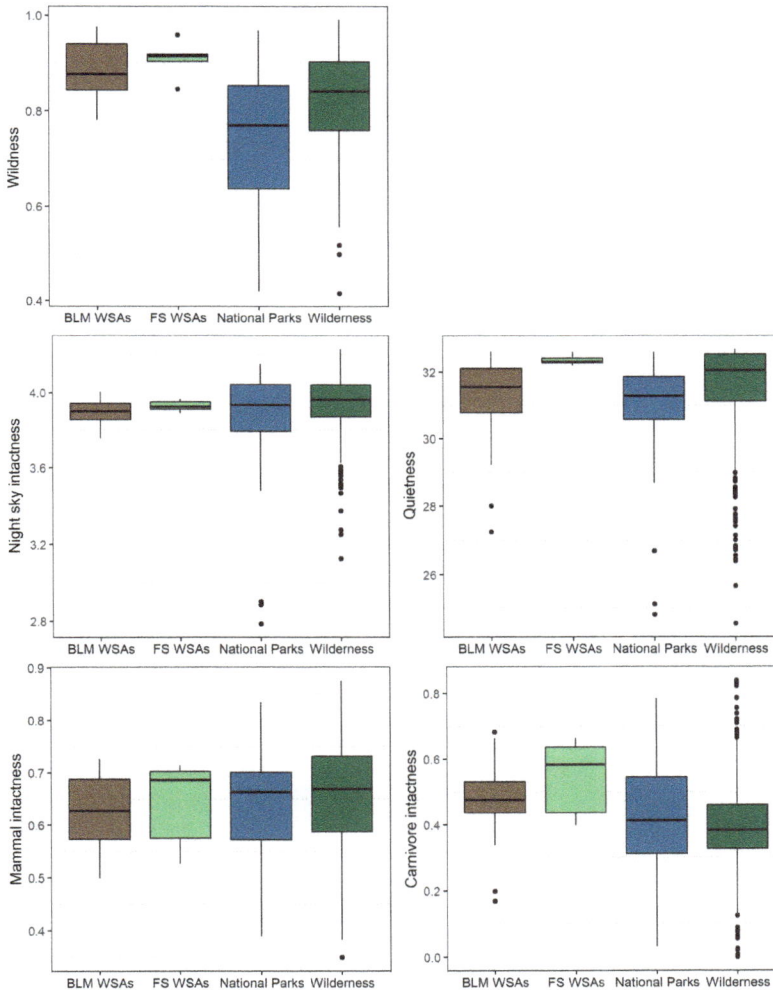

Figure 3. Boxplots of values used to assess wildland quality of Wilderness Study Areas targeted for release administered by the Bureau of Land Management (BLM) and U.S. Forest Service (FS) compared to the national parks and wilderness areas. Upper and lower bounds of boxplots represent the 75th and 25th percentiles, the dark lines in the box are the median values, and 'whiskers' are the 95th and 5th percentiles. Filled circles beyond the whiskers are values from units outside the 95th and 5th percentiles.

Night sky intactness estimates of targeted FS and BLM WSAs were similar to levels in national parks and wilderness areas (Figure 3). One of the targeted FS WSAs (Big Snowies) and six of the BLM WSAs (Twin Coulee, Bridge Coulee, Wales Creek, Dog Creek South, Stafford, and Axolotl Lakes) where characterized by darker night skies than half of all national parks and wilderness areas combined (Figure S2). Landscape quietness estimates in the targeted FS WSAs were higher compared to national parks and wilderness areas, while quietness in targeted BLM WSAs were similar to parks and wilderness areas (Figure 3). All five targeted FS WSAs and seven targeted BLM WSAs (East Fork Blacktail Deer Creek, Twin Coulee, Axolotl Lakes, Wales Creek, Hoodoo Mountain, Blacktail Mountains, and Seven Blackfoot) experienced quieter conditions than half of all national parks and wilderness areas combined (Figure S3).

Mammal intactness was similar among targeted WSAs and national parks and wilderness areas (Figure 3). Three of the five targeted FS WSAs (Blue Joint, Sapphire, and West Pioneer) and 11 BLM WSAs had higher mammal intactness values than half of national parks and wilderness areas combined (Figure S4). On average, carnivore intactness tended to be higher in targeted WSAs than national parks and wilderness areas (Figure 3). Estimates of carnivore intactness in all targeted FS WSAs and 20 BLM WSAs were greater than half of all national parks and wilderness areas. In fact, three targeted FS WSAs and two BLM WSAs had carnivore estimates greater than 90% of all national parks and wilderness areas (Figure S5).

4. Discussion

The Wilderness Study Areas targeted for demotion in recently introduced legislation to the U.S. Congress are comparable to nationally-significant protected areas with respect to their wildland quality. In some cases, individual units targeted for release represent some of the most valuable areas with respect to the wildland qualities compared to nationally-celebrated parks and wilderness areas. While legislation proposing demotion of the conservation status (i.e., "release") of these WSAs argues that these 29 units are unsuitable for wilderness-level protection, this national comparison suggests otherwise. These 29 areas may be just as valuable—and in some cases, based on some metrics, more valuable—than existing national parks and wilderness areas.

My assessment of "wildness" using Theobald's map of human modification shows the 29 areas to be comparable to the wildest national parks and wilderness areas in the country. As wildlands continue to be developed globally and nationally [2,5], it is critical that we include assessments of wildness in conservation prioritization [11]. Wildness and a low degree of human modification are also associated with high degree of permeability for moving organisms and processes [16,28,29]. Protecting lands that maintain a high degree of wildness should inform future conservation globally [6] and nationally [11], and may provide benefits to species under climate change [30].

Dark skies and quiet landscapes may be considered components of wildness, but here I evaluate them as separate qualities for the comparison between targeted WSAs and existing national parks and wilderness areas. Dark night skies and relatively quiet conditions indicate that these lands are important areas to maintain these qualities both for their ecological [18,19] and recreational value. As with wildness, the targeted WSAs are comparable to national parks and wilderness areas and, in some cases, maintain darker night skies and quieter conditions than national parks and wilderness areas combined. Protecting the remaining lands that maintain darks skies and quiet landscapes may offer benefits to species sensitive to these impacts, while also maintaining high quality lands for wildland recreation.

Finally, intactness of assemblages of species has only recently been quantified and mapped at large spatial extents [20,23,31,32]. Here, I focused on mapping the proportion of mammals and mammal carnivores that still occur of that which occurred before pre-European local extinctions [20]. Lands that maintain a high degree of "historical fidelity" in their species composition are considered more biologically intact [33]. Few places in the contiguous United States still maintain the full complement of species that existed before extinctions following Euro-American colonization. The targeted WSAs in

Montana are comparable to existing national parks and wilderness areas with respect to the intactness of mammal species. However, the targeted WSAs tended to be more highly intact with respect to the carnivores of conservation concern included here. Three of the five Forest Service WSAs were more intact than 90% of all national parks and wilderness areas combined. These areas still maintain grizzly bear habitat and at least occasional occupancy and are home to other mammal carnivores such as wolverines, grey wolves, mountain lions, and lynx. Few other areas host such an intact assemblage of carnivores.

The value of these WSAs in sustaining wildlands, dark skies, quiet landscapes, and relatively intact mammal assemblages should be considered when evaluating the suitability of these lands for elevated levels of protection. Indeed, before demoting these lands and opening them for commercial timber extraction, energy development, and intensified motorized recreation, a national perspective on their value, such as the one I present here, is needed. Local assessment will always be an important step in conservation planning, but increasingly we have datasets and tools available to place lands into a more national perspective to assess their value at that scale. Federal lands should be managed with such a perspective. Assessments such as the one I present here will not be the deciding factor in the future of these 29 WSAs, but such an assessment should provide important perspective on their value.

5. Conclusions

Federal legislation has been introduced to demote the conservation status of 29 Wilderness Study Areas (WSAs) in Montana. Three decades ago, these WSAs were deemed unsuitable for conservation status promotion by the Forest Service and BLM. However, the proposed legislation would actually demote their conservation status and the protections afforded by their existing designations. This demotion could result in increased commercial development in these areas, including timber harvesting, energy exploration and development, road-building, and motorized recreation. I have shown that these areas maintain a high degree of wildland quality when compared with existing national parks and wilderness areas in the contiguous United States. A similar analysis that places local evaluations into a national perspective should be conducted when similar policies or bills are proposed that would demote the conservation status of lands.

Supplementary Materials: The following are available online at http://www.mdpi.com/2073-445X/7/2/69/s1.

Funding: This research received no external funding.

Acknowledgments: Thanks to Connor Bailey for compiling and sharing data on Wilderness Study Areas. Barb Cestero and Scott Brennan provided comments and reviews that improved the paper.

Conflicts of Interest: The author declares no conflict of interest.

References

1. Venter, O.; Sanderson, E.W.; Magrach, A.; Allan, J.R.; Beher, J.; Jones, K.R.; Possingham, H.P.; Laurance, W.F.; Wood, P.; Fekete, B.M.; et al. Sixteen years of change in the global terrestrial human footprint and implications for biodiversity conservation. *Nat. Commun.* **2016**, *7*, 1–11. [CrossRef] [PubMed]
2. Theobald, D.M.; Zachmann, L.J.; Dickson, B.G.; Gray, M.E.; Albano, C.M.; Landau, V.; Harrison-Atlas, D. *The Disappearing West: Description of the Approach, Data, and Analytical Methods Used to Estimate Natural Land Loss in the Western U.S.*; Conservation Science Partners: Truckee, CA, USA, 2016.
3. Aplet, G.H. On the nature of wildness: Exploring what wilderness really protects. *Denver Law Rev.* **1999**, *76*, 347–367.
4. Martin, J.-L.; Maris, V.; Simberloff, D.S. The need to respect nature and its limits challenges society and conservation science. *Proc. Natl. Acad. Sci. USA* **2016**, *113*, 6105–6112. [CrossRef] [PubMed]
5. Watson, J.E.M.; Shanahan, D.F.; Marco, M.; Di Allan, J.; Laurance, W.F.; Sanderson, E.W.; Mackey, B.; Venter, O. Catastrophic declines in wilderness areas undermine global environment targets. *Curr. Biol.* **2016**, *26*, 1–6. [CrossRef] [PubMed]

6. Kormos, C.F.; Badman, T.; Jaeger, T.; Bertzky, B.; Merm, R.; Van Osipova, E.; Shi, Y.; Larsen, P.B. *World Heritage, Wilderness, and Large Landscapes and Seascapes*; International Union for Conservation of Nature: Gland, Switzerland, 2017.

7. Allan, J.R.; Venter, O.; Watson, J.E.M. Temporally inter-comparable maps of terrestrial wilderness and the Last of the Wild. *Sci. Data* **2017**, *4*, 1–8. [CrossRef] [PubMed]

8. Cantú-Salazar, L.; Gaston, K.J. Very Large Protected Areas and Their Contribution to Terrestrial Biological Conservation. *BioScience* **2010**, *60*, 808–818. [CrossRef]

9. Gaston, K.J.; Jackson, S.F.; Cantú-Salazar, L.; Cruz-Piñón, G. The ecological performance of protected areas. *Annu. Rev. Ecol. Evolut. Syst.* **2008**, *39*, 93–113. [CrossRef]

10. Aycrigg, J.L.; Groves, C.; Hilty, J.A.; Scott, J.M.; Beier, P.; Boyce, D.A.; Figg, D.; Hamilton, H.; Machlis, G.; Muller, K.; et al. Completing the system: Opportunities and challenges for a national habitat conservation system. *BioScience* **2016**, *66*, 774–784. [CrossRef]

11. Belote, R.T.; Dietz, M.S.; Jenkins, C.N.; McKinley, P.S.; Irwin, G.H.; Fullman, T.J.; Leppi, J.C.; Aplet, G.H. Wild, connected, and diverse: Building a more resilient system of protected areas. *Ecol. Appl.* **2017**, *27*, 1050–1056. [CrossRef] [PubMed]

12. U.S. Department of the Interior. *The Federal Land Policy and Management Act, as Amended*; Bureau of Land Management and Office of the Solicitor, Ed.; U.S. Department of the Interior: Washington, DC, USA, 2001.

13. Belote, R.T.; Irwin, G.H. Quantifying the National Significance of Local Areas for Regional Conservation Planning: North Carolina's Mountain Treasures. *Land* **2017**, *6*, 35. [CrossRef]

14. Noss, R.F.; Platt, W.J.; Sorrie, B.A.; Weakley, A.S.; Means, D.B.; Costanza, J.; Peet, R.K. How global biodiversity hotspots may go unrecognized: Lessons from the North American Coastal Plain. *Divers. Distrib.* **2015**, *21*, 236–244. [CrossRef]

15. USA National Park Service. *Compilation of the Administrative Policies for the National Parks and National Monuments of Scientific Significance (Natural Area Category)*; U.S. Government Publishing Office: Washington, DC, USA, 1970; 147p.

16. Belote, R.T.; Dietz, M.S.; McRae, B.H.; Theobald, D.M.; McClure, M.L.; Irwin, G.H.; McKinley, P.S.; Gage, J.A.; Aplet, G.H. Identifying corridors among large protected areas in the United States. *PLoS ONE* **2016**, *11*, e0154223. [CrossRef] [PubMed]

17. Theobald, D.M. A general model to quantify ecological integrity for landscape assessments and US application. *Landsc. Ecol.* **2013**, *28*, 1859–1874. [CrossRef]

18. Longcore, T.; Rich, C. Ecological light pollution. *Front. Ecol. Environ.* **2004**, *2*, 191–198. [CrossRef]

19. Shannon, G.; McKenna, M.F.; Angeloni, L.M.; Crooks, K.R.; Fristrup, K.M.; Brown, E.; Warner, K.A.; Nelson, M.D.; White, C.; Briggs, J.; et al. A synthesis of two decades of research documenting the effects of noise on wildlife. *Biol. Rev.* **2016**, *91*, 982–1005. [CrossRef] [PubMed]

20. Faurby, S.; Svenning, J.C. Historic and prehistoric human-driven extinctions have reshaped global mammal diversity patterns. *Divers. Distrib.* **2015**, *21*, 1155–1166. [CrossRef]

21. Tricker, J.; Landres, P.; Dingman, S.; Callagan, C.; Stark, J.; Bonstead, L.; Fuhrmann, K.; Carver, S. *Mapping Wilderness Character in Death Valley National Park*; Natural Resource Report NPS/DEVA/NRR—2012/503; National Park Service, U.S. Department of the Interior: Fort Collins, CO, USA, 2012.

22. Tricker, J.; Landres, P.; Chenoweth, J.; Hoffman, R.; Scott, R. *Mapping Wilderness Character in Olympic National Park Final Report*; U.S. Department of Agriculture, Forest Service, Rocky Mountain Research Station, Aldo Leopold Wilderness Research Institute: Missuola, MT, USA, 2013.

23. Newbold, T.; Hudson, L.N.; Hill, S.L.L.; Contu, S.; Gray, C.L.; Scharlemann, J.P.W.; Börger, L.; Phillips, H.R.P.; Sheil, D.; Lysenko, I.; et al. Global patterns of terrestrial assemblage turnover within and among land uses. *Ecography* **2016**, *39*, 1151–1163. [CrossRef]

24. Aplet, G.; Thomson, J.; Wilbert, M. Indicators of wildness: Using attributes of the land to assess the context of wilderness. In Proceedings of the Wilderness Science in a Time of Change, Missoula, MT, USA, 23–27 May 1999; McCool, S.F., Cole, D.N., Borrie, W.T., O'Laughlin, J., Eds.; Rocky Mountain Research Station: Fort Collins, CO, USA, 2000; pp. 89–98.

25. Monahan, W.B.; Gross, J.E.; Svancara, L.K.; Philippi, T. *A Guide to Interpreting NPScape Data and Analyses*; Natural Resource Technical Report NPS/NRSS/NRTR—2012/578; National Park Service, U.S. Department of the Interior: Fort Collins, CO, USA, 2012.

26. Mennitt, D.; Sherrill, K.; Fristrup, K. A geospatial model of ambient sound pressure levels in the contiguous United States. *J. Acoust. Soc. Am.* **2014**, *135*, 2746–2764. [CrossRef] [PubMed]

27. Nelson, L.; Kinseth, M.; Flowe, T. *Explanatory Variable Generation for Geospatial Sound Modeling Standard Operating Procedure*; Natural Resource Report NPS/NRSS/NRR—2015/936; National Park Service, U.S. Department of the Interior: Fort Collins, CO, USA, 2015.

28. Theobald, D.M.; Reed, S.E.; Fields, K.; Soulé, M. Connecting natural landscapes using a landscape permeability model to prioritize conservation activities in the United States. *Conserv. Lett.* **2012**, *5*, 123–133. [CrossRef]

29. Anderson, M.G.; Clark, M.; Sheldon, A.O. Estimating climate resilience for conservation across geophysical settings. *Conserv. Biol.* **2014**, *28*, 959–970. [CrossRef] [PubMed]

30. Martin, T.G.; Watson, J.E.M. Intact ecosystems provide best defence against climate change. *Nat. Clim. Chang.* **2016**, *6*, 122–124. [CrossRef]

31. Newbold, T.; Hudson, L.N.; Arnell, A.P.; Contu, S.; Palma, A.; De Ferrier, S.; Hill, S.L.L.; Hoskins, A.J.; Lysenko, I.; Phillips, H.R.P.; et al. Has land use pushed terrestrial biodiversity beyond the planetary boundary? A global assessment. *Science* **2016**, *353*, 288–291. [CrossRef] [PubMed]

32. Newbold, T.; Hudson, L.N.; Hill, S.L.L.; Contu, S.; Lysenko, I.; Senior, R.A.; Börger, L.; Bennett, D.J.; Choimes, A.; Collen, B.; et al. Global effects of land use on local terrestrial biodiversity. *Nature* **2015**, *520*, 45–50. [PubMed]

33. Scholes, R.J.; Biggs, R. A biodiversity intactness index. *Nature* **2005**, *434*, 45–49. [CrossRef] [PubMed]

land

MDPI

Perspective

Fairness and Transparency Are Required for the Inclusion of Privately Protected Areas in Publicly Accessible Conservation Databases

Hayley S. Clements [1,*], Matthew J. Selinske [2,3], Carla L. Archibald [4,5], Benjamin Cooke [3], James A. Fitzsimons [6,7], Julie E. Groce [8], Nooshin Torabi [3] and Mathew J. Hardy [2,3]

[1] Centre for Complex Systems in Transition, Stellenbosch University, Stellenbosch 7600, South Africa
[2] Australian Research Council Centre of Excellence for Environmental Decisions, RMIT University, Melbourne, VIC 3000, Australia; matthew.selinske@rmit.edu.au (M.J.S.); mat.hardy@rmit.edu.au (M.J.H.)
[3] Centre for Urban Research, School of Global, Urban and Social Studies, RMIT University, Melbourne, VIC 3001, Australia; ben.cooke@rmit.edu.au (B.C.); nooshin.torabi@rmit.edu.au (N.T.)
[4] Australian Research Council Centre of Excellence for Environmental Decisions, University of Queensland, St. Lucia, QLD 4072, Australia; c.archibald@uq.edu.au
[5] School of Earth and Environmental Sciences, University of Queensland, St. Lucia, QLD 4072, Australia
[6] The Nature Conservancy, P.O. Box 57, Carlton South, VIC 3053, Australia; jfitzsimons@tnc.org
[7] School of Life and Environmental Sciences, Deakin University, Burwood, VIC 3125, Australia
[8] School of Biological Sciences, Monash University, Clayton, VIC 3800, Australia; julie.groce@monash.edu
* Correspondence: clementshayley@gmail.com; Tel.: +27-21-808-2704

Received: 7 July 2018; Accepted: 11 August 2018; Published: 13 August 2018

Abstract: There is a growing recognition of the contribution that privately-owned land makes to conservation efforts, and governments are increasingly counting privately protected areas (PPAs) towards their international conservation commitments. The public availability of spatial data on countries' conservation estates is important for broad-scale conservation planning and monitoring and for evaluating progress towards targets. Yet there has been limited consideration of how PPA data is reported to national and international protected area databases, particularly whether such reporting is transparent and fair (i.e., equitable) to the landholders involved. Here we consider PPA reporting procedures from three countries with high numbers of PPAs—Australia, South Africa, and the United States—illustrating the diversity within and between countries regarding what data is reported and the transparency with which it is reported. Noting a potential tension between landholder preferences for privacy and security of their property information and the benefit of sharing this information for broader conservation efforts, we identify the need to consider equity in PPA reporting processes. Unpacking potential considerations and tensions into distributional, procedural, and recognitional dimensions of equity, we propose a series of broad principles to foster transparent and fair reporting. Our approach for navigating the complexity and context-dependency of equity considerations will help strengthen PPA reporting and facilitate the transparent integration of PPAs into broader conservation efforts.

Keywords: Convention on Biological Diversity; Aichi Target 11; conservation planning; protected area reporting; equity framework; private land conservation; privacy

---

1. Introduction

Protected areas remain a core global strategy for curbing the current biodiversity extinction crisis [1]. Under the Convention on Biological Diversity (CBD), signatory governments have committed to conserve at least 17% of their terrestrial and 10% of their marine environments by 2020 through "ecologically representative" protected area networks (Aichi Target 11) [2]. Despite significant expansion in the global protected area estate in the past two decades, many countries are predicted to fall short of this target, particularly in terms of ecological representation [3,4]. The majority of the world's reported protected areas are owned and managed by governments [1], with their efficacy and extent constrained in many cases by limited governmental resources and competing priorities [5]. The capacity of these protected areas to protect representative samples of biodiversity is further limited by often historical biases in their locations towards higher elevations, steeper slopes, and less productive portions of the landscape [4,6]. In many countries, the majority of land is privately owned, particularly in highly productive areas, and some of these contain threatened and/or under-represented ecosystems [7–10]. Increasingly, privately protected areas (PPAs) are being used to conserve biodiversity on private land, complementing government-owned protected area estates [8,11–14]. The increase in the number of, and area covered by, PPAs around the world in recent decades [15,16] poses unique opportunities and challenges for monitoring, managing, and expanding the protected area estate.

The potential for PPAs to contribute to conservation has been emphasized by the International Union for Conservation of Nature (IUCN) World Commission on Protected Areas (IUCN-WCPA) [17], and in some countries, PPAs are counted towards international conservation targets (e.g., Aichi Target 11 [15]). PPA information (e.g., geospatial data, property name, management authority) is included within some national and international protected area databases, such as the World Database on Protected Areas (WDPA) [16]. These databases provide information on the distribution of a country's complete protected area estate, which is important for systematic conservation planning, monitoring, and evaluation [18,19] and enables transparent assessments of progress towards international conservation targets (e.g., [20]). While the inclusion of PPAs in national and international databases is beneficial from a conservation perspective, the public accessibility of many of these databases raises questions about whether current approaches of including PPA information are fair and transparent to the PPA owners.

Defined as protected areas under private governance [15], PPAs encompass a diverse set of private property conservation arrangements (e.g., conservation easements, conservation covenants, and land stewardship agreements). In many countries, PPAs are commonly owned by individuals or families (hereafter referred to as "landholders"), and in some instances these landholders live on and/or derive a primary income from their land [15]. These individual or family PPA landholders (as opposed to non-government organizations, NGOs, who own PPAs) are the focus of this paper. The capacity and motivation of these landholders for managing their PPAs can vary [21], as can their awareness of the obligations of owning protected land [22]. Whilst some PPAs are established and managed with public funding (e.g., incentives programs), others rely on private funding or independent action by landholders [23]. This diversity in landholder motivations, management approaches, and residential or financial dependence on PPAs suggests there may be potential differences in landholder preferences regarding the sharing of PPA information. Thus, the inclusion of PPA information in conservation databases warrants considerations beyond those required for protected areas on public land.

However, there has been limited consideration of PPA data reporting processes, and it remains unclear what PPA information is collated, who has access to this information, whether landholders are aware of reporting procedures, and how their perspectives are accounted for [24–26]. Examining these questions is timely given the current development of PPA best-practice guidelines by the IUCN (referred to in [16]) and the 2020 deadline for achieving the CBD Aichi Targets. The 2016 IUCN World Conservation Congress approved a resolution on supporting PPAs, which calls on IUCN members to "include privately protected areas that meet the requirements of IUCN Protected Area Standards when reporting about protected area coverage and other related information, including to the World Database on Protected Areas (WDPA) and to the CBD, *in collaboration and agreement with the owners of such areas*" (emphasis added) [17] (p. 2). A discussion on PPA data reporting is also timely given the General Data Protection Regulation that became enforceable in the European Union earlier this year (a region in which PPA numbers are increasing [14,27,28]).

Acknowledging the benefits of compiling PPA databases, we explore in this paper current international and several national procedures for reporting PPA data and the extent to which landholders are informed of these procedures. We focus on the three countries with the highest numbers of reported PPAs in the WDPA: Australia, South Africa, and the United States of America (USA) [16]. Given the apparent need for transparency and fairness in PPA reporting, particularly for PPAs owned by individuals (as opposed to NGOs), we apply an equity framework to identify the potential tensions and implications of including PPA data in publicly accessible databases. This broad framework leaves room for the diversity of contexts in which PPAs are administered. As such, our aim is to prompt reflection against equity principles in a manner that acknowledges this context diversity rather than offer a rigid tool for universal application. Finally, we synthesize the insights into a series of principles to help policy makers navigate potential issues and promote equitable reporting of PPA data. We emphasize the importance of navigating these issues to ensure effective integration of PPAs into conservation planning, management, and monitoring where agreed by the owners of such areas.

2. Current International and National PPA Data Management Procedures

Debate around providing access to personal or sensitive data spans many domains, including big data, e-health, and law enforcement (e.g., [29,30]). One prominent example within conservation is whether to publish the location of threatened species. Some researchers argue that location data should be kept confidential given the risks of poaching [31], while others promote open-access to enable effective conservation planning and management [32,33]. In the case of PPAs, the value of comprehensive protected area databases for conservation planning and management may come with risks to the landholders who make available their property information. For example, concerns have been raised by conservation organizations and landholders that publishing the location of a PPA may encourage trespassing or be used by property developers to identify undervalued land [24,26]. The need for the data owner's consent to share their information is also a common feature in international data-sharing policies [34]. It is thus important to explore whether and how current international and national protected area reporting processes navigate these considerations.

The importance of international collaboration between state and private actors for successful conservation initiatives, including protected areas, is increasingly discussed [35–37]. In reporting protected area data, this collaboration typically takes the form of national governments collating protected area (including in some instances PPA) information from a variety of sources (e.g., national and subnational government agencies, NGOs) and reporting this information to the United Nations Environmental Program-World Conservation Monitoring Centre, which curates the WDPA [38] (Figure 1). NGOs can also submit protected area data directly to the WDPA subsequent to data verification by the WCPA [16]. Unless otherwise specified, this data is made freely available online where it is used for a variety of purposes, including conservation research, the development of conservation indicators and targets (e.g., Sustainable Development Goals Indicator 15.1.2 and Aichi Target 11), and reporting on conservation progress (e.g., Protected Planet Report) [38]. The WDPA also accepts data with restrictions on use and dissemination. If PPA data is considered sensitive by the data provider, it can be used by WDPA managers for analyses but not shared further [16]. The WDPA requires, at a minimum, the protected area name, management authority, and geographic location [38]. Ideally, management plans and geospatial data of protected area boundaries are also provided. Data is only accepted into the WDPA after the provider has signed a contributor agreement, stipulating whether the data can be shared publicly and verifying that the "relevant stakeholders and rights-holders" have agreed to the provision of the data [38] (p. 60). However, for PPA data provided to the WDPA, there is limited information regarding the extent to which relevant landholders are aware of, and agree with, the inclusion of their data [16]. There is therefore a need to examine how PPA data is collected, managed, and reported within countries (Figure 1).

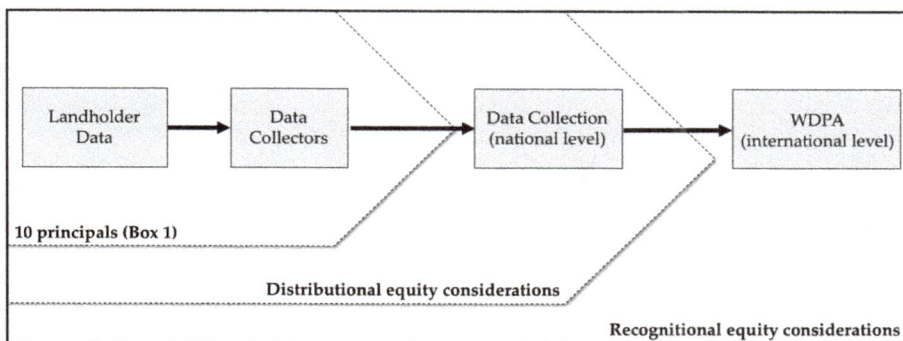

Figure 1. Conceptual flow of privately protected area (PPA) data from local (landholder) to international (World Database on Protected Areas; WDPA) reporting levels. Recognitional equity considerations are important at all levels, while distributional and procedural equity considerations are most applicable to the PPA data collection and collation decisions, rules, and responsibilities established within a country (Table 1). The ten principles described in Box 1 are intended to guide equitable data collection by the local, state, or regional organizations that oversee the PPA agreements.

In their recent review of 17 countries, Stolton et al. (2014) [15] found that 12 countries had national databases of their PPA estates, though the majority were incomplete. Here we illustrate the diversity in PPA reporting using three examples: Australia, South Africa, and the USA. These three countries currently report the greatest number of PPAs to the WDPA (together accounting for 87% of all PPAs in the WDPA in 2017 [16]), and therefore represent good examples for considering reporting procedures across three diverse continents. These are not intended to be in-depth case studies but are rather illustrations of current PPA reporting issues.

In Australia, conservation covenants and private reserves owned by NGOs are the key mechanisms for establishing PPAs [25,39]. Conservation covenants are binding agreements between landholders and an authorized agency (e.g., state-level government, NGOs), established to protect natural features in perpetuity, sometimes incentivized through the provision of financial support or management guidance [25]. The PPA estate in Australia has seen impressive expansion in recent years, increasing the biodiversity representation and connectivity of the National Reserve System [19,25]. While Australia has a publicly available national protected area database that has shown good progress in its inclusion of PPAs (Collaborative Australian Protected Area Database; CAPAD), covenants are designated under state legislation and only some states and agencies provide data to the national database [25]. Factors impeding covenant reporting include privacy concerns about revealing property locations and a lack of coordination among stakeholders (e.g., state governments versus the Australian government) [25,40]. While many interviewed landholders viewed inclusion of their covenanted land in the national database positively, others had a negative perspective towards inclusion due to concerns that their contribution may lessen the government's responsibility for meeting national protected area targets on publicly-owned land (i.e., additionality does not take place through state protected area expansion because governments count PPAs that were being conserved anyway towards their targets) [41]. For some PPAs, there is transparency that landholders' property information will be included in national databases upon signing agreements (e.g., new PPAs purchased by NGOs with funds from the Australian Government's National Reserve System Program and new covenants signed under the Tasmanian Private Forest Reserve Program in exchange for financial incentives) [25]. For other conservation covenants that are currently reported nationally, it is unclear whether landholders are made aware that information about their properties may be included as PPAs in national and international databases. We are aware that some programs have not sought explicit permission to include this data in the CAPAD.

In South Africa, the Department of Environmental Affairs has a legislative mandate to maintain a publicly available register of South Africa's conservation estate [42] and an advanced legislative system for formally recognizing PPAs, which have become the focus of protected area expansion efforts [43]. South Africa's primary tool for expanding its estate of protected areas on privately owned land is the national biodiversity stewardship initiative [44]. Biodiversity Stewardship Agreements are established through a contract between the landholder and the provincial conservation agency, and data-reporting policies are not evident in the national contract template [45]. However, the establishment of a formal PPA requires a public gazettal process [45], suggesting landholders are likely to be aware of the public availability of their property's name, land use, and geospatial data.

The establishment of most private conservation areas in the USA is negotiated by organizations that either purchase land directly or specialize in conservation easements (agreements between landholders and organizations regarding land use restrictions to achieve conservation in exchange for payment, tax benefits, or development permits [23]). The publicly accessible National Conservation Easement Database has successfully aggregated data on easements held by thousands of conservation organizations (NGOs, state and federal governments), but many organizations have declined to provide data [26]. Concern for landholder privacy is reported as a primary deterrent for those not reporting as well as prior agreements with landholders not to share locations and fear that this data could be used by developers to identify undervalued properties [24,26]. Although the USA does not formally recognize a unified national protected area system and has no PPA definition, an impressive 8731 USA PPAs are reported in the WDPA—the most of any country [46].

These examples highlight some of the diversity both within and between countries in the extent of, processes associated with, and transparency regarding the reporting of PPA data. This diversity reflects complex governance and legislative arrangements involving landholders, NGOs, and governmental agencies that operate at local through to national scales. Based on the examples above, it is not always clear whether affected landholders are made aware of data management policies and whether they are provided the opportunity to state their preference regarding being labeled as a PPA and included in national and international databases and towards conservation targets. While some landholders viewed inclusion in national protected area systems positively [41], several studies have noted landholder concerns regarding the public sharing of their information ranging from privacy risks to a reluctance to negate governments of their conservation responsibilities [26,41]. The lack of reporting transparency and landholder concerns around PPA data sharing are not specific to the three focal countries. During the compilation of a PPA database for Mexico, for example, concerns were raised by some landholders that the misuse of their PPA information by others could lead to instances of blackmail [16,47]. When these concerns are weighed against the importance of sharing this information for broader societal benefits, such as effective conservation planning, potential tensions emerge between private and public good [24]. These tensions need to be navigated fairly and transparently.

3. Issues of Equity around PPA Data Inclusion in Publicly Accessible Databases

Questions of fairness and transparency in PPA reporting suggest the need to engage with the concept of equity, which is broadly defined as the fair and just treatment of individuals or groups within society [48]. The consideration of equity in conservation can be regarded as both fundamental (i.e., it is inherently right) and outcome-based (i.e., it can assist in achieving effective long-term conservation) [49]. For example, perceptions of unfairness amongst communities affected by conservation policies can lead to increased costs for conservation programs [50]. There is increasing recognition that conservation policy has often neglected issues of equity regarding people affected by policy prescriptions [51]. Equity brings a focus to questions of legitimate process, participant buy-in, increased accountability, and transparent compliance for conservation [50,51].

Table 1. Considerations of three dimensions of equity [48–50] in the inclusion of privately protected area (PPA) data (e.g., geospatial data, property name, management authority) in publicly available databases and towards national and international conservation targets. These considerations are relevant to the organizations that engage directly with landholders, facilitate PPA agreements, and collect PPA data ('data collectors'; Figure 1). Some considerations are also relevant to the national organizations that collate this information and decide whether to contribute it to international databases.

Dimensions of Equity	Level of Relevance	Considerations
Procedural: equitable involvement and inclusion of all stakeholder groups in rule-making and decisions	Data collectors	• If PPA inclusion in publicly accessible databases and/or towards targets is mandatory for a program or country, are efforts made to ensure landholders understand the reporting process and its implications when they enter into a PPA agreement, and have agreed to this? • Who is responsible for making landholders aware of their obligations and rights, and for communicating any changes in national or international policies to landholders (e.g., the organization engaging with PPA landholders, or the regional or national government)?
	National	• To what extent should landholders be involved in decisions about what data is reported to national and international databases? Is it important that landholders feel their preferences have been considered? • Are national and international reporting rules and agreements consistent with those that landholders sign when establishing PPAs on their properties or do they place extra levels of obligation not previously agreed to? • Who owns the PPA data (e.g., the landholder, the conservation organization, the government), and is this recognized formally in PPA agreements? If additional parties provide funding or support in establishing or managing a PPA, does that entitle them to (partial) ownership of the data and thus a say in decisions about the data's accessibility?
Distributional: equitable distribution of costs, benefits, rights, responsibilities, and risk within and among groups from present and future generations	Data collectors	• What data security risks are associated with storing and sharing PPA information, and how are they being managed by the organizations involved in PPA reporting? (e.g., sharing PPA data could cause trespassing issues or create a target for developers or marketers [26]). • How are equity issues addressed as the PPA changes ownership? (e.g., an original owner may have agreed to share information, yet a subsequent owner does not; see [22]).
	National	• What are the landholders' costs, risks, and responsibilities of inclusion relative to the public benefits derived from their inclusion in publicly accessible databases? • Is there a risk that the transaction costs associated with reporting could impact rates of PPA establishment and ongoing landholder satisfaction, thus influencing public conservation benefits [52]? • If public funds are used in the process of establishing PPAs (e.g., identifying, incentivizing, capacity building [53]) should landholders be required to allow data related to their properties to be used for national conservation reporting [24,26]? • Does the use of PPAs in delivering 'common good' conservation outcomes (e.g., knowledge of conservation progress and improved ability to plan strategically for conservation at a landscape scale [54]) compel data sharing? Does this apply in situations where landholders finance the costs of PPA establishment and management? • Does the reporting of PPA data that contributes to national conservation estates add to or substitute for government responsibility for meeting international targets? (e.g., could PPAs reduce the responsibilities of governments for meeting international protected area targets through public land management/investment [26,41]?).

Table 1. *Cont.*

Dimensions of Equity	Level of Relevance	Considerations
	Data collectors	• Have reporting and communication processes considered the diversity in landholders' awareness of the consequences of PPA reporting (e.g., [22])? • How are geographical, political and program-specific considerations incorporated into the reporting process? (e.g., in the USA, different programs are subject to different laws that influence inclusion requirements and privacy concerns [26]).
Recognitional: equitable respect for knowledge systems, values, social norms, and rights of all stakeholders, and consideration of the diversity of institutional and political settings.	National	• Are all institutions and levels of government involved in reporting processes considered trustworthy by landholders and international conservation bodies, and does this influence landholders' willingness to share data [52,55]? • Has consideration been made for landholders' motivations and how these might influence willingness to share information? Engaging with motives can be vital for PPA success [55,56]; some motivations may lead to support for data-sharing (e.g., motivation to contribute to national conservation efforts [41]), while others may result in a lack of support for data-sharing (e.g., motivation for privacy [26]). • Does reporting consider land tenure and cultural differences (e.g., individualistic versus collective ownership, history of land tenure, differences between first and subsequent generation PPA landholders [22]). • To what extent could PPA reporting influence conflict over governance arrangements and contested property rights? (e.g., property rights and governance processes can be fluid and unstable depending on institutional, community and individual power dynamics [24]).

The importance of equitable processes in the establishment and management of protected areas is emphasized by the CBD under Aichi Target 11, which states that indigenous and local communities "should equitably share in the benefits arising from protected areas and should not bear inequitable costs" [2]. Here we highlight the relevance of equity issues beyond protected area establishment and management to protected area data reporting when private landholders are involved. This is particularly relevant considering that a new post-2020 global Strategic Plan for Biodiversity will be negotiated over the coming years and new targets will be set that will require equity considerations. Following from recent work on equity in conservation contexts [48–50], we use three dimensions of equity—procedural, distributional, and recognitional—to unpack potential fairness considerations and tensions associated with including PPAs in publicly accessible databases. We frame these considerations as a series of questions to facilitate reflection on reporting processes for different organizations (Table 1). We include questions that are (1) most relevant to the organizations that engage with landholders, facilitate PPA agreements, and collect PPA information, and (2) also relevant to organizations that collate this information at a national level (see Figure 1). The questions listed are not intended to be prescriptive, nor do we assume they are all-encompassing given the likely range of equity issues in different PPA contexts. The questions offer prompts for considering how organizations can respond to the issues raised in a manner that is applicable to their context and governance arrangements.

Procedural equity refers to the equitable involvement and inclusion of all stakeholder groups in rule-making and decisions [48]. Questions of procedural equity for PPA data reporting relate to how decisions are made regarding whether to include PPAs in publicly accessible databases and towards international targets and the extent to which different landholders can participate in these decisions (Table 1). Distributional equity refers to the equitable distribution of costs, benefits, rights, responsibilities, and risk within and among groups from present and future generations [48] and is thus associated with how the benefits and costs for public and private stakeholders involved in PPA reporting processes are distributed (Table 1). Finally, recognitional equity relates to the equitable respect for knowledge systems, values, social norms, and rights of all stakeholders and consideration of the diversity of institutional and political settings [48], calling for consideration of the diversity of PPA landholders and the environments in which PPAs operate (Table 1). Questions of distributional, procedural, and recognitional equity are likely to be important for guiding decisions, rules, and responsibilities established by organizations collecting, managing, and sharing PPA data within a country (Figure 1). In addition, the relevance of recognitional equity considerations may extend to the international organizations aggregating and sharing this data (Figure 1).

4. Navigating Equity Considerations: Principles for Fair and Transparent PPA Data Reporting

Organizations face significant challenges in ensuring that PPA reporting processes are fair and transparent (challenges of capacity, complex governance arrangements, etc. [15,26]). However, potentially larger challenges to long-term conservation may arise if reporting is not transparent. An equitable reporting process is important for developing and maintaining organizational legitimacy and ensuring landholders feel supported—two factors that can influence landholder motivation to participate in conservation activities and thus have a notable impact on conservation outcomes [50,53]. Equitable reporting is also imperative for maintaining the quality of PPA data and fulfilling the WDPA requirement that the data contributors "have the rights, permissions and authority" to report data [38] (p. 59). The retraction of non-consented data from the WDPA (in line with the UNEP-WCMC protocol) could have direct consequences for the ability of conservation practitioners, policy makers and researchers to effectively plan and monitor conservation efforts across landscapes and at national and international scales. While arrangements that ensure the fair inclusion of PPAs in public databases will be context-specific (Table 1), we propose a set of broad principles that could guide the consideration of equity in a consistent and transparent process across a range of contexts (Box 1).

Box 1. Ten principles for the fair and transparent inclusion of PPA data in publicly accessible databases (adapted from Greenleaf (2012) [34]).

1.	**Data collection**—only adequate and relevant data is collected, it is not excessive in relation to stated purpose, and it is collected fairly, lawfully, and with the landholder's full knowledge and consent.
2.	**Data quality**—data is accurate and kept up-to-date.
3.	**Purpose specification**—at the time of collection, the data collector provides clear information about what data is being collected and for what purpose.
4.	**Notice of rights**—at the time of collection, the data collector provides clear information about their practices and policies, and the choices available to the landholder.
5.	**Limited use**—data is stored and used for specific and clearly defined purposes for no longer than is required.
6.	**Data security**—data is protected from risks (e.g., loss, unauthorized access) by reasonable security safeguards.
7.	**Openness**—there is a policy of openness about any changes made to the practices and policies with respect to data.
8.	**Access**—landholders are able to enquire and receive confirmation about what data, relating to them or their property, has been collected and is stored.
9.	**Correction**—landholders are able to challenge data relating to them or their property and, if incorrect, have the data erased, rectified or amended.
10.	**Accountability**—data controllers are accountable for complying with measures above and must ensure that the recipient agency or organization will protect the information in the same manner.

The collection, management, and reporting of privately-owned data is a widespread challenge that has led to the development of privacy laws in at least 87 countries [34] and, most recently, the General Data Protection Regulation that became enforceable in the European Union in early 2018. Drawing from several regional agreements on data privacy from around the world, Greenleaf (2012) collated ten "global" principles for transparent data reporting [34]. The widespread geographic use of these principles thus makes them a useful starting point for considering fair and transparent reporting processes regarding PPAs around the world. We present these ten principles (Box 1) as considerations for organizations involved in the collection, management, and reporting of PPA data (Figure 1). While these organizations generally have the power to influence procedural equity (which is the focus of many of the proposed principles), the intent is to account for recognitional and distributional equity through such procedures.

It is imperative that transparency around PPA data procedures begins during discussions with landholders interested in establishing conservation agreements for their property and that this transparency is maintained for prospective buyers of properties already under a conservation agreement. The organization should ensure landholders are made aware of what information is to be collected, how it will be managed and distributed, by whom, and for what purposes (P1-4; Box 1). In some contexts (e.g., South Africa), national reporting is legally mandated and consent thus entails making landholders aware of their obligations. In other contexts, landholders may be able to decide what (if any) of their information is included in national databases. The process of obtaining consent should consider the recognitional aspects of equity (e.g., landholders' motivations and awareness of responsibilities) as well as distributional aspects (e.g., whether it is fair to provide landholders with options around data-sharing if public funds have been used to establish the PPA; Table 1). Irrespective of the context and nature of consent, organizations collecting and reporting PPA information need to have clear processes for ensuring landholders are made aware of their rights, obligations, and options.

Transparency of data management practices and the security of the data collected are key considerations regarding PPA data reporting (P5-7; Box 1). Managers of PPA data need to be cognizant of the risks associated with sharing this data (distributional equity). These risks are likely to vary between different landholders (recognitional equity), and consideration should be given to how these risks can be mitigated through data-management processes. Data-protection policies suggest that data should be used only to fulfill the purposes of collection and other compatible purposes [34]. These purposes need to be made explicit, particularly regarding whether they justify conservation organizations sharing PPA information to national and international databases. In some instances, a tiered data management approach may be most appropriate whereby a broad set of relevant data could be collected by the conservation agency with the full consent of the landowner, with just a subset of this data then shared with the national government, to meet national reporting needs while protecting more sensitive data. The national government could share a subset of this data with WDPA to meet international reporting requirements. An initiative similar to the Indigenous and Community Conserved Areas (ICCA) registry could be developed for PPAs. The ICCA registry was initiated in response to data-reporting concerns by communities and enables the communities themselves to choose what data is made publicly available [57].

In addition to the organization's responsibilities towards data collection and security, landholders whose personal data has been collected should be able to enquire about, have access to, and ensure accuracy of that data (P8-10; Box 1). It is unclear to what extent landholders can currently access and correct their personal information in national and international databases. Finally, the organizations who are collecting and reporting this data—the first point of contact for the landholders—should be the ones accountable for complying with data sharing protocols, such as these ten principles, and should gain assurances from the recipient organizations about data accuracy and protection (Figure 1).

We recognize that operationalizing these principles will have practical implications for conservation organizations, many of which are already operating in a resource-constrained environment. Given the diversity of PPA mechanisms and contexts, the implications of implementing these principles will vary but may, for example, require PPA programs to review the wording of their PPA agreements, data administration and security, current reporting procedures, and internal guidelines. Amongst this, programs will need to take into consideration a variety of issues, such as accommodating landholder preferences in both existing and future PPA agreements and managing these as PPAs change ownership (Table 1).

While these principles are presented for PPAs specifically, they are equally applicable for other effective area-based conservation measures (OECMs) on private land, which also contribute to global conservation targets [44]. Further, these principles are also worth considering in other conservation contexts that involve the public reporting of data pertaining to non-public stakeholders, such as carbon and biodiversity offset programs. Organizations that follow these principles would be well-placed to meet the WDPA requirements that data is shared "with the free, prior and informed consent of communities and/or indigenous peoples involved in the management, governance or ownership of the sites described in the dataset" [38] (p. 60).

While we have focused largely on recommended procedures going forward, we are aware that in some jurisdictions, PPA data is already being reported at national and international levels without the express permission of the landholders. In line with the 2016 World Conservation Congress motion on PPAs [17], we stress that IUCN members and other organizations reporting PPA information to international databases should do so "in collaboration and agreement with the owners of such areas" and that the WDPA review PPAs already in the database to ensure such agreements are evident. We call on national governments, the WDPA, and PPA programs to work together to implement the principles outlined above to achieve equitable PPA reporting.

5. Conclusions

Considerations of equity are gaining increasing attention in conservation, including protected area establishment and management [58]. Here we have illustrated that such considerations need to extend to protected area data reporting processes when private landholders are involved. It is important to consider procedural, distributional, and recognitional dimensions of equity, including questions of landholder consent regarding data-sharing, the distribution of costs and benefits of data-sharing between private individuals and the public good, and recognition of diverse landholder motivations and tenure arrangements (Table 1). We have offered a set of ten broad principles to help organizations navigate the complexity and context-dependency of equity considerations for PPA data reporting (Box 1). With the growing number and extent of PPAs around the world, there is increasing recognition that conservation planning, management, monitoring, and evaluation would greatly benefit from the inclusion of these privately-owned properties. Wherever the reporting of PPA data is required, and deemed appropriate, we stress the importance of providing fair and transparent reporting processes. This will facilitate effective and integrated conservation efforts and rigorous assessments of progress at national and international levels.

Author Contributions: All authors contributed to the following: Conceptualization, Writing—Original Draft Preparation, and Writing—Review & Editing.

Funding: This research was funded by a Claude Leon Fellowship (HC); Australian Postgraduate Awards (CA, JG); the Australian Research Council Centre of Excellence for Environmental Decisions grant number CE11001000104, funded by the Australian Government (CA, MH, MS); and The Nature Conservancy (JF).

Acknowledgments: We thank Brent Mitchell and two anonymous reviewers for valuable feedback on an earlier version of this paper.

Conflicts of Interest: The authors declare no conflict of interest.

References

1. UNEP-WCMC; IUCN. *Protected Planet Report 2016*; UNEP-WCMC: Cambridge, UK; Gland, Switzerland, 2016.
2. CBD. *Convention on Biological Diversity's Strategic Plan for 2020*; CBD: Montreal, QC, Canada, 2010.
3. Butchart, S.H.M.; Clarke, M.; Smith, R.J.; Sykes, R.E.; Scharlemann, J.P.W.; Harfoot, M.; Buchanan, G.M.; Angulo, A.; Balmford, A.; Bertzky, B.; et al. Shortfalls and solutions for meeting national and global conservation area targets. *Conserv. Lett.* **2015**, *8*, 329–337. [CrossRef]

4. Venter, O.; Magrach, A.; Outram, N.; Klein, C.J.; Marco, M.D.; Watson, J.E.M. Bias in protected-area location and its effects on long-term aspirations of biodiversity conventions. *Conserv. Biol.* **2018**, *32*, 127–134. [CrossRef] [PubMed]

5. Watson, J.E.M.; Dudley, N.; Segan, D.B.; Hockings, M. The performance and potential of protected areas. *Nature* **2014**, *515*, 67–73. [CrossRef] [PubMed]

6. Joppa, L.N.; Pfaff, A. High and far: Biases in the location of protected areas. *PLoS ONE* **2009**, *4*, e8273. [CrossRef] [PubMed]

7. Norton, D.A. Conservation biology and private land: Shifting the focus. *Conserv. Biol.* **2000**, *14*, 1221–1223. [CrossRef]

8. Gallo, J.A.; Pasquini, L.; Reyers, B.; Cowling, R.M. The role of private conservation areas in biodiversity representation and target achievement within the Little Karoo region, South Africa. *Biol. Conserv.* **2009**, *142*, 446–454. [CrossRef]

9. Fitzsimons, J.; Wescott, G. The role and contribution of private land in Victoria to biodiversity conservation and the protected area system. *Aust. J. Environ. Manag.* **2001**, *8*, 142–157. [CrossRef]

10. Knight, A.T. Private lands: The neglected geography. *Conserv. Biol.* **1999**, *13*, 223–224. [CrossRef]

11. Shanee, S.; Shanee, N.; Monteferri, B.; Allgas, N.; Pardo, A.A.; Horwich, R.H. Protected area coverage of threatened vertebrates and ecoregions in Peru: Comparison of communal, private and state reserves. *J. Environ. Manag.* **2017**, *202*, 12–20. [CrossRef] [PubMed]

12. Pegas, F.D.V.; Castley, J.G. Private reserves in Brazil: Distribution patterns, logistical challenges, and conservation contributions. *J. Nat. Conserv.* **2016**, *29*, 14–24. [CrossRef]

13. Von Hase, A.; Rouget, M.; Cowling, R.M. Evaluating private land conservation in the Cape lowlands, South Africa. *Conserv. Biol.* **2010**, *24*, 1182–1189. [CrossRef] [PubMed]

14. Manolache, S.; Nita, A.; Ciocanea, C.M.; Popescu, V.D.; Rozylowicz, L. Power, influence and structure in Natura 2000 governance networks. A comparative analysis of two protected areas in Romania. *J. Environ. Manag.* **2018**, *212*, 54–64. [CrossRef] [PubMed]

15. Stolton, S.; Redford, K.H.; Dudley, N. *The Futures of Privately Protected Areas*; IUCN: Gland, Switzerland, 2014.

16. Bingham, H.; Fitzsimons, J.A.; Redford, K.H.; Mitchell, B.A.; Bezaury-Creel, J.; Cumming, T.L. Privately protected areas: Advances and challenges in guidance, policy and documentation. *Parks* **2017**, *23.1*, 13–28. [CrossRef]

17. IUCN. *WCC-2016-Res-036-EN Supporting Privately Protected Areas*; IUCN: Gland, Switzerland, 2016; Available online: https://portals.iucn.org/library/node/46453 (accessed on 12 August 2018).

18. Margules, C.R.; Pressey, R.L. Systematic conservation planning. *Nature* **2000**, *405*, 243–253. [CrossRef] [PubMed]

19. Fitzsimons, J.A.; Wescott, G. The role of multi-tenure reserve networks in improving reserve design and connectivity. *Landsc. Urban Plan.* **2008**, *85*, 163–173. [CrossRef]

20. Tittensor, D.P.; Walpole, M.; Hill, S.L.L.; Boyce, D.G.; Britten, G.L.; Burgess, N.D.; Butchart, S.H.M.; Leadley, P.W.; Regan, E.C.; Alkemade, R.; et al. A mid-term analysis of progress toward international biodiversity targets. *Science* **2014**, *346*, 241–248. [CrossRef] [PubMed]

21. Fitzsimons, J.A.; Carr, C.B. Conservation covenants on private land: Issues with measuring and achieving biodiversity outcomes in Australia. *Environ. Manag.* **2014**, *54*, 606–616. [CrossRef] [PubMed]

22. Stroman, D.A.; Kreuter, U.P. Perpetual conservation easements and landowners: Evaluating easement knowledge, satisfaction and partner organization relationships. *J. Environ. Manag.* **2014**, *146*, 284–291. [CrossRef] [PubMed]

23. Kamal, S.; Brown, G. Conservation on private land: A review of global strategies with a proposed classification system. *J. Environ. Plan. Manag.* **2015**, *58*, 576–597. [CrossRef]

24. Olmsted, J.L. The invisible forest: Conservation easement databases and the end of clandestine conservation of natural lands. *Law Contemp. Probl.* **2011**, *74*, 51–82.

25. Fitzsimons, J.A. Private protected areas in Australia: Current status and future directions. *Nat. Conserv.* **2015**, *10*, 1–23. [CrossRef]

26. Rissman, A.R.; Owley, J.; L'Roe, A.W.; Morris, A.W.; Wardropper, C.B. Public access to spatial data on private-land conservation. *Ecol. Soc.* **2017**, *22*, 24. [CrossRef]

27. Rafa, M. Spain. In *The Futures of Privately Protected Areas*; Stolton, S., Redford, K.H., Dudley, N., Eds.; IUCN: Gland, Switzerland, 2014; pp. 92–94.

28. Heinonen, M. Finland. In *The Futures of Private Protected Areas*; Stolton, S., Redford, K.H., Dudley, N., Eds.; IUCN: Gland, Switzerland, 2014; pp. 70–74.
29. Goldenfein, J. Police photography and privacy: Identity, stigma and reasonable expectation. *UNSW Law J.* **2013**, *36*, 256.
30. Hoffman, S. Citizen science: The law and ethics of public access to medical big data. *Berkeley Technol. Law J.* **2015**, *30*, 1741–1806.
31. Lindenmayer, D.B.; Scheele, B. Do not publish. *Science* **2017**, *356*, 800–801. [CrossRef] [PubMed]
32. Lowe, A.J.; Smyth, A.K.; Atkins, K.; Avery, R.; Belbin, L.; Brown, N.; Budden, A.E.; Guru, S.; Hardie, M.; Smits, J.; et al. Publish openly but responsibly. *Science* **2017**, *357*, 141–142. [CrossRef] [PubMed]
33. Tulloch, A.I.T.; Auerbach, N.; Avery-Gomm, S.; Bayraktarov, E.; Butt, N.; Dickman, C.R.; Ehmke, G.; Fisher, D.O.; Grantham, H.; Holden, M.H.; et al. A decision tree for assessing the risks and benefits of publishing biodiversity data. *Nat. Ecol. Evol.* **2018**, *2*, 1209–1217. [CrossRef] [PubMed]
34. Greenleaf, G. The influence of European data privacy standards outside Europe: Implications for globalization of Convention 108. *Int. Data Priv. Law* **2012**, *2*, 68–92. [CrossRef]
35. Lemos, M.C.; Agrawal, A. Environmental governance. *Annu. Rev. Environ. Resour.* **2006**, *31*, 297–325. [CrossRef]
36. Nita, A.; Ciocanea, C.M.; Manolache, S.; Rozylowicz, L. A network approach for understanding opportunities and barriers to effective public participation in the management of protected areas. *Soc. Netw. Anal. Min.* **2018**, *8*, 1–11. [CrossRef]
37. Bodin, Ö. Collaborative environmental governance: Achieving collective action in social-ecological systems. *Science* **2017**, *357*, eaan1114. [CrossRef] [PubMed]
38. UNEP-WCMC. *World Database on Protected Areas User Manual 1.5*; UNEP-WCMC: Cambridge, UK, 2017.
39. Fitzsimons, J.A. Private Protected Areas? Assessing the suitability for incorporating conservation agreements over private land into the National Reserve System: A case study of Victoria. *Environ. Plan. Law J.* **2006**, *23*, 365–385.
40. Hardy, M.J.; Fitzsimons, J.A.; Bekessy, S.A.; Gordon, A. Exploring the permanence of conservation covenants. *Conserv. Lett.* **2017**, *10*, 221–230. [CrossRef]
41. Fitzsimons, J.A.; Wescott, G. Perceptions and attitudes of land managers in multi-tenure reserve networks and the implications for conservation. *J. Environ. Manag.* **2007**, *84*, 38–48. [CrossRef] [PubMed]
42. Fourie, N. The South African database on Protected and Conserved areas (SAPAD)—Realising the objectives of the SDI Act and custodianship. In *Geomatics Indaba Proceedings 2015—Stream 2*; EE Publishers: Muldersdrift, South Africa, 2015; pp. 88–99.
43. DEA; SANBI. *National Protected Area Expansion Strategy for South Africa*; DEA: Springfield, VA, USA, 2009.
44. Mitchell, B.A.; Fitzsimons, J.A.; Stevens, C.M.D.; Wright, D.R. PPA or OECM? Differentiating between privately protected areas and other effective area-based conservation measures on private land. *Parks* **2018**, *24 (Special Issue)*, 49–60. [CrossRef]
45. DEA. *Biodiversity Stewardship Guidelines*; Department of Environmental Affairs: Pretoria, South Africa, 2009.
46. IUCN & UNEP-WCMC. *The World Database on Protected Areas (WDPA), September 2017*; IUCN: Cambridge, UK, 2017. Available online: https://www.iucn.org/theme/protected-areas/our-work/world-database-protected-areas (accessed on 12 August 2018).
47. Bezaury-Creel, J.E.; Ochoa-Ochoa, L.M.; Torres-Origel, J.F. *Base de Datos Geográfica de las Reservas de Conservación Privadas y Comunitarias en México—Versión 2.1 Diciembre 31, 2012*; The Nature Conservancy: Mexico City, Mexico, 2012.
48. McDermott, M.; Mahanty, S.; Schreckenberg, K. Examining equity: A multidimensional framework for assessing equity in payments for ecosystem services. *Environ. Sci. Policy* **2013**, *33*, 416–427. [CrossRef]
49. Law, E.A.; Bennett, N.J.; Ives, C.D.; Friedman, R.; Davis, K.J.; Archibald, C.; Wilson, K.A. Equity trade-offs in conservation decision making. *Conserv. Biol.* **2018**, *32*, 294–303. [CrossRef] [PubMed]
50. Pascual, U.; Phelps, J.; Garmendia, E.; Brown, K.; Corbera, E.; Martin, A.; Gomez-Baggethun, E.; Muradian, R. Social equity matters in payments for ecosystem services. *BioScience* **2014**, *64*, 1027–1036. [CrossRef]
51. Dawson, N.; Martin, A.; Danielsen, F. Assessing equity in protected area governance: Approaches to promote just and effective conservation. *Conserv. Lett.* **2018**, *11*, e12388. [CrossRef]

52. Torabi, N.; Mata, L.; Gordon, A.; Garrard, G.; Wescott, W.; Dettmann, P.; Bekessy, S.A. The money or the trees: What drives landholders' participation in biodiverse carbon plantings? *Glob. Ecol. Conserv.* **2016**, *7*, 1–11. [CrossRef]

53. Selinske, M.J.; Cooke, B.; Torabi, N.; Hardy, M.J.; Knight, A.T.; Bekessy, S.A. Locating financial incentives among diverse motivations for long-term private land conservation. *Ecol. Soc.* **2017**, *22*, 7. [CrossRef]

54. Cooke, B.; Moon, K. Aligning "public good" environmental stewardship with the landscape-scale: Adapting MBIs for private land conservation policy. *Ecol. Econ.* **2015**, *114*, 152–158. [CrossRef]

55. Cooke, B.; Langford, W.T.; Gordon, A.; Bekessy, S. Social context and the role of collaborative policy making for private land conservation. *J. Environ. Plan. Manag.* **2012**, *55*, 469–485. [CrossRef]

56. Selinske, M.J.; Coetzee, J.; Purnell, K.; Knight, A.T. Understanding the motivations, satisfaction, and retention of landowners in private land conservation programs. *Conserv. Lett.* **2015**, *8*, 282–289. [CrossRef]

57. ICCA. Indigenous and Community Conserved Areas Registry. Available online: http://www.iccaregistry.org/ (accessed on 7 July 2018).

58. Franks, P.; Booker, F.; Roe, D. *Understanding and Assessing Equity in Protected Area Conservation*; IEED Issue Paper; IEED: London, UK, 2018; ISBN 9781784315559.

land

MDPI

Perspective

Understanding the Biodiversity Contributions of Small Protected Areas Presents Many Challenges

Robert F. Baldwin * and Nakisha T. Fouch

Department of Forestry and Environmental Conservation, Clemson University, Clemson, SC 29634, USA;
nfouch@g.clemson.edu
* Correspondence: baldwi6@clemson.edu; Tel.: +1-864-656-1976

Received: 27 September 2018; Accepted: 18 October 2018; Published: 20 October 2018

Abstract: Small protected areas dominate some databases and are common features of landscapes, yet their accumulated contributions to biodiversity conservation are not well known. Small areas may contribute to global biodiversity conservation through matrix habitat improvement, connectivity, and preservation of localized ecosystems, but there is relatively little literature regarding this. We review one database showing that the average size of nearly 200,000 protected areas in the United States is ~2000 ha and the median is ~20 ha, and that small areas are by far the most frequent. Overall, 95% and 49% of the records are less than the mean (1648 ha) and median (16 ha), respectively. We show that small areas are prevalent features of landscapes, and review literature suggesting how they should be studied and managed at multiple scales. Applying systematic conservation planning in a spatially hierarchical manner has been suggested by others and could help insure that small, local projects contribute to global goals. However, there are data and financial limitations. While some local groups practice ecosystem management and conservation planning, they will likely continue to protect what is "near and dear" and meet site-based goals unless there is better coordination and sharing of resources by larger organizations.

Keywords: conservation landscapes; scale of assessment; conservation planning

1. Introduction

There is wide agreement that the biodiversity crisis is best met with a carefully planned system of large conservation areas within well-connected networks [1]. However, humanity did not act in a concerted, globally-minded fashion to produce a functional system that protects biodiversity. Rather, what exists on the landscape is a distribution of conserved parcels with histories related as much to aesthetics and socioeconomic convenience as to systematic scientific goals [2,3].

There has been much research examining the effectiveness of the protected areas' estate. There is a preponderance of literature concluding that protected areas' coverage is inadequate, spatial distributions are biased, edge effects are eroding interior conditions, and that there is poor governance resulting in alienation of local human populations [4–6]. There is less understanding of how these negative conditions might be ameliorated by very local actions, resulting in a matrix of improved landscape conditions and more engaged conservation constituents [7,8]. An area of research that needs attention is the landscape-level function and roles played by the plethora of very small protected areas, although marine protected areas have received more attention [9]. In fact, there is little descriptive information revealing the dominance by frequency across the land surface of small areas. To that end, we will briefly review one dataset—the Protected Areas Database of the United States—as to size distributions, and then discuss broader conservation issues concerning small areas. Throughout, we make recommendations for further research.

The Protected Areas Database of the United States V1.4 (PADUS) has 194,518 parcels for the continental US, including designations as local as county parks and conservation easements, as well as large national parks and forests [10]. The PADUS is maintained by the United States Geological Survey's National Gap Analysis Project in order to organize and assess the management status of elements of biodiversity projection, i.e., "protected areas". The database includes all known public areas and voluntarily provided, private protected areas (e.g., conservation easements). The PADUS has versions between 2005 and 2016, currently at V1.4 (https://gapanalysis.usgs.gov/padus/data/metadata/).

We derived simple summary statistics of the PADUS area attributes and displayed them. The median area of parcels in the PADUS was 16 ha, the mean was 1648 ha (SD = 33,207.1), and the frequency distribution is skewed strongly right (Figure 1). Small areas were nonuniformly distributed but present in every major ecoregion (Figure 2). There are 95,116 parcels <16 ha and 185,751 <1648 ha in size, representing 49% and 95% of the count, respectively. However much they dominated by count, the smallest areas (<16 ha) and those <1648 ha represented 0.13% and 4% of the total area, respectively. Acknowledging possible errors of omission and commission in the dataset (see below), these descriptive analyses serve to illustrate the point that while the area covered is dominated by largeness, by number, most areas are not large. This presents unique research and management problems.

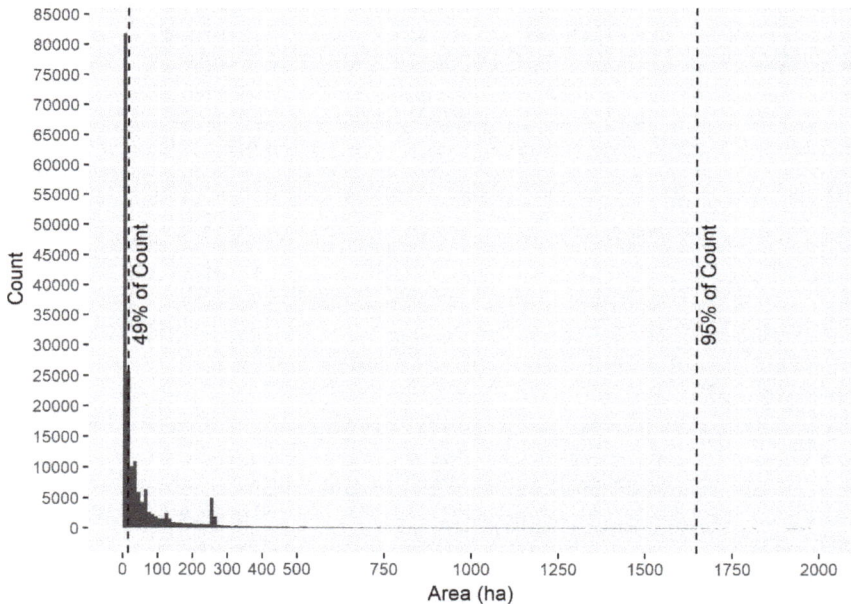

Figure 1. Frequency distribution of protected area parcel sizes in the United States. The *x* axis is truncated at 2000 ha in order to show the distribution on the *y* axis without transforming the data. Bin count is 10 ha. Bars represent the percentage of total records smaller than the mean (95%) and median areas (49%). R version 3.4.2.

As noted in their metadata, PADUS is subject to errors of omission and commission. Some of the errors they report are due to overlap—topology—and they mention "slivers" specifically. We observed that some of the smallest polygons in the dataset were geoprocessing artifacts, including slivers [11]. These were typically very small, but the smallest records in the PADUS were a relatively minor portion. For example, 2.4% are <0.1 ha in size. On the other hand, many very small polygons are legitimate conserved parcels. Many records (27,338 or 14.1%) were less than 1 ha and distinct parcels, as illustrated in Figure 3. While many of the small records have attributes indicating they are conservation easements or small parks, others may not be real or their spatial attributes may be inaccurate; investigating the characteristics of these parcels is an important research question, because while most are conserved areas, others may have inaccurate spatial information and influence the size distribution. Another source of error could be lack of clarity over what constitutes a separate protected area in an ownership sense, as distinct from a polygon sense. For example, the database has some records that are split polygons, so that one continuous protected area is represented by two records.

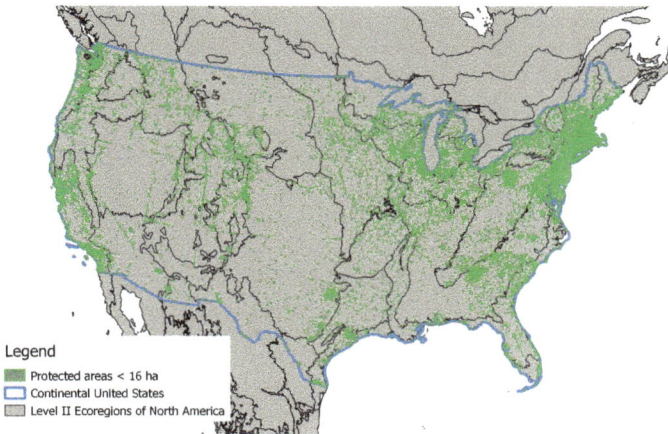

Figure 2. Small protected areas distributed across the continental United States. Shown in green on the map are the protected areas <16 ha, which is the median parcel size for the area shown. Data from PADUS V1.4 Combined Feature Class. QGIS V3.2.

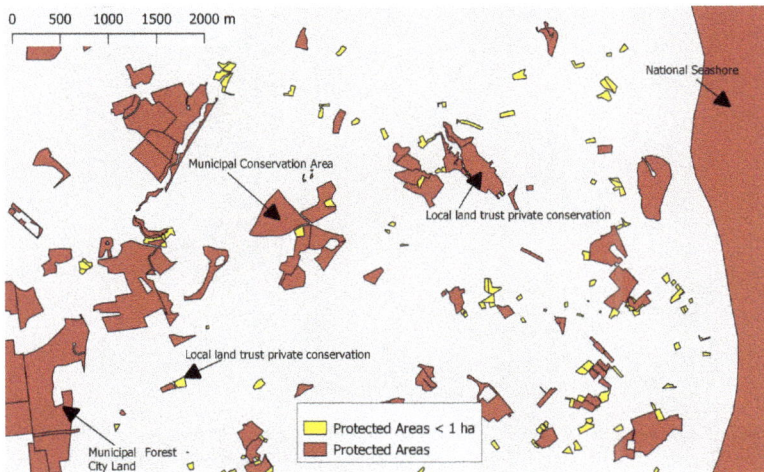

Figure 3. A portion of Cape Cod, USA, showing extremely small conserved parcels <1 ha, in yellow, in a matrix of slightly larger projects typical of those discussed in this paper. A selection of parcels is labeled to show the types of conservation lands in this landscape. Even the <1 ha areas have "conservation" designations in the database, i.e., are not data anomalies. To what extent small protected areas perform landscape-level biodiversity functions is largely unknown. QGIS V 3.2.

The PADUS is a massive undertaking managing hundreds of thousands of polygons generated by a plethora of sources, and errors are to be expected. The PADUS metadata states that the project reports spatial data as they are received. A thorough study of the PADUS should quantify the extent of errors, how they may influence analyses, compare representation of small areas in the PADUS with other protected areas datasets such as the World Database on Protected Areas (WDPA), and investigate spatial and designation rules for how distinct protected areas are defined.

Since the PADUS is built from contributions from states and others who keep public land records, and relies on voluntary submission for private areas, it is probably incomplete. There is evidence that it disproportionately underrepresents smaller parcels. The National Conservation Easement Database (NCED 2018) has more conservation easements recorded for the Continental US than does PADUS. NCED contains over 158,168 records, while in the continental PADUS, only 54,590 are categorized as easements. According to its documents, the NCED itself is incomplete. It relies on voluntary submission by land trusts, many of whom do not want to share to a large public database due to privacy and other concerns [12]. More research is needed on the structure of datasets that serve spatial data on protected areas in regards to smaller areas, and errors of commission and omission, before a firm knowledge of size distribution of protected areas can be obtained, and this needs to be undertaken on a global scale [13].

2. What Is "Small"?

For this review, we consider "small" based on the observed distribution in the PADUS. A "small protected area" is at the lower end of the range, between 16 ha (the US median) and 1648 ha (the US mean) in size. By this definition, then, small areas would be from tens to hundreds of hectares in size, rather than thousands of hectares. Areas in the range of tens to hundreds of hectares are often the focus of local conservation projects (Figures 3 and 4).

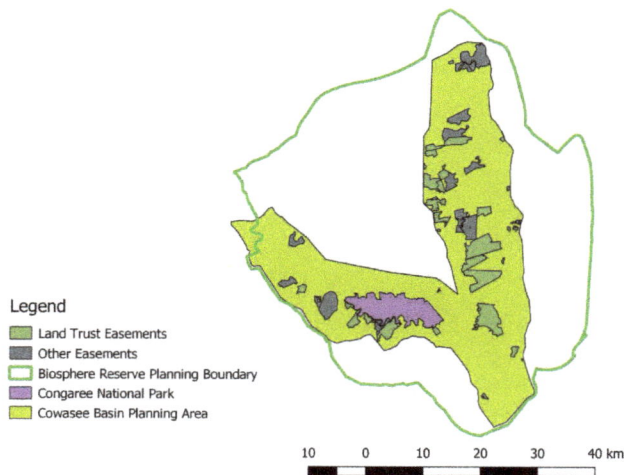

Legend
- Land Trust Easements
- Other Easements
- Biosphere Reserve Planning Boundary
- Congaree National Park
- Cowasee Basin Planning Area

10 0 10 20 30 40 km

Figure 4. A landscape in South Carolina, USA showing small conservation lands within two overlapping planning projects, an International Biosphere Reserve, and the COWASEE Basin Task Force. Conservation actions within these areas are coordinated to meet landscape-level objectives. QGIS V3.2.

Figure 4 shows the Congaree–Wateree River region of South Carolina, USA. Numerous conserved parcels ranging from 4 to 6179 ha in size are represented, established by public and private entities for conservation purposes. Also shown are two overlapping planning areas illustrating landscape context as mentioned below. This is an example of a landscape in which there are many small protected areas, and in which ecosystem management and systematic conservation planning are practiced, to a degree.

3. Potential Importance of Small Areas

Considerable cost is incurred in establishing and managing small protected areas [14]. Aside from their sheer number and costs of acquisition, smaller areas may cost more to manage than larger ones due to the challenges of having management tasks distributed. On the other hand, smaller areas arise from local concerns and may serve local purposes better than larger, more distant areas. Small parks and private conserved lands are typically generated and managed at the local scale—as in counties, states, cities, land trusts, landowners, and indigenous people. These include the bulk of conservation easements, local parks, historical and archaeological sites, private reserves, wildlife refuges, and community forests [15,16]. One drawback is that some would not be considered by international bodies to be areas whose primary purpose is biodiversity conservation. If they maintain natural land cover and are unable to be developed, they nonetheless function to exclude conversion to development. Furthermore, they often have specific conservation goals, and these include those related to biodiversity [17,18].

Despite their prevalence and potential benefit, there is very little known about how smaller protected areas function to provide biodiversity protection that may complement large, public protected areas and help address global issues [19,20]. The abundance of small protected areas—at least in the continental United States—underscores the need for a better understanding of their collective and individual function and how future conservation efforts may be oriented towards achieving larger landscape goals. Unfortunately, the literature on these issues is not robust. Much research on small protected areas remains to be done and applied to conservation planning, design, acquisition, and management.

4. Landscape-Level Views of Small Areas

A large body of literature shows that as islands of habitat become smaller and more isolated, they lose local populations of species [21]. Parks and protected areas, when fragments of larger once-connected landscapes, follow these island-biogeographic trends [22]. Ecological isolation increases if the conserved area is in a matrix of intensive land uses decreasing permeability and dispersal opportunities and increasing edge penetration to the core [6]. As a consequence, smaller areas in human-dominated matrices with intensive land uses are expected to have reduced biodiversity conservation function than larger, more well-connected areas.

As a whole, perhaps because of the importance of large size, the literature on protected areas is focused on those that are the largest. For example, globally protected areas are biased towards locations where they will have the least effect of reducing land conversion [4]. That study, and many others like it, investigate the coarse-scale pattern using minimum grain sizes of 1 km^2 or greater. The number and types of studies focusing on the smallest areas appears to be increasing, and the potential importance of grassroots conservation, private land, and multiscale conservation becomes more evident [23]. In the United States, the rapid increase in the number of conservation easements has given rise to studies of their effectiveness [19,24]. There are many state and county parks which may provide meaningful habitat protection. For example, the life histories of some wetland animals may be accommodated by protected areas considered to be small, as in state parks in the hundreds of hectares range [25]. Others have focused on how the scale of conservation can fit the scale of the ecosystem, community, policy environment, or process of concern [26]. Vernal pool systems in the Northeastern United States may be significantly protected by local conservation actions [27]. In some parts of the world, community-owned and managed forests are localized socioecological systems that may have greater conservation outcomes and more local support than traditional protected areas [15].

Studies focused on how populations and communities function across time and space provide insight into how small protected areas may function and how their internal processes may be influenced by events occurring outside their boundaries. Populations and communities are not isolated in space; conditions at a varying distances can influence response variables such as occupancy, population size, and community composition and structure at a site [28,29]. Dispersal and migration are common foci of these studies, as they are mechanisms connecting populations, and can provide insight into how to design conservation landscapes. Some of these movements, such as for reptiles and amphibians, are highly local in scale (e.g., hundreds of meters), yet still require landscape-level thinking for conservation purposes [30]. In fact, using landscape-level analyses to configure habitats in conservation plans at local scales for pond-breeding amphibians has been a productive area of research [31,32]. Small protected areas may be important for climate change adaptation, such as when considering the need to provide habitats during range shifts through human-dominated landscapes [33].

Studies that incorporate multiple spatial scales may help a manager of a small parcel to understand how the management of adjacent properties within specified distances influences conditions observed at the site. For example, forest cover is important at certain thresholds and distances [34], loss of wetland density at landscape scales influences local diversity [35], butterfly metapopulations depend on a spatially and temporally dynamic system of habitat patches [36], and turtles follow an array of routes over large areas and are strongly influenced by road traffic within that landscape [37]. A small reserve may have wildlife populations that may be perceived as resident, but routinely exchange wildlife with surrounding properties.

Some small protected areas serve very specific and well-known functions in a process that covers a much larger area. For example, bird migratory stopovers and seabird breeding sites are very spatially specific, and small protected areas have been established to cover those functions. Cape May, New Jersey, is a place where waterfowl, raptors, shorebirds, and songbirds congregate during spring migration [38]. Cape May National Wildlife Refuge is about 4900 ha in size and is a key location in the Atlantic Flyway. L'île Bonaventure-et-du-Rocher Percé in eastern Canada was protected in 1919 as a 1361-ha Migratory Bird Sanctuary, hosting thousands of breeding pairs of seabirds, including about half of the Northern Gannets breeding in the Atlantic [39]. The Western Hemisphere Shorebird Reserve Network has collated information on 102 locations with sizes ranging from several hundred to tens of thousands of hectares from the tip of South America to Alaska, and provides scientific resources to aid in their conservation (https://www.whsrn.org/).

When small protected areas are embedded in human-dominated landscapes, they are more susceptible to ecological isolation than larger protected areas in those same areas. Urbanization creeps out from metropolitan areas and causes rural lands to transition to higher density residential landscapes with attendant roads, industries, and services [40]. Land values rise during this process, and while there may be more funds and incentives available for conservation deals on private land, these deals are competing with the real estate market [41]. Consequently, most land transitions in the United States are from rural to residential, resulting in the increased isolation of the remaining natural lands. At the same time, small conservation areas embedded in urbanizing areas may provide disproportionately high conservation values [42] and benefits for human health and wellbeing [43].

5. Ecosystem Management for Small Areas

Ecosystem management is a way of organizing thinking about individual protected areas and private parcels in relation to the spatial and temporal scales at which ecosystem processes occur, as opposed to within political or jurisdictional boundaries. Its management roots were in the Pacific Northwest in the late 20th century, where there were significant endangered species management issues and social conflicts resulted [44]. Accommodating dispersal, gene flow, disturbance, migration, and other ecological processes requires working across management units, ownerships, and jurisdictions. For example, conserving migratory movements of ungulates may be dependent on highly localized natural features that occur outside protected areas, practices of landowners such as fencing, differing game laws across political boundaries, and departments of transportation who are responsible for wildlife underpasses and overpasses [45,46]. Such tasks will be more challenging in landscapes where there are lots of smaller areas.

Natural disturbance regimes help to maintain biological diversity, but are difficult to manage across boundaries of protected areas [47]. Managing a landscape-level disturbance, such as fire, is a challenge for protected areas as neighboring landowners are influenced. Because of perimeter-to-area ratios, such boundary-related tasks will be more difficult the smaller the protected area is. Nonetheless, fire management is frequently practiced on small reserves. For example, even in the urban setting of the Hitchcock Woods in South Carolina (850 ha), the longleaf pine systems are maintained by repeated burning. Much education is undertaken with local residents. However, knowledge of how to manage fire in local socioeconomic systems is rarely published [48]. Thus, there is much more knowledge available than is in present in the scientific literature and can be accessed via websites, workshops, and so on. Smaller protected areas may be vulnerable to large disturbances, as a hurricane or pathogen could wipe out a large percentage of its contents. For example, old growth forests on a 2180-ha portion of the Cape Romain National Wildlife Refuge in South Carolina were decimated—for one pine species, almost total losses—after Hurricane Hugo in 1989 [49]. The smaller the protected area, the more likely a single disturbance event will change the entire ecosystem. If this happens to a small protected area that is isolated by development, problems are compounded.

Ecosystem management may be more important for smaller protected areas due to their isolation and abundance and complexities of ownerships. A landscape with many protected areas with different managers and owners is a challenge in cross-boundary cooperation [50]. For one thing, smaller areas are more likely to have ecological processes that transcend boundaries. One approach to cooperation, examined for two very small parcels, was to identify the owners responsible for landscape features important to ecosystem function, and then develop a plan for the neighbors to accommodate movements of animals using habitats on each parcel. Such an approach required mutual understanding of the ecological and conservation issues and agreement on values [51]. Another example is the Southern Low Country ACE Basin longleaf partnership that manages the longleaf system which involves frequent burning (https://www.longleafalliance.org/soloace).

The region shown in Figure 4 has two overlapping planning areas: an International Biosphere Reserve and a landscape-level alliance or Task Force. Both planning areas share members, for example, from the National Park Service, Congaree Land Trust, and state agencies. The idea is to increase cooperation to achieve landscape-level conservation goals. The hope is that even though the protected areas are small, their collective impact may be increased by better management of un-conserved lands, while preserving local resource-based economies. Similarly, community conserved areas (CCAs) are a global phenomenon that provide biodiversity services while satisfying the needs of local and indigenous peoples. They are not very well understood in terms of their overall contributions to biodiversity conservation, but due to their extensive coverage (as much as 11% of global forests), are an example of how relatively small, local projects can accumulate effects [52]. These approaches can be facilitated by landscape-level cooperatives and alliances, especially if they respect human dimensions. For example, in governance of conservation networks, recognizing and trading leadership roles based on what is needed to accomplish particular goals is better than traditional, top-down management [53].

6. Systematic Conservation Planning for Small Areas

Systematic conservation planning is a rapidly growing field in spatial ecology that attempts to allocate conservation actions across the landscape based on patterns of representation, size and configuration, and threat [54]. Systematic conservation planning uses spatial datasets in complex, mathematically-driven layering operations to identify which parts of the landscape meet conservation goals. Many spatially extensive analyses and localized projects have been completed, such as that presented in [55]. One of the weaknesses of conservation planning is the lack of relevance to local decision-making when there is a mismatch between grain size and data quality and the scale at which decision-makers need information [56,57].

Local conservationists seeking to prioritize small parcels may not be able to develop models comparable to those produced at greater spatial extents. Spatial data is often less available at the fine grain sizes relevant to local conservation than at the coarse grain sizes relevant for large-landscape projects, resulting in maps and models for systematic planning that show broad patterns, but are less relevant locally. Biodiversity data may consist of points with high accuracy, but the coverage of points is often nonuniform and biased towards survey routes or other observation patterns. Resulting species distribution maps, which are often the basis for conservation plans, are influenced by these data limitations [58]. Smaller conservation organizations often do not have the technical capacity or funding to invest in new, fine-grain data for spatial planning. Even if they engage in systematic conservation planning, and many do, they are forced to use coarser-grain data. Grain size and extent of analysis can influence model results, resulting in locations prioritized for conservation that are influenced by the accuracy of the data [59]. Resolving questions of scale is an important area of research if spatial conservation planning is to be relevant for the smallest protected areas.

Despite these challenges, systematic conservation planning has the potential to greatly improve the biodiversity impact of new small protected areas. Until the advent of these methods, many conservation projects were focused on protecting single populations of rare and endemic plants and particular ecosystems (e.g., cliffs and caves), regardless of spatial context. The Nature Conservancy is one of the largest private conservation organizations in the world. It is a leader in conservation planning, yet has numerous smaller reserves in the US that represent an occurrence of a rare species or important localized ecosystem. Their Places We Protect website provides background information on these, for example, the 202-ha Loverens Mill Cedar Swamp in New Hampshire (https://www.nature.org/en-us/get-involved/how-to-help/places-we-protect/). The Nature Conservancy was a leader in forming the field of systematic conservation planning, attempting to "scale up" from their site-based approach to larger and more connected areas [60]. Their approach now integrates the small, important sites with landscape-level patterns and processes, focused on resilience to climate change [61].

Given the preponderance of small protected areas, those engaged in systematic conservation planning may need to be more focused on how to improve site selection at the local scale. It appears that at least in some areas of the United States, land trusts and others in conservation are protecting many areas that are close to where they live and work [20]. It is unclear to what degree these decisions are being driven by spatial conservation priorities, or by opportunities and aesthetics, and local knowledge about site conditions.

Initiating more systematic conservation planning at local scales for small areas will necessitate better databases that are more fine-scale. Since land use and land cover data are the basis of most conservation planning models, developing more localized land cover data should be a top priority [62]. Fine-scale biodiversity data can be improved through USGS GAP species distribution models, the Global Biodiversity Information Facility (GBIF), and proprietary sources such as NatureServe. As stated earlier, more complete representation of very small areas in protected areas databases including PADUS, NCED, and the World Database on Protected Areas (WDPA) will help.

Local land use decision-making usually occurs in the context of development, but conservation planners can get involved and bring biodiversity concerns into that context [63]. Such planning projects should be done within a spatially hierarchical framework (e.g., landscapes–ecoregions), so that what is accomplished locally will have relevance when viewed in greater spatial contexts. Larger regional analyses may point to localities that have biodiversity importance at an ecoregional level (e.g., species geographic ranges, across climate gradients, and large land use transitions). Within those localities prioritized as being important at a coarser-grain level, local conservation groups should plan using finer-grained data. Funding agencies should recognize that planning and data development require support at every scale.

7. Summary

More research is needed on the spatial distribution and landscape-level functions of small protected areas. At least in the United States, small areas dominate the frequency distribution and may have greater collective impact on conservation than their individual sizes suggest. Larger areas still account for most of the coverage, but their function may be enhanced by a matrix-improving impact of small sites, which may also serve to represent localized natural phenomena. Scattered across the landscape at varying densities, small areas may provide landscape-level functions that enhance biodiversity conservation. Better landscape-level management and planning at local scales could improve their current function and spatial distribution of additional areas. Datasets are incomplete and may have inaccuracies, so should be improved so that the above hypotheses may be tested.

Considerable effort is expended to protect and manage small areas, but not enough is known about how they contribute to global biodiversity goals. Even the smallest protected areas may protect localized ecosystems, collectively improve landscape-level conditions such as permeability and matrix quality, and may be important to human community health and wellbeing. While systematic conservation planning has made great scientific strides and is being applied at most scales, data availability and accuracy, financial restrictions, and lack of access to technology may limit its application to small projects. More research and data development are needed to ensure that the cumulative effect of small areas is a net gain for global biodiversity conservation.

7.1. Recommendations for Research

- Examine effects of small protected areas on landscape-level patterns and processes, including local representation of regional diversity, improvement of matrix quality, metapopulation and community services, stopover sites, and network centrality.
- Examine protected areas dataset completeness and topology errors for small areas.
- Develop fine-grain biological and economic data to be used in local-scale systematic conservation planning.
- Examine social and economic barriers that keep partners who own, manage, or monitor small protected areas—e.g., land trusts, local governments, indigenous peoples—from participating in larger landscape efforts.
- Conduct economic research on contributions of small areas to ecosystem services, including water purification and supply, biodiversity, and flood storage.
- Research how small protected areas may contribute to resiliency following large-scale disturbances and contribute to climate change adaptation.
- Examine role of small protected areas in local communities and how they contribute to social, physical, and psychological health and wellbeing.

7.2. Recommendations for Management

- Manage for ecosystem processes such as dispersal and disturbance that extend outside property boundaries.
- Identify the neighboring landowners that accommodate ecosystem processes influencing the site; collaborate across boundaries.
- Provide mapped data on small areas to public databases so it can be used in conservation planning.
- Contribute time and resources to protecting other parcels in your landscape to form functional networks.
- Participate in watershed-level activities to enhance water quality.
- Participate in large landscape efforts to restore natural disturbance regimes.
- Scale up: represent the locality in larger landscape conservation efforts.
- Educate stakeholders about the importance of participating in larger efforts.

Author Contributions: Conceptualization, R.F.B.; Methodology, R.F.B. and N.F.; Software, R.F.B. and N.F.; Formal Analysis, R.F.B. and N.F.; Data Curation, R.F.B.; Writing—Original Draft Preparation, R.F.B.; Writing—Review & Editing, R.F.B. and N.F.; Visualization, R.F.B. and N.F.; Supervision, R.F.B.; Funding Acquisition, R.F.B.

Funding: This research was funded by The National Science Foundation, grant number 1518455, and the Margaret H. Lloyd-SmartState Endowment for South Carolina.

Acknowledgments: We thank Congaree Land Trust, Congaree National Park, Cowasee Basin Task Force, the Department of Forestry and Environmental Conservation at Clemson University, and colleagues on the Clemson-Conservation Easement project (NSF 1518455) for their support.

Conflicts of Interest: The authors declare no conflict of interest.

References

1. Noss, R.F.; Dobson, A.P.; Baldwin, R.F.; Beier, P.; Davis, C.R.; dellaSala, D.A.; Francis, J.; Locke, H.; Nowak, K.; Lopez, R.; et al. Bolder thinking for conservation. *Conserv. Boil.* **2012**, *26*, 1–4. [CrossRef] [PubMed]
2. Margules, C.R.; Pressey, R.L. Systematic conservation planning. *Nature* **2000**, *405*, 243–253. [CrossRef] [PubMed]
3. Jenkins, C.N.; Van Houtan, K.S.; Pimm, S.L.; Sexton, J.O. US protected lands mismatch biodiversity priorities. *Proc. Natl. Acad. Sci. USA* **2015**, *112*, 5081–5086. [CrossRef] [PubMed]
4. Joppa, L.N.; Pfaff, A. High and far: Biases in the location of protected areas. *PLoS ONE* **2009**, *4*, e8273. [CrossRef] [PubMed]
5. DeFries, R.; Karanth, K.K.; Pareeth, S. Interactions between protected areas and their surroundings in human-dominated tropical landscapes. *Boil. Conserv.* **2010**, *143*, 2870–2880. [CrossRef]
6. Woodroffe, R.; Ginsberg, J.R. Edge effects and the extinction of populations inside protected areas. *Science* **1998**, *280*, 2126–2128. [CrossRef] [PubMed]
7. Powell, R.B.; Cuschnir, A.; Peiris, P. Overcoming governance and institutional barriers to integrated coastal zone, marine protected area, and tourism management in Sri Lanka. *Coast. Manag.* **2009**, *37*, 633–655. [CrossRef]
8. McRae, B.H.; Hall, S.A.; Beier, P.; Theobald, D.M. Where to restore ecological connectivity? Detecting barriers and quantifying restoration benefits. *PLoS ONE* **2012**, *7*, e52604. [CrossRef] [PubMed]
9. Gurney, G.G.; Pressey, R.L.; Ban, N.C.; Álvarez-Romero Jorge, G.; Jupiter, S.; Adams, V.M. Efficient and equitable design of marine protected areas in Fiji through inclusion of stakeholder-specific objectives in conservation planning. *Conserv. Boil.* **2015**, *29*, 1378–1389. [CrossRef] [PubMed]
10. USGS. *Protected Areas Database of the United States (PAD-US), Version 1.4 Combined Feature Class*; U.S. Department of the Interior: U.S. Geological Survey, 2016.
11. Lipscomb, D.J.; Baldwin, R.F. Geoprocessing solutions developed while calculating Human Footprint statistics for zones representing protected areas and adjacent lands at the continent scale. *Math. Comput. For. Nat. Resour. Sci.* **2010**, *2*, 72–78.
12. Rissman, A.R.; Owley, J.; L'Roe, A.W.; Morris, A.W.; Wardropper, C.B. Public access to spatial data on private-land conservation. *Ecol. Soc.* **2017**, *22*, 24. [CrossRef]
13. Clements, H.; Selinske, M.; Archibald, C.; Cooke, B.; Fitzsimons, J.; Groce, J.; Torabi, N.; Hardy, M. Fairness and Transparency Are Required for the Inclusion of Privately Protected Areas in Publicly Accessible Conservation Databases. *Land* **2018**, *7*, 96. [CrossRef]
14. Armsworth, P.R.; Cantú-Salazar, L.; Parnell, M.; Davies, Z.G.; Stoneman, R. Management costs for small protected areas and economies of scale in habitat conservation. *Boil. Conserv.* **2011**, *144*, 423–429. [CrossRef]
15. Porter-Bolland, L.; Ellis, E.A.; Guariguata, M.R.; Ruiz-Mallén, I.; Negrete-Yankelevich, S.; Reyes-García, V. Community managed forests and forest protected areas: An assessment of their conservation effectiveness across the tropics. *For. Ecol. Manag.* **2012**, *268*, 6–17. [CrossRef]
16. Brewer, R. *Conservancy: The Land Trust Movement in America*; Dartmouth College, University of Press of New England: Hanover, NH, USA, 2003; p. 364.
17. Dudley, N.; Parrish, J.D.; Redford, K.H.; Stolton, S. The revised IUCN protected areas management categories: The debate and ways forward. *Oryx* **2010**, *44*, 485–490. [CrossRef]
18. Merenlender, A.M.; Huntsinger, L.; Guthey, G.; Fairfax, S.K. Land trusts and conservation easements: Who is conserving what for whom? *Conserv. Boil.* **2003**, *18*, 65–75. [CrossRef]

19. Rissman, A.R.; Merenlender, A.M. The conservation contributions of conservation easements: Analysis of the San Francisco Bay Area Protected Lands Database. *Ecol. Soc.* **2008**, *13*, 25. [CrossRef]

20. Baldwin, R.F.; Leonard, P.B. Interacting social and environmental predictors for the spatial distribution of conservation lands. *PLoS ONE* **2015**, *10*, e0140540. [CrossRef] [PubMed]

21. Hunter, M.L.; Gibbs, J.P. *Fundamentals of Conservation Biology*, 3rd ed.; Blackwell Publishing: Oxford, UK, 2007.

22. Newmark, W.D. Extinction of mammal populations in Western North American National Parks. *Conserv. Boil.* **1995**, *9*, 512–526. [CrossRef]

23. Hilty, J.; Merenlender, A.M. Studying biodiversity on private lands. *Conserv. Boil.* **2003**, *17*, 132–137. [CrossRef]

24. Kiesecker, J.M.; Comendant, T.; Grandmason, E.; Gray, E.M.; Hall, C.; Hilsenbeck, R.; Kareiva, P.; Lozier, L.; Naehu, P.; Rissman, A.R.; et al. Conservation easements in context: A quantitative analysis of their use by The Nature Conservancy. *Front. Ecol. Environ.* **2007**, *5*, 125–130. [CrossRef]

25. Pitt, A.L.; Howard, J.H.; Baldwin, R.F.; Baldwin, E.D.; Brown, B.L. Small Parks as Local Social–Ecological Systems Contributing to Conservation of Small Isolated and Ephemeral Wetlands. *Nat. Areas J.* **2018**, *38*, 237–249. [CrossRef]

26. Calhoun, A.J.K.; Jansujwicz, J.S.; Bell, K.P.; Hunter, M.L. Improving management of small natural features on private lands by negotiating the science–policy boundary for Maine vernal pools. *Proc. Natl. Acad. Sci. USA* **2014**, *111*, 11002–11006. [CrossRef] [PubMed]

27. Hunter, M.L. Valuing and conserving vernal pools as small-scale ecosystems. In *Science and Conservation of Vernal Pools in Northeastern North America*; Calhoun, A.J.K., deMaynadier, P.G., Eds.; CRC Press: New York, NY, USA, 2008; pp. 1–10.

28. Hanski, I. Metapopulation dynamics. *Nature* **1998**, *396*, 41–49. [CrossRef]

29. Leibold, M.A.; Holyoak, M.; Mouquet, N.; Amarasekare, P.; Chase, J.M.; Hoopes, M.F.; Holt, R.D.; Shurin, J.B.; Law, R.; Tilman, D.; et al. The metacommunity concept: A framework for multi-scale community ecology. *Ecol. Lett.* **2004**, *7*, 601–613. [CrossRef]

30. Semlitsch, R.D. Differentiating migration and dispersal processes for pond-breeding amphibians. *J. Wildl. Manag.* **2008**, *72*, 260–267. [CrossRef]

31. Compton, B.W.; McGarigal, K.; Cushman, S.A.; Gamble, L. A resistant-kernal model of connectivity for amphibians that breed in vernal pools. *Conserv. Boil.* **2007**, *21*, 788–799. [CrossRef] [PubMed]

32. Harper, E.B.; Rittenhouse, T.A.G.; Semlitsch, R.D. Demographic consequences of terrestrial habitat loss for pool-breeding amphibians: Predicting extinction risks associated with inadequate size of buffer zones. *Conserv. Boil.* **2008**, *22*, 1205–1215. [CrossRef] [PubMed]

33. Nuñez, T.; Lawler, J.J.; McRae, B.H.; Pierce, D.J.; Krosby, M.R.; Kavanagh, D.M.; Singleton, P.H.; Tewksbury, J.J. Connectivity planning to facilitate species movements in response to climate change. *Conserv. Boil.* **2013**, *27*, 407–416. [CrossRef] [PubMed]

34. Homan, R.N.; Windmiller, B.S.; Reed, J.M. Critical thresholds associated with habitat loss for two vernal pool-breeding amphibians. *Ecol. Appl.* **2004**, *14*, 1547–1553. [CrossRef]

35. Gibbs, J.P. Importance of small wetlands for the persistence of local populations of wetland-associated animals. *Wetlands* **1993**, *13*, 25–31. [CrossRef]

36. Hanski, I.; Thomas, C.D. Metapopulation dynamics and conservation: A spatially explicit model applied to butterflies. *Boil. Conserv.* **1994**, *68*, 167–180. [CrossRef]

37. Beaudry, F.; deMaynadier, P.G.; Hunter, M.L. Identifying road mortality threat at multiple spatial scales for semi-aquatic turtles. *Boil. Conserv.* **2008**, *141*, 2550–2563. [CrossRef]

38. Niles, L.J.; Burger, J.; Clark, K.E. The Influence of Weather, Geography, and Habitat on Migrating Raptors on Cape May Peninsula. *Condor* **1996**, *98*, 382–394. [CrossRef]

39. Chardine, J.W.; Rail, J.-F.; Wilhelm, S. Population dynamics of Northern Gannets in North America, 1984–2009. *J. Field Ornithol.* **2013**, *84*, 187–192. [CrossRef]

40. Theobald, D.M. Placing exurban land-use change in a human modification framework. *Front. Ecol. Environ.* **2004**, *2*, 139–144. [CrossRef]

41. Bell, K.P.; Irwin, E.G. Spatially explicit micro-level modelling of land use change at the rural-urban interface. *Agric. Econ.* **2002**, *27*, 217–232. [CrossRef]

42. Baldwin, R.F.; deMaynadier, P.G. Assessing threats to pool-breeding amphibian habitat in an urbanizing landscape. *Boil. Conserv.* **2009**, *142*, 1628–1638. [CrossRef]

43. Larson, L.R.; Jennings, V.; Cloutier, S.A. Public parks and wellbeing in urban areas of the United States. *PLoS ONE* **2016**, *11*, e0153211. [CrossRef] [PubMed]

44. Grumbine, R.E. What is ecosystem management? *Conserv. Boil.* **1994**, *8*, 27–38. [CrossRef]

45. Berger, J. The last mile: How to sustain long-distance migration in mammals. *Conserv. Boil.* **2004**, *18*, 320–331. [CrossRef]

46. Trombulak, S.C.; Frissell, C.A. Review of ecological effects of roads on terrestrial and aquatic communities. *Conserv. Boil.* **2000**, *14*, 18–30. [CrossRef]

47. Baker, W.L. The landscape ecology of large disturbances in the design and management of nature reserves. *Landsc. Ecol.* **1992**, *7*, 181–194. [CrossRef]

48. Neel, L. *The Art of Managing Longleaf*; University of Georgia Press: Athens, GA, USA, 2010; p. 211.

49. Conner, W.H.; Mixon, W.D.; Wood, G.W. Maritime forest habitat dynamics on Bulls Island, Cape Romain National Wildlife Refuge, SC, following Hurricane Hugo. *For. Ecol. Manag.* **2005**, *212*, 127–134. [CrossRef]

50. Chester, C.C. *Conservation across Borders: Biodiversity in an Interdependent World*; Island Press: Washington, DC, USA, 2006.

51. Colbert, N.; Baldwin, R.F. A developer-initiated conservation plan for pool-breeding amphibians in Maine, USA: A case study. *J. Conserv. Plan.* **2011**, *7*, 27–38.

52. Kothar, A. Community conserved areas: Towards ecological and livelihood security. *Parks* **2006**, *16*, 3–13.

53. Imperial, M.T.; Ospina, S.; Johnston, E.; O'Leary, R.; Thomsen, J.; Williams, P.; Johnson, S. Understanding leadership in a world of shared problems: Advancing network governance in large landscape conservation. *Front. Ecol. Environ.* **2016**, *14*, 126–134. [CrossRef]

54. Baldwin, R.F.; Trombulak, S.C.; Leonard, P.B.; Noss, R.F.; Hilty, J.A.; Possingham, H.P.; Scarlett, L.; Anderson, M.G. The Future of Landscape Conservation. *BioScience* **2018**, *68*, 60–63. [CrossRef] [PubMed]

55. Leonard, P.; Baldwin, R.F.; Hanks, D. Landscape-scale conservation design across biotic realms: Sequential integration of aquatic and terrestrial landscapes. *Sci. Rep.* **2017**, *7*, 14556. [CrossRef] [PubMed]

56. Guerrero, A.M.; McAllister, R.R.J.; Corcoran, J.; Wilson, K.A. Scale Mismatches, Conservation Planning, and the Value of Social-Network Analyses. *Conserv. Boil.* **2013**, *27*, 35–44. [CrossRef] [PubMed]

57. Theobald, D.M.; Spies, T.; Kline, J.D.; Maxwell, B.; Hobbs, N.T.; Dale, V.H. Ecological support for rural land-use planning. *Ecol. Appl.* **2005**, *15*, 1906–1914. [CrossRef]

58. Franklin, J. *Mapping Species Distributions: Spatial Inference and Prediction*; Cambridge University Press: Cambridge, UK, 2009; p. 320.

59. Leonard, P.B.; Duffy, E.B.; Baldwin, R.F.; McRae, B.H.; Shah, V.B.; Mohapatra, T.K. GFLOW: Software for modelling circuit theory-based connectivity at any scale. *Methods Ecol. Evol.* **2017**, *8*, 519–526. [CrossRef]

60. Groves, C.; Jensen, D.; Valutis, L.L.; Redford, K.H.; Shaffer, M.; Scot, J.M.; Baumgartner, J.V.; Higgins, J.V.; Beck, M.W.; Anderson, M.G. Planning for biodiversity conservation: Putting conservation science into practice. *Bioscience* **2002**, *52*, 499–512. [CrossRef]

61. Anderson, M.G.; Clark, M.; Sheldon, A.O. Estimating climate resilience for conservation across geophysical settings. *Conserv. Boil.* **2014**, *28*, 959–970. [CrossRef] [PubMed]

62. Theobald, D. A General-Purpose Spatial Survey Design for Collaborative Science and Monitoring of Global Environmental Change: The Global Grid. *Remote Sens.* **2016**, *8*, 813. [CrossRef]

63. Theobald, D.M.; Hobbs, R.J.; Bearly, T.; Zack, J.A.; Shenk, T.; Riebsame, W.E. Incorporating biological information in local land-use decision-making: Designing a system for conservation planning. *Landsc. Ecol.* **2000**, *15*, 35–45. [CrossRef]

land

MDPI

Article
Setting and Implementing Standards for Management of Wild Tigers

M. K. S. Pasha [1,*], Nigel Dudley [2,3], Sue Stolton [2], Michael Baltzer [1], Barney Long [4], Sugoto Roy [5], Michael Belecky [1], Rajesh Gopal [6] and S. P. Yadav [6]

[1] WWF Singapore, 354 Tanglin Road, Tanglin Block, Tanglin International Centre, Singapore 247672, Singapore; mbaltzer@wwfnet.org (M.Ba.); mbelecky@wwf.sg (M.Be.)

[2] Equilibrium Research, 47 The Quays, Spike Island, Cumberland Road, Bristol BS1 6UQ, UK; nigel@equilibriumresearch.com (N.D.); sue@equilibriumresearch.com (S.S.)

[3] School of Geography, Planning and Environmental Management, University of Queensland, Brisbane, QLD 4072, Australia

[4] Global Wildlife Conservation, 500 N Capital of Texas Hwy Building 1, Suite 200, Austin, TX 78746, USA; blong@globalwildlife.org

[5] International Union for Conservation of Nature, Global Species Programme, Rue Mauverney 28, 1196 Gland, Switzerland; Sugoto.Roy@iucn.org

[6] Global Tiger Forum, 200, Jor Bagh Road, Third Floor (Near Jor Bagh Metro Station), New Delhi 110003, India; rajeshgopal.sg.gtf@gmail.com (R.G.); spyadavifs@gmail.com (S.P.Y.)

* Correspondence: kpasha@wwf.sg

Received: 30 June 2018; Accepted: 27 July 2018; Published: 31 July 2018

Abstract: Tiger numbers have collapsed so dramatically that conservationists are adopting a strategy of securing populations in priority conservation landscapes. This includes improving management effectiveness in these sites. The Conservation Assured | Tiger Standards (CA | TS) are designed to help ensure effectiveness and provide a benchmark against which to measure progress. CA | TS is a distillation of best practice and a roadmap to management effectiveness, linking management to expert-driven standards covering all aspects of management, including those which are tiger-specific (monitoring, maintenance of prey, control of poaching). Sites are audited against a set of standards and if met, are accredited as CA | TS Approved. We describe CA | TS in the context of tiger conservation, describe the evolution and philosophy of the system and consider its application across the tiger range, before drawing on lessons learned from 5 years of development. Important benefits include the independence of CA | TS from existing governmental or NGO institutions, the emphasis on regional governance and the existence of active support groups. Conversely, the participatory approach has slowed implementation. CA | TS remains more attractive to well managed sites than to sites that are struggling, although building capacity in the latter is its key aim. The close connections between people working on tiger conservation make some aspects of independent assessment challenging. Finally, if CA | TS is to succeed in its long term aims, it needs to go hand in hand with secure and adequate funding to increase management capacity in many tiger conservation areas.

Keywords: Tiger; conservation standards; protected area management; management effectiveness; accreditation; conservation assured

1. Introduction

Global tiger (*Panthera tigris*) population size has fallen by over 95% since the turn of the 20th century—down from perhaps 100,000 to roughly 3900 individuals (Wolf and Ripple, 2017) [1]. Furthermore, tigers have lost over 93% of their historic range (Walston et al., 2010) [2]. There are now only 13 recognised Tiger Range Countries (TRCs: Bangladesh, Bhutan, Cambodia, China, India, Indonesia, Lao PDR, Malaysia, Myanmar, Nepal, Russia, Thailand, and Vietnam), of which only 10 still have wild tiger populations (WWF, 2015 [3], see Figure 1). Many other countries that formerly held tigers, such as South Korea and Kazakhstan, have lost their populations altogether. Tigers have been all but driven to extinction by poaching, illegal trade (Stoner and Pervushina, 2013) [4], habitat fragmentation, deforestation and diminishing prey (Damania et al., 2008) [5].

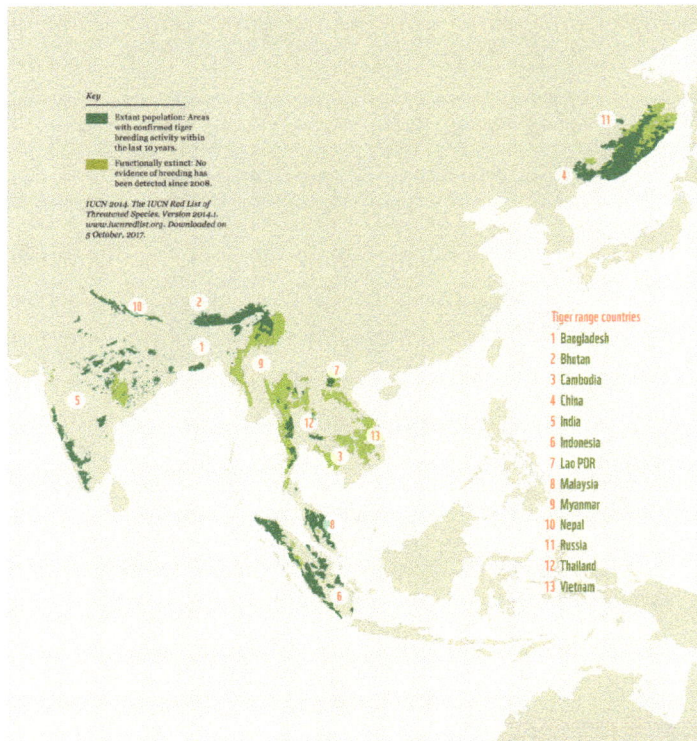

Figure 1. Distribution Range Wild Tigers.

In response to a serious threat of extinction of wild tigers, in 2010 the conservation community committed to doubling the global wild tiger population by 2022 (known as TX2) at a major 'tiger summit' in St. Petersburg (Global Tiger Initiative, 2010) [6]. This global goal was supported by all 13 TRCs and a global political infrastructure was established to coordinate and monitor action towards this goal. Many governments, donors and non-governmental organisations committed additional funds and personnel to focus on tiger conservation, such as German support for the Integrated Tiger Habitat Conservation Programme (Roy et al., 2016) [7], and there were high profile statements of support from global leaders.

Tigers require large areas of forest to survive. Historically, conservation strategies have focused on the designation of protected areas. There is clear evidence that site protection should remain the cornerstone of conservation for the vast majority of threatened amphibians, reptiles, birds and mammals. Quantitative analysis provides strong evidence that well-managed protected areas are an effective method of site protection (Geldmann et al., 2018) [8]. In the face of a worsening conservation status for tigers, an identified priority action is to secure stable tiger populations within identified protected areas, as a source of landscape-scale expansion (Walston et al., 2010) [2].

However, designation of protected areas alone is not enough. Global analysis has shown that many protected areas are currently not effectively managed (Leverington et al., 2010) [9], and the problem of 'paper parks' is well noted (e.g., Brun et al., 2015) [10]. Recent research confirms many management shortcomings in tiger conservation areas (Conservation Assured, 2018) [11–13].

Therefore, although the number of tiger conservation areas (TCAs) has increased over the past few years, a commensurate increase in tiger population will only occur if TCAs attain effective management. Concrete, measurable steps to improve management effectiveness in tiger reserves are a critical component of tiger conservation strategies. In turn, there is need for clear guidance on how best to manage tigers, address wider issues such as human-wildlife conflict, and develop agreed ways of verifying the effectiveness of tiger conservation.

In this paper, we discuss a set of verifiable management standards known as Conservation Assured I Tiger Standards (CA I TS). We examine how these differ from existing management effectiveness evaluation (MEE) processes, the development and application of CA I TS, application to date, lessons learned, and the scope for widening the approach.

2. Moving from Management Effectiveness to Management Standards

Over the last 20 years, our understanding of what constitutes good management of protected areas has increased considerably. More than 40 management effectiveness (MEE) data collection tools have been developed. These have been based largely on principles from the IUCN World Commission for Protected Areas. Hockings et al. (2006) [14] describe the mechanisms and rationale for MEE, along with many of the methods used world-wide. By 2013, the date of the last global survey, the IUCN Protected Area Management Effectiveness database held over 10,000 records and 90 countries had met the Convention on Biological Diversity (CBD) 2010 target of carrying out MEE in at least 30 per cent of their protected areas (Coad et al., 2013) [15]. An MEE system has also been developed explicitly for tiger protected areas in India called the MEETR—Management Effective Evaluation for Tiger Reserves (Mathur et al., 2014) [16].

MEE tools identify when things are going wrong, but they do not necessarily provide concrete steps for improvement, nor do they set standards against which to measure results and document change. The CBD identified the need for such standards in its 2004 Programme of Work on Protected Areas (CBD, 2004) [17]. Standards can, among other things, define key terminology; determine which data, methods and approaches are most favourable; and specify monitoring needs and the reporting format. They provide clear guidance on minimum requirements for effective management. There is good evidence that standards improve adoption of best practices, learning, performance-based awards, and the quality of results (Polasky, 2015) [18]. Standards also help donors to assess if investments are used effectively (Ferraro and Patanayak, 2006; McCarthy, 2012) [19,20]; for both institutional and private donors (Bennett et al., 2015) [21].

MEE tools, although useful for evaluating management objectives, offer less in terms of measuring management against best practice. CA I TS was designed to fill this gap and to provide species-specific conservation standards to help drive and measure progress towards improved tiger conservation (Conservation Assured, 2018a) [12]. It does not replace existing MEE tools; indeed, the implementation of site-based MEE is one of the standards of management included within the broader aims of CA I TS.

3. Methods

CA I TS was developed after an exhaustive literature review and stakeholders' consultation including protected area managers in TRCs and tiger experts from around the world. The approach drew from experience with establishing voluntary certification systems for natural resources such as food and timber, but went considerably further in terms of setting detailed standards for management. Standards were developed in accordance with the ISEAL Principles for Credible and Effective Sustainability Standards Systems (ISEAL, 2013) [22]. Workshops were conducted in India, Nepal, Bhutan and Malaysia, to draw up and refine the draft standards, which were field-tested and subject to further review. The system was designed to have applicability across all TRCs, covering varied geographical, cultural and ecological needs. The CA I TS standards are reviewed periodically as best practice evolves.

Governance (see Figure 2) begins with a broad CA I TS Partnership, which includes tiger and conservation management experts from around the world. An International Executive Committee ensures good governance and equivalence of implementation across TRCs. CA I TS functions through National Committees comprising government, NGO, experts and academic representatives, ideally embedded within an existing policy and institutional framework. The CA I TS Council brings representatives from National Committees together to share experiences; and is represented on the International Executive Committee. The CA I TS management team ensures technical and financial viability and the CA I TS support group (a wide range of non-governmental and intergovernmental organizations) provides support to TRCs undertaking CA I TS.

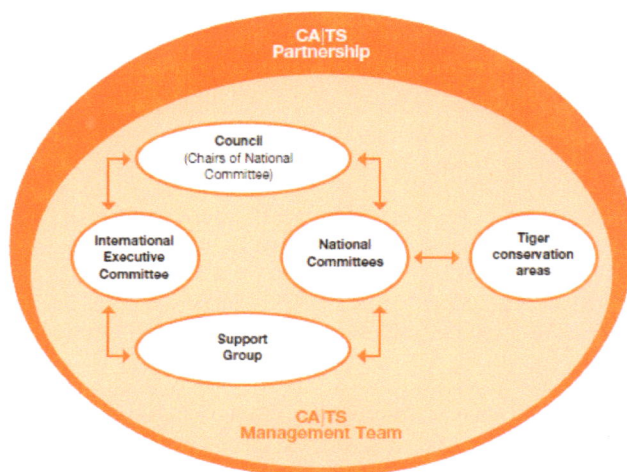

Figure 2. Conservation Assured I Tiger Standards (CA I TS) management structure.

CA I TS has a hierarchical structure; seven *'pillars'* covering different management issues, 17 *elements* (see Figure 3), subdivided into *standards*, for which *criteria* have been laid down (management actions required and a list of actions taken). For each standard, four values have been provided for reporting on compliance: Standard exceeded, standard achieved, standard mainly achieved and standard not achieved.

Protected areas and other tiger conservation areas first become CA I TS Registered and then go through an accreditation process to judge whether they meet CA I TS Approved status. CA I TS Registration demonstrates commitment to achieve best practice and identifies gaps in current management, providing a strategy for managers and a clear indication of needs to donors and potential supporters. For some sites CA I TS Registration may initiate a several year process of management intervention and improvement to reach the CA I TS Approved status; in these cases, the standards and criteria can be used as a gap analysis and planning tool to identify management actions which need to be put into place to reach Approved status.

PILLAR	ELEMENT
CONSERVATION ASSURED	
IMPORTANCE AND STATUS	1. Social, cultural and biological significance 2. Area design 3. Legal status, regulation and compliance
MANAGEMENT	4. Management planning 5. Management plan/system implementation 6. Management processes 7. Staffing (full-time and part-time) 8. Infrastructure, equipment and facilities 9. Sustainability of financial resources 10. Adaptive management (feedback loop)
COMMUNITY	11. Human–wildlife conflict (HWC) 12. Community relations 13. Stakeholder relationships
TOURISM	14. Tourism and interpretation *Note: this standard is only applicable for areas with major tourism objectives*
PROTECTION	15. Protection
TIGER STANDARDS	
HABITAT MANAGEMENT	16. Habitat and prey management
TIGER POPULATIONS	17. Tiger populations

Figure 3. CA I TS pillars and elements.

The CA I TS assessment is an iterative process which starts with a manager's self-assessment of the site against the CA I TS standards and criteria. The assessment is then reviewed and refined by the National Committee. An independent reviewer also reviews the entire CA I TS process before a final decision on status is made. Once the national process is complete and the National Committee recommends that a site has reached CA I TS Approved status, the International Executive Committee provides a final check to ensure parity between countries. Training is provided for all those taking part in CA I TS to ensure effective and consistent application. Approval lasts for 3 years, when a streamlined review process takes place. If there are significant changes in management or circumstances the area may need to update and resubmit its dossier to the National Committee and International Executive Committee.

4. Results

Since its launch, CA I TS has achieved broad-based support and obtained commitment to implement from all TRCs. Presently it is being implemented across 60 sites in seven TRCs (Bhutan, Bangladesh, China, India, Malaysia, Nepal and Russia), where National Committees have been established. Three sites have attained CA I TS Approved status (Chitwan National Park, Nepal, Sikhote-Alin Nature Reserve, Russia and Lansdowne Forest Division, India) and currently (as of June 2018) 18 dossiers are pending final approval. CA I TS has also been used for assessing the readiness of tiger reintroduction in Cambodia (Gray, et al., 2017) [23].

CA I TS can also measure progress in implementing international conventions, multilateral treaties, global initiatives and national conservation plans (Conservation Assured, 2014) [24]. CA I TS is partnered with IUCN's Green List of Protected and Conserved Areas, an initiative to identify and promote protected areas reaching excellent standards of management (IUCN 2013) [25]. It will help governments and their partners to meet the CBD's Strategic Plan for Biodiversity, particularly Targets 11, 12, (Strategic Goal C: To improve the status of biodiversity by safeguarding ecosystems, species and genetic diversity); 14 (Strategic Goal D: Enhance the benefits to all from biodiversity and ecosystem services) and 19 (Strategic Goal E: Enhance implementation through participatory planning, knowledge management and capacity building). Within the Convention on International Trade in Endangered Species of Wild Fauna and Flora (CITES), CA I TS helps state parties to fulfil commitments under 16.68(b) to 'provide information on incidents of poaching' (CITES, 2016) [26]. It also helps to evaluate anti-poaching plans, to fulfil the 'Zero Poaching Initiative' approach, formally endorsed in the most recent CITES Conference of Parties (Decision 17.225).

5. Experience with Existing CA I TS Approved Sites

As noted above, three sites have so far been awarded the CA I TS Approved status. Each site is unique and this variety has helped the CA I TS Partnership learn more about the CA I TS approach. A summary of the assessment and findings is given for each site below.

Nepal: Nepal has been actively engaged in the development of CA I TS from the beginning and nominated the first site to be CA I TS Registered. The assessment of Chitwan National Park (CNP) for CA I TS was conducted in 2013/2014 and was facilitated by WWF Nepal and the Department of National Parks & Wildlife Conservation. Field survey, consultations and meetings were conducted to verify that management met the CA I TS criteria. Based on this evidence, a dossier was submitted to the CA I TS National Committee of Nepal for evaluation and feedback. The feedback provided by the committee was utilized to finalize the document, which was followed by an independent review of the CA I TS process and outputs by an expert not involved in the CA I TS assessment.

Despite the pressures that CNP faces, including the constant threat from poachers and high demands from tourism, the site has seen an increasingly effective management and protection regime put in place. Overall CNP demonstrated high conformity to CA I TS and, therefore, qualified to be a CA I TS Approved Site. The CA I TS National Committee of Nepal concluded that the dossier developed to indicate the achievement of the CA I TS criteria shows that CNP is managed to the standards set out in CA I TS (Figure 4). In particular, CNP has a well-developed governance structure to include local people in management; active and supportive buffer zone communities; and benefit sharing is enshrined in law. In terms of community relationship management, as well as tiger monitoring and management, the site exceeds the CA I TS criteria (and thus highlights best practices). All the standards under the tourism, protection and habitat management pillars achieved the desired level of conformity, Standard Achieved. In a few areas of management not all the criteria were fully met (i.e., were assessed as Standard Mainly Achieved) but in these cases activities are underway to resolve outstanding issues and in no case do unfulfilled criteria represent a major impediment to tiger conservation. (Indeed, the Standard Mainly Achieved assessment can be a way of donors identifying priorities for investment.) For example, field posts needed to be improved through better provision and maintenance of clean drinking water (absent from some posts), solar power and infrastructure; however, there was insufficient budget available for these activities in the 2012–2016 management plan but fundraising was on-going at the time of the assessment and additional funding was secured after CA I TS Approval. When making the CA I TS Approved decision, the International Executive Committee requested the site to report on progress towards these field post improvements in 2016, particularly regarding safe drinking water at guard posts. WWF Nepal, as a focal agency facilitating CA I TS in Nepal worked with the National Park, Army and Best Paani Pvt Ltd. to provide bio-sand filtration systems, which were installed in 30 guard posts in September 2017, providing clean water for around 450 people.

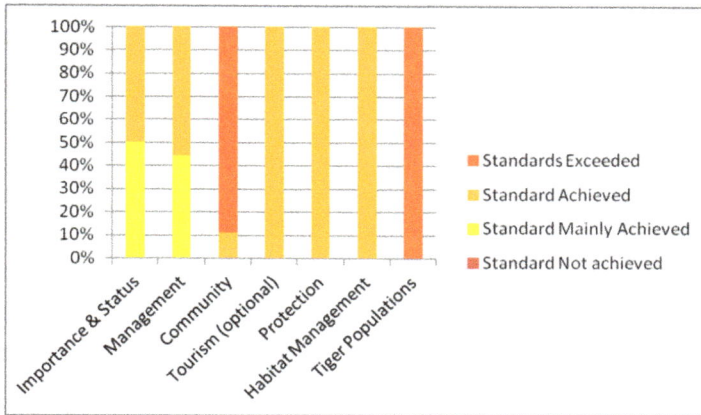

Figure 4. Overview of conformity of the 17 elements under seven pillars of CA I TS (see Figure 2) of CA I TS for Chitwan National Park.

The finalized dossier for CNP was submitted to the National Committee for its approval, which was achieved in October 2014. The Dossier was then passed to the CA I TS International Executive Committee for final approval. After discussion and some further information requirements from CNP (see above) the site was accredited as CA I TS Approved in January 2015. Since approval, Nepal has suffered a disastrous earthquake and flooding. After each disaster, updates on the impacts for tiger conservation in Chitwan were prepared by the protected area and National Committee. As the first site to receive CA I TS Approved status, Chitwan is also the first site to undertake the newly developed re-approval process, which is on-going (Conservation Assured, 2018a) [12].

Russia: Sikhote-Alin Nature Reserve (SANR) in Russia's Far East forms part of the easternmost habitat complex of tigers and has a unique combination of ecosystems. Tigers are found in low densities across a wide landscape, of which the SANR is only a small part. SANR had a tiger population of between 13–20 tigers in 2015 and is a source for populations for surrounding areas.

SANR was nominated as the first Russian site to take on CA I TS and the site was registered in December 2014. The field assessment began in 2015 and was aided by local and regional staff of WWF and the Wildlife Conservation Society. The National Committee (referred to in this case as the Jurisdictional Committee as it only representative of the small part of Russia where tigers are found) was established in March 2015. The assessment process went through several iterations after feedback from the Jurisdictional Committee, independent reviewers and CA I TS management team. After approval from the Jurisdictional Committee and International Executive Committee, CA I TS Approved status was conferred on SANR in July 2015.

SANR has a strong management system and a history of continuous monitoring of tigers, using the same protocols, going back 80 years. Annual plans for administration, finance, protection and infrastructure are linked to 3-year plans, and the management system has a robust auditing mechanism.

The CA I TS Dossier for SANR highlights both best practices and minor gaps in management practice (Figure 5). The results show that standards under the tiger population and habitat management pillar of CA I TS have exceedingly high levels of conformity, i.e., Standard Exceeded. All standards under the protection pillar have achieved the desired level of conformity, Standard Achieved. As in Nepal, in a few places where standards were assessed as Mainly Achieved, activities are underway to resolve outstanding issues and in no case does an unfulfilled criterion represent a major impediment to tiger conservation. For example, SANR has tourism potential. This could be a contributing factor to the local economy and could improve community livelihoods. At present, there is no tourism management plan in the reserve; however, the Amur Tiger Centre has a major goal to develop ecological tourism focused on the tiger in Russia and a project is being developed to raise interest in tigers amongst people living in European Russia, and thus to encourage tourism in the tiger range. SANR has now developed a plan to target ecological tourism aimed at the local community and which encourages volunteers and students to engage with management. Eco-trails have been built to give an experience of the wildlife and habitat along with an interpretation centre. An Ecotourism Club has also been setup locally to create more awareness about wildlife conservation and protection in Sikhote-Alin.

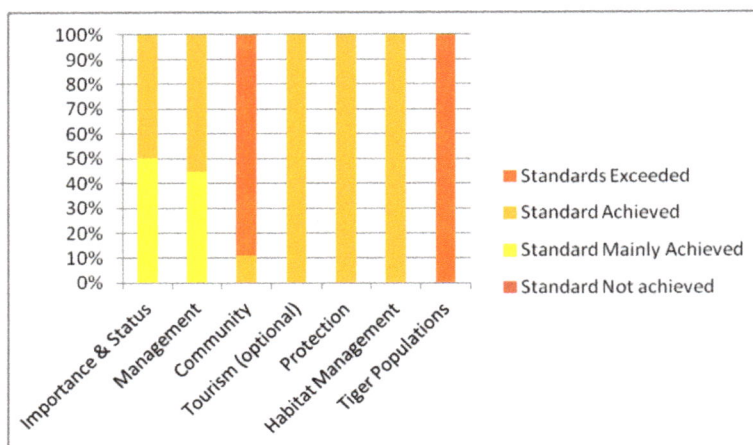

Figure 5. Overview of conformity of the 17 elements under seven pillars of CA I TS (see Figure 2) of CA I TS for Sikhote-Alin Nature Reserve.

India: Project Tiger was launched in 1973 in response to a serious decline in tiger numbers. Focused on the development and management of tiger reserves, the project has clearly had a beneficial impact on safeguarding wild tigers in the country (Johnsingh and Goyal, 2005) [27]. However, there are many important tiger populations and habitats outside tiger reserves, which have tended to receive less support and management focus. It was agreed that CA I TS would initially focus on these areas for its roll-out in India and the CA I TS National Committee in India drew up a preliminary list of nine forest reserves for the registration and implementation of CA I TS in September 2015.

Lansdowne Forest Division was the first area to become CA I TS Approved. Lansdowne is in the Pauri Garhwal district of Uttarakhand state in Northern India. It provides a crucial link between Rajaji and Corbett Tiger Reserves and is an important tiger and wildlife habitat in the western part of theTerai Arc Landscape. One-hundred-and-fifty-six tigers were camera trapped in March 2015 across 84 locations in Lansdowne; 23 were identified, of which 10 were females.

The CA I TS dossier was developed over 2016. After independent review and revision, the site was approved by the National Committee and reviewed by the International Executive Committee in May 2017. CA I TS Approved status was conferred after the submission of some additional material from the site.

The CA I TS pillars of community, protection, habitat management and tiger populations were all assessed as Standard Achieved (Figure 6). It was noted that a very competent protection strategy had been put in place over the last 7–8 years by the division manager who, amongst other things, had developed an intelligence-based protection system. Thus, although there is a lot of poaching pressure around the tiger reserves in the area, the Landsdowne tiger population is stable and has been growing over the last few years. Excellence (i.e., Standard Exceeded) was noted in both the importance and status and management pillars; however, some areas of management were also assessed as

Standards Mainly Achieved, as staffing and funding gaps had been a problem at the site. However, the incorporation of Landsdowne into CA | TS leveraged further government funding to the site, allowing steps to fill staff capacity and infrastructure gaps. Tourism, which is a voluntary standard in CA | TS as not all tiger sites have tourists, has not been fully in developed in Landsdowne. However, there are plans to develop tourism further, to take pressure off the neighbouring tiger reserves and provide additional community benefits.

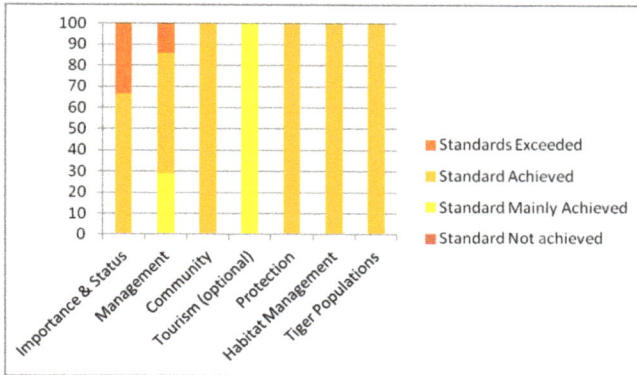

Figure 6. Overview of conformity of the 17 elements under seven pillars of CA | TS (see Figure 2) of CA | TS for Lansdowne Forest Division.

6. Discussion: Lessons Learned from Five Years Developing the CA | TS Standards

The development of CA | TS has been a multi-year project, involving many different actors across a range of countries. CA | TS differs from many conventional environmental certification systems, because of its focus on capacity-building and the lack of an immediate market-based mechanism (such as a price premium for a certified product). As a result, there has been a steep learning curve amongst those involved. The following section summarises some of the lessons learned to date.

Although a number of existing organisations have been closely involved in the establishment of the system, particularly WWF, which has provided substantial logistical and financial support, CA | TS has from the start been an independent body. The assumption is that any single NGO, or any government body, will not have the independence, or at least will not be perceived as having the independence, to run an accreditation system. The result, successful to date, has been that governments, NGOs and international bodies such as UN agencies can all sit around the table as equal partners.

Regional governance is seen as critical to success, so that the CA | TS Council is chaired by a government representative (currently from Bhutan), which makes it easier for governments to engage directly in the process and has resulted in all 13 TRCs becoming actively engaged. This has already led, for example, to cross border exchanges between staff to share experiences and learn different approaches and techniques. It has also strengthened links with other international processes such as the Global Tiger Forum (GTF) and the St Petersburg Declaration. The existence of an active Support Group, including several NGOs, UNDP, GTF and IUCN, has resulted in enough funds and assistance to ensure that the process has continued. Furthermore, cooperation with the IUCN Green List, a global and more generally protected area accreditation system that has been under development during the same period (IUCN 2013) [25], has helped in the development of some of the processes.

However, implementation has taken longer than originally planned, in part because of the inevitable time-lags involved in running a participatory process. This is however problematic in light of the short time period remaining to achieve TX2 and the urgency of wider tiger conservation efforts. Perhaps unsurprisingly, it has proven easier to get active involvement from those countries already doing well in terms of tiger conservation than from those where capacity is low and effectiveness compromised, yet the latter are the primary target of the CA I TS initiative. The countries of Southeast Asia in particular have generally weaker tiger conservation systems (Conservation Assured, 2018) [11–13] and at the same time have been less consistently engaged with CA I TS. Further efforts, from governments, NGOs and donor agencies, remain important if CA I TS is not to be simply a reward for good management rather than also a tool for improving management.

While there has been good buy-in from TRCs, there are still challenges in getting CA I TS fully integrated with national systems and national Tiger Action Plans. Many of these had been completed by the time that the CA I TS standard was in place and this, therefore, needs to be retrofitted into existing plans, if it is to become an integral part of everyday conservation and not a stand-alone project. Furthermore, the wide involvement of many stakeholders has made the system much stronger, but it also brings problems in terms of independent peer-review. Given the relatively small world of tiger conservation, most national experts suitable to reviews a site for CA I TS registration or approval will likely know many of the people involved in management, or themselves be involved in CA I TS. While there are policies in place to avoid direct conflict of interest, and the global International Executive Committee provides a check, this will inevitably remain an issue for any system of protected area verification, particularly in smaller countries.

Finally, the underlying philosophy of CA I TS driving increased management effectiveness is only going to work in the weaker TCAs if there is associated funding to build capacity. To date, this has been acquired on a fairly ad hoc basis, drawing on a surge of global support for tiger conservation. A longer term approach is needed and a business plan has been drawn up to this effect (Conservation Assured, 2018b) [13]. As well as setting out plans for the effective and efficient functioning of CA I TS at scale, the plan outlines an 'Advancement Programme' aimed at boosting the capacity of the TCAs to improve their management. This will require a very significant investment of funds either by the TRC governments or from donor sources. The plan suggests a new collective fund to provide support to the TCAs as the best method for financing these needs and the CA I TS Support Group has been discussing options for the development of such a fund (Conservation Assured, 2018b) [13].

7. Conclusions—CA I TS and Into the Future

The CA I TS vision is that wild tigers have spaces to live and breed safe from threat resulting in increased populations and recovery of range. To achieve this vision, CA I TS' two goals are that: (1) The adoption and implementation of CA I TS ensures tiger habitats are effectively conserved, well-managed and ecologically connected to maintain, secure and recover viable populations; and (2) CA I TS demonstrates and promotes best practice in protected area management in Asia. Whether achieving these goals will entail a long-term accreditation system, which will persist for decades, or whether CA I TS is more of a medium term capacity building tool whose usefulness will diminish over time is still open to question.

It is not simply the number of species that define the health of a landscape (Pollock et al., 2017) [28]. Tigers range across more than 50 habitats, with huge differences in elevation, temperature and ecosystem, so that CA I TS contributes to issues far broader than tiger conservation. Furthermore, in the context of severe declines in many flagship species (Butchart et al., 2010) [29], including particularly those targeted by the wildlife trade such as rhinoceros, elephant and pangolin, the CA I TS methodology could be modified for other species reliant on site-based conservation. The five conservation pillars of CA I TS are directly transferable; specific standards would need to be developed in accordance with other species' biology, protection and habitat management. Discussions are ongoing on developing Conservation Assured standards for several species groups such as rhinos, pangolins, river dolphins and some plants. Integrating the Conservation Assured philosophy into wider conservation discourse is the next priority in the development of this approach.

Author Contributions: Writing-Original Draft Preparation: M.K.S.P., N.D. and S.S.; Writing-Review & Editing: M.B. (Michael Baltzer), B.L., S.R., M.B. (Michael Belecky), R.G. and S.P.Y.

Funding: This research received no external funding.

Acknowledgments: We would like to thank all the Tiger Range Governments (TRCs) and the CA I TS Partnership for their support in rolling out CA I TS. We are also grateful to all those sitting on the National Committees in Nepal and India and Jurisdictional Committees in Russia, site managers, reviewers and the CA I TS International Executive Committee members who have invested their time to work on the assessments and review of the three CA I TS Approved sites presented in the paper. Special thanks to Dmitry Gorshkov, Yury Darman, Sergey Aramilev and Dale Miquelle from Russia; DVS Khati, Djananjay Mohan and Nitish Kumar from India; Bishwa Nath Oli, Maheshwar Dkahal, Diwakar Chapagain, Shyam Bajimaya and all the teams in WWF Russia, India and Nepal for facilitating the CA I TS implementation and site assessments. We are thankful to the State Forest Department of Uttarakhand, for being the first State to roll out CA I TS in India. The CA I TS Support Group has been a strong force for driving the CA I TS globally, we are thankful to all of them. The GTF has also played a very strong role in the adoption of CA I TS across TRCs and advocating the benefits of CA I TS for long-term tiger conservation, we thank them for their support.

Conflicts of Interest: The authors declare no conflict of interest.

References

1. Wolf, C.; Ripple, W.J. Range contractions of the world's large carnivores. *R. Soc. Open Sci.* **2017**, *4*, 170052. [CrossRef] [PubMed]

2. Walston, J.; Robinson, J.G.; Bennett, E.L.; Breitenmoser, U.; Da Fonseca, G.A.B.; Goodrich, J.; Gumal, M.; Hunter, L.; Johnson, A.; Karanth, K.U.; et al. Bringing the tiger back from the brink—The six per cent solution. *PLoS Biol.* **2010**, *8*, e1000485. [CrossRef] [PubMed]

3. WWF. Priority Species. 2015. Available online: http://tigers.panda.org/tiger-facts/ (accessed on 27 July 2018).

4. Stoner, S.S.; Pervushina, N. *Reduced to Skin and Bones Revisited: An Updated Analysis of Tiger Seizures from 12 Tiger Range Countries (2000–2012)*; Traffic: Kuala Lumpur, Malaysia, 2013.

5. Damania, R.; Seidensticker, S.; Whitten, T.; Sethi, G.; MacKinnon, K.; Kiss, A.; Kushlin, A. *A Future for Wild Tigers*; World Bank: Washington, DC, USA, 2008.

6. Global Tiger Initiative (GTI). The St. Petersburg Declaration on Tiger Conservation. 2010. Available online: cmsdata.iucn.org/downloads/st_petersburg_declaration_english.pdf (accessed on 9 January 2018).

7. Roy, S.; Vie, J.C.; Gelsi, T. The Integrated Tiger Habitat Conservation Programme—Progress to date. *Cat News* **2016**, *63*, 35–36.

8. Geldmann, J.; Coad, L.; Barnes, M.; Craigie, I.D.; Woodley, S.; Balmford, A.; Brooks, T.M.; Hockings, M.; Knights, K.; Mascia, M.B.; et al. A global analysis of management capacity and ecological outcomes in protected areas. *Conserv. Lett.* **2018**, *11*, e12434. [CrossRef]

9. Leverington, F.; Costa, L.L.; Pavese, H.; Lisle, A.; Hockings, M. A Global Analysis of Protected Area Management Effectiveness. *Environ. Manag.* **2010**, *46*, 685–698. [CrossRef] [PubMed]

10. Brun, C.; Cook, A.R.; Lee, J.S.H.; Wich, S.A.; Pin Koh, L.; Carrasco, L.R. Analysis of deforestation and protected area effectiveness in Indonesia: A comparison of Bayesian spatial models. *Glob. Environ. Chang.* **2015**, *31*, 285–295. [CrossRef]

11. Conservation Assured. *Safe Havens for Wild Tigers: A Rapid Assessment of Management Effectiveness against the Conservation Assured Tiger Standards*; Conservation Assured: Singapore, 2018.

12. Conservation Assured. *CA I TS Manual Version 2.0*; Conservation Assured: Singapore, 2018.

13. Conservation Assured. *Conservation Assured I Tiger Standards 2018–2022*; Business Plan; Conservation Assured: Singapore, 2018.

14. Hockings, M.; Stolton, S.; Leverington, F.; Dudley, N.; Courrau, J. *Evaluating Effectiveness: A Framework for Assessing Management Effectiveness of Protected Areas*, 2nd ed.; IUCN: Gland, Switzerland; Cambridge, UK, 2006. Available online: https://portals.iucn.org/library/efiles/documents/pag-014.pdf (accessed on 27 July 2018).

15. Coad, L.; Leverington, F.; Burgess, N.D.; Cuadros, I.C.; Geldmann, J.; Mathews, T.R.; Mee, J.; Nolte, C.; Stoll-Kleemann, S.; Vansteelant, N. Progress towards the CBD protected area management effectiveness targets. *PARKS* **2013**, *19*, 13–24. [CrossRef]

16. Mathur, V.B.; Gopal, R.; Yadav, S.P.; Negi, H.S. *Management Effectiveness Evaluation of Tiger Reserves*; Technical Manual No. WII-NTCA/01/2010 pp 21; Revised and Updated Version; WII-NTCA/01/2014 pp 25; National Tiger Conservation Authority and Wildlife Institute of India: New Dehli, India, 2014.

17. Convention on Biological Diversity (CBD). *Programme of Work on Protected Areas (UNEP/CBD/COP/7/21)*; Secretariat of the Convention on Biological Diversity: Montreal, QC, Canada, 2004.

18. Polasky, S.; Tallis, H.; Reyes, B. Setting the bar: Standards for ecosystem services. *Proc. Nat. Acad. Sci. USA* **2015**, *112*, 7356–7361. [CrossRef] [PubMed]
19. Ferraro, P.J.; Pattanayak, S.K. Money for nothing? A call for empirical evaluation of biodiversity conservation investments. *PLoS Biol.* **2006**, *4*, e105. [CrossRef] [PubMed]
20. McCarthy, D.P.; Donald, P.F.; Scharlemann, J.P.W.; Buchanan, G.M.; Balmford, A.; Green, J.M.H.; Bennun, L.A.; Burgess, N.D.; Fishpool, L.D.; Garnett, S.T.; et al. Financial costs of meeting global biodiversity conservation targets: Current spending and unmet needs. *Science* **2012**, *338*, 946–949. [CrossRef] [PubMed]
21. Bennett, J.R.; Maloney, R.; Possingham, H.P. Biodiversity gains from efficient use of private sponsorship for flagship species conservation. *Proc. R. Soc. B Biol. Sci.* **2015**, *282*, 1–7. [CrossRef] [PubMed]
22. ISEAL. *Principles for Credible and Effective Sustainability Standards*; ISEAL: London, UK, 2013.
23. Gray, T.N.E.; Crouthers, R.; Ramesh, K.; Vattakaven, J.; Borah, J.; Pasha, M.K.S.; Lim, T.; Phan, C.; Singh, R.; Long, B.; et al. A framework for a ssessing readiness for tiger Panthera tigris reintroduction: A case study from eastern Cambodia. *Biodivers. Conserv.* **2017**, *26*, 2383–2399. [CrossRef]
24. Conservation Assured. *Conservation Assured | Tiger Standards: A Multifunctional Protected Area Management Tool to Aid Implementation of International Conventions, Multilateral Treaties, Global Initiatives & National Action*; Conservation Assured: Petaling Jaya, Malaysia, 2014.
25. IUCN. The IUCN Red List of Threatened Species. IUCN Green List Boosts Partnership with a New Tiger Conservation Initiative at the Asia Parks Congress. 2013. Available online: http://www.iucnredlist.org/news/iucn-green-list-boosts-partnership-with-a-new-tiger-conservation-initiative-at-the-asia-parks-congress (accessed on 27 July 2018).
26. CITES. Convention on International Trade in Endangered Species of Wild Fauna and Flora (CITES). In Proceedings of the Seventeenth Meeting of the Conference of the Parties, Johannesburg, South Africa, 24 September–5 October 2016; CoP17 Doc. 60.1. Available online: https://cites.org/sites/default/files/eng/cop/17/WorkingDocs/E-CoP17-60-01.pdf (accessed on 27 July 2018).
27. Johnsingh, A.J.T.; Goyal, S.P. Tiger Conservation in India: The Past, Present and the Future. *Indian For.* **2005**, *131*, 1279–1296.
28. Pollock, L.J.; Thuiller, W.; Jetz, W. Large conservation gains possible for global biodiversity facets. *Nature* **2017**, *546*, 141–157. [CrossRef] [PubMed]
29. Butchart, S.H.M.; Walpole, M.; Collen, B.; Van Strein, A.; Scharlemann, J.P.W.; Almond, R.E.A.; Baillie, J.E.M.; Bomhard, B.; Brown1, C.; Bruno, J. Global biodiversity: Indicators of recent declines. *Science* **2010**, *328*, 1164–1168. [CrossRef] [PubMed]

Article

Assessing Local Indigenous Knowledge and Information Sources on Biodiversity, Conservation and Protected Area Management at Khuvsgol Lake National Park, Mongolia

Christopher McCarthy [1,*], Hitoshi Shinjo [1], Buho Hoshino [2] and Erdenebuyan Enkhjargal [3]

[1] Graduate School of Global Environmental Studies, Kyoto University, Kyoto 606-8501, Japan; shinhit@kais.kyoto-u.ac.jp

[2] Graduate School of Dairy Sciences, Rakuno Gakuen University, Hokkaido 069-0836, Japan; aosier@rakuno.ac.jp

[3] Graduate School of Global Studies, Doshisha University, Kyoto Prefecture 602-8580, Japan; erdenebuyan.enkhjargal@gmail.com

* Correspondence: cmccarth@ucsd.edu

Received: 4 September 2018; Accepted: 11 October 2018; Published: 11 October 2018

Abstract: Indigenous knowledge about biodiversity and conservation is valuable and can be used to sustainably manage protected areas; however, indigenous communities continue to be marginalized due to the belief that their values and behaviors do not align with the overarching mission of conservation. This paper explores the extent of local knowledge and awareness of biodiversity, conservation and protected area management of indigenous communities at Khuvsgol Lake National Park, Mongolia. We investigate current levels of biodiversity awareness and explore perceptions toward conservation values and park management governance. Most respondents had a high awareness of existing biodiversity and held positive attitudes toward nature conservation and protected areas; however, insufficient knowledge of park rules and low levels of trust between local residents and park authorities may undermine conservation objectives in the long run. We identify an unequal share of economic benefits from tourism and preferential treatment toward elite business owners as a source of conflict. Limited information channels and poor communication between local residents and park authorities are also a source for low-level participation in conservation activities. Leveraging the increasing use of information communication technology, such as mobile phones, can serve as a new mechanism for improved information sharing and transparent reporting between local communities, conservationists and protected area authorities.

Keywords: protected areas; biodiversity; conservation; protected area management; information communication technology; Mongolia

---

1. Introduction

National parks and protected areas have become the most effective strategy to conserve and protect biodiversity and natural ecosystems [1]. As of 2017, more than 240,000 protected areas exist covering a total of 15 percent of the world's terrestrial area [2]. Under the Convention on Biodiversity and Aichi Biodiversity Targets the world's governments have pledged to increase the number of terrestrial protected areas to more than 17 percent by 2020 (Convention on Biodiversity 2010). However, some experts are calling for even bolder action to preserve at least 50 percent of terrestrial area globally [3,4]. While the international community's decision to increase protected lands is encouraging, those managing protected areas often lack the adequate resources to effectively manage and enforce park rules and regulations, resulting in inefficient conservation efforts—nearly one-third of the world's protected areas are susceptible to intense human pressure [5]. Furthermore, large numbers of these protected areas, which are home to some of the highest levels of biodiversity in the world, fully or partially overlap the traditional lands of indigenous peoples [6]. The governance authorities and arrangements of these protected lands often do not recognize local inhabitants' traditional knowledge, practices, means of livelihoods, and collective tenure, which can create inequalities that undermine the values that long-term and sustained conservation is based [7].

It has long been believed that problems like pollution, deforestation, species extinction, and soil degradation have been due to local, indigenous misuse of natural resources [8]. Although the need to reduce human impacts on biodiversity has been widely acknowledged, research has found that local people do value, utilize and efficiently manage their environments, suggesting that local involvement is the first and most important line of defense in protecting biodiversity [9]. Since 1994 there has been some acknowledgement by the international community to recognize indigenous peoples' rights and ownership of protected areas, beginning with the World Conservation Union and later the World Commission on Protected Areas [10–12]. Nevertheless, while these efforts suggest a shift from past conservation philosophy, little has been done beyond the declarations and guidelines made on the international stage to ensure that indigenous peoples are included in the decision making and co-management of protected lands [13]. In the face of these challenges, the need for new channels of communication that promote the transparent exchange of information between conservationists, park authorities and indigenous peoples has been established [5,14,15]. Recent studies have identified information communication technologies (ICT) as an emerging tool for conservation in Africa while some development agencies like the World Bank have reported preliminary success implementing information sharing campaigns via mobile phone short message service (SMS) for remote communities in Asia [16,17]. Unlike television and radio, the proliferation of mobile phones, specifically smart phones, provides an opportunity for information exchange and feedback loops between groups and does not have the same cost barrier or need for reliable access to an electricity grid [15].

Two questions addressed in this paper aim to elucidate the generally accepted theory that indigenous communities are aware of their environment, and conservation goals can be better achieved if local communities are engaged and included in the decision-making process. A unique aspect of this study aims to identify current information sources and new channels for information exchange as a means to encourage participation of local and indigenous communities in protected area governance. We ask: (1) What is the current level of local awareness of biodiversity, conservation and protected area management at Khuvsgol Lake National Park (KLNP) in Mongolia and; (2) What are the information sources, usage and usage patterns of information communication technologies within indigenous KLNP households?

To understand the issues addressed in the study we turn to Mongolia, a country home to some of the world's largest remaining wild areas that support a vast and diverse group of native flora and fauna. However, in the face of a rapidly growing mining and tourism industry, conservation of wildlife and traditional pastoral livelihoods are being threatened [18]. In Mongolia, conservation policies are under the direction of the Ministry of Nature, Environment and Tourism, which oversees 64 protected areas covering approximately 21 million hectares or 14% of Mongolia's total terrestrial area [19]. Many of these protected areas overlap with local and indigenous communities and conflicts between public administration officers, park authorities and local inhabitants are occurring with greater frequency [20]. At KLNP the Ministry of Nature, Environment and Tourism has overseen park management and conservation activities since the park was designated as a protected area in 2004. For the most part, conservation law in Mongolia is consistent with international practice; however, there is a significant exception in the law that promotes: (1) the participation of local people and communities in protected area establishment, planning and management; and (2) the sharing of benefits from protected areas with local people [21]. Despite this, government efforts to expand tourism in the Khuvsgol Lake region have increased rapidly with KLNP receiving national and international attention, and while interest for tourism grows development has fallen short of integrating local communities into the planning and decision-making process and providing an equal share of benefits from the tourism sector.

The Setting

Located in northern Mongolia, along the Russian border, at the foot of the eastern Sayan Mountains, Khuvsgol Lake National Park is designated as a protected area (IUCN Category II) for its ecological importance including natural beauty, high biological richness (794 plant species, 369 animal species, 258 migratory birds), pristine water resources and unique historic and cultural values. The region lies 1645 m above sea level and covers more than 8300 square kilometers. KLNP is classified by the Ministry of Nature, Environment and Tourism as a "strictly protected area". The region is home to a large number of vulnerable and endangered wildlife and fish species including argali, elk, musk deer, sable, Siberian marmot as well as burbot, grayling, lenok and perch. Despite the protected status of KLNP a variety of challenges continue to exist, including illegal logging, illegal mining, commercial fishing, unregulated development, poor sanitation and water quality, and litter [22].

The namesake of the park, two million-year-old Khuvsgol Lake, is one of seventeen ancient lakes in the world, and is considered to be one of the most pristine fresh water sources on the planet. More than 136 km long, 35 km wide and 262 m deep the lake contains 4% of global fresh water and 70% of Mongolia's fresh water resources [23].

Present-day inhabitants of KLNP are mainly settled and mobile pastoral households including the Tsaatan, a community of reindeer herders living in the northern part of the reserve [24]. Originally from bordering Tuva Republic, the Tsaatan are one of the world's last remaining groups of nomadic reindeer herders. Currently, some 40 reindeer households live within the boundaries of KLNP. In addition, nearly 200 mobile pastoral families live at KLNP grazing their herds of sheep, goat, cow, yak and horse in seasonal pasture areas. In recent years, the park has attracted many new residents owing largely to the rapidly growing tourism sector. Within the interior of KLNP there are 2 major districts and a number of sub districts, a few of which have become permanent and semi-permanent tourist areas. Districts included within park boundaries are subject to park regulations; however, laws are often unevenly enforced. Many of the districts within KLNP are some of the poorest in Mongolia [25]. Total population of KLNP has increased from 12,000 in 2001 to nearly 16,000 in 2017 [26].

KLNP has seen a rapid increase in tourist numbers bringing with it profound changes in the local economy. Many mobile pastoral households which had traditionally relied on animal husbandry now make some of their income from the seasonal tourism sector. Common jobs include horse guides, tourist camp operators and sellers of indigenous arts and craft souvenirs. Local communities; however, have limited capacity or resources to fully access tourism-related benefits with most of the business going to the elite tourist camp operators, which can provide higher quality service at lower cost. We find that outside camp operators account for 62 of 82 (76 percent), of all tourist camps at KLNP in 2017 as opposed to 10 percent in 2000. For many pastoral residents, livestock grazing will continue to be a principal livelihood, yet these benefits are in decline due to a growing encroachment of tourism camps onto traditional grazing areas.

From 2010 to 2014, annual tourist visits to KLNP rose from 11,000 to 60,000 [25], largely due to improved road access and reduced visa restrictions. While, the government has targeted KLNP as a key region for development, tourism-related expansion occurs in the absence of planning and there is little to no coordination between KLNP administration, communities, tour operators and tourism facilitators. Uncontrolled sewage and litter from tourism is threatening lake water quality and exerting additional pressure on grazing areas depended on by mobile pastoral households [27]. Recent studies have even reported the presence of high-levels of microplastic pollution within the lake [28].

In carrying out the functions of the park, KLNP is staffed by a small number of administrators and rangers, officially 1 per 32 square kilometers. Interviews with park administrators established that rangers are tasked to enforce park rules and interact with local community members; however, outreach and interaction is infrequent. Park regulations are not made available or published for local inhabitants to see, leading to an information imbalance that often results in uneven enforcement of park policies. The importance of transparent governance cannot be overstated, as we found it to play a key role in the lack of trust between local community members and park authorities. Furthermore, park rangers lack the adequate training to evaluate human activity as being acceptable or unacceptable based on traditional and cultural values. Examples include citing local inhabitants for harvesting of biomass for cooking and heating, fishing for non-commercial purposes, and grazing of animals in traditional pasture areas. While these activities are officially illegal by Mongolian law, park authorities' unwillingness to recognize local knowledge about the environment may contribute to the marginalization of local citizenry at KLNP.

2. Methods

Thirty households within KLNP were asked 60 open and close-ended questions designed to assess knowledge on biodiversity awareness, conservation and protected area management within KLNP, including beliefs on environmental protection and conservation, identification of important biodiversity species, and knowledge on rules, regulation and management (Table S1). A unique aspect of this study, the survey also explored information sources, technology adoption and mobile phone use within local and indigenous households.

The survey was carried out during two field seasons: June–September 2016 and July–August 2017 in the district of Jankhai 28 km north of Khatgal, the administrative center (Figure 1) of KLNP. The study area was selected given its high number of settled and mobile pastoral households. It also serves as a major tourist destination for its proximity to the lake and road access and has seen a massive increase in tourism related development in recent years. In total, nearly 200 households were living within Jankhai at the time of the study. To minimize selection bias, a snowball technique sampling method was employed with one person being chosen at random and after asked to identify a friend or acquaintance that could participate in the survey. This sampling method was selected for its reliability and convenience. Of the 200 households, 70 individuals from 30 unique settled and mobile pastoral households were approached due to time constraints and the large distances between households. Of these 30 that were approached, 100% participated in the survey. Data input, coding, classification, categorization and analysis were conducted using STATA 13.0.

A description of the demographic data of participants at KLNP can be found in Table 1. We define settled households as those living in a fixed structure who do not participate in pastoralism, these households own few to no animals and rely less on the environment for their source of livelihood. Mobile pastoral households, on the contrary, dwell in the traditional Mongolian 'ger', a round wooden framed structure surrounded by felt, and depend heavily on the environment to earn a living. Many of these families own a variety of animals including sheep, goat, yak, cow and horse and move seasonally to different grazing areas. We make the distinction between settled and mobile pastoral households in order to compare the latter groups that practice more traditional indigenous livelihoods to the former who are more sedentary and therefore more likely to be involved in conventional occupations and have greater access to information and knowledge resources. Our study sample is representative of regional and national averages [25].

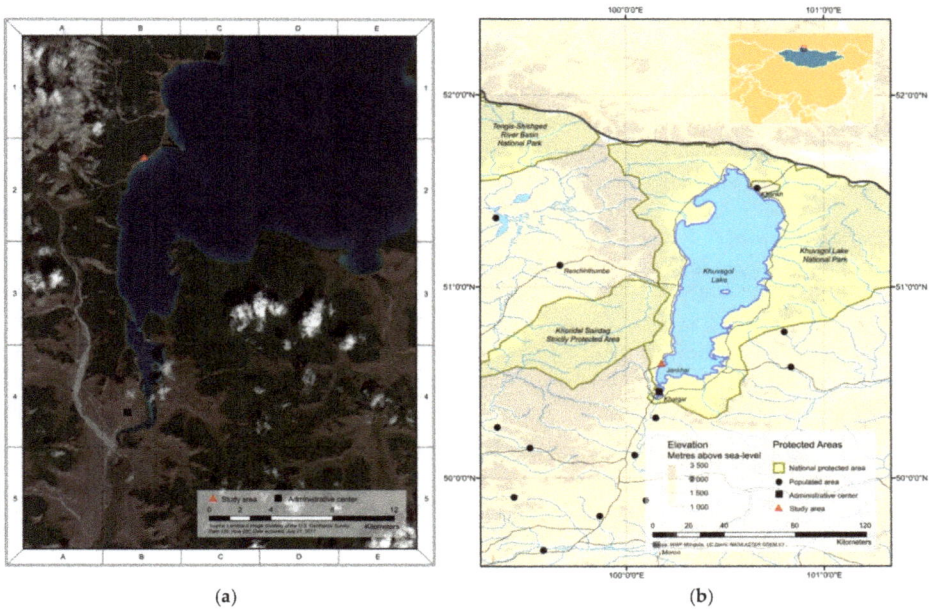

(a)   (b)

Figure 1. (**a**) Landsat-8 image (Path 135, Row 026; Date acquired 21 July 2017) of study area (2B) and administrative center of KLNP (4B); (**b**) Map showing environs of KLNP.

Table 1. Demographic data of participants at KLNP.

Characteristics	Group	Total
Gender	Female	41 (59%)
	Male	29 (41%)
Age (years)	8–19	16 (23%)
	20–29	19 (27%)
	30–39	6 (9%)
	40–49	10 (14%)
	50–59	10 (14%)
	≥60	9 (13%)
Education	Primary school	40 (57%)
	High school	11 (16%)
	University	2 (3%)
	Vocational training	2 (3%)
Occupation	Herder	31 (44%)
	Fisherman	2 (3%)
	Construction	5 (7%)
	Tourism	9 (12%)
	Teacher	3 (4%)
	Student	10 (14%)
	Pensioner	10 (14%)
Dwelling	Mobile pastoralist	51 (73%)
	Settled	19 (27%)

3. Results

3.1. Biodiversity Awareness

In order to understand the level of biodiversity awareness among local residents we asked respondents to identify species within KLNP from photographs and, if identified correctly, state whether they believed the animal was considered endangered, vulnerable, or at no risk (Table 2). Responses indicated that mobile pastoralists have a high degree of knowledge on vulnerable and endangered species in KLNP. Even argali and red deer which graze deep into the Sayan mountains away from pastoral zones were recognized as being animals found within park boundaries and highly endangered. Settled participants reported a similar degree of awareness for protected and endangered species. Many participants also noted that the Khuvsgol Grayling, commonly cited by government agencies as being overfished by local residents, is highly vulnerable and therefore deserving of protection; however, subsistence fishing should be permitted. When questioned how a species became endangered over hunting was the most commonly cited answer among both pastoral and settled households.

Table 2. Levels of awareness for key and vulnerable species at KLNP.

Species	Demographic		Total
	Pastoralist	Settled	
Argali (*Ovis ammon*)	33 (64%)	12 (63%)	45 (64%)
Musk Deer (*Moschus moschiferus*)	44 (86%)	16 (84%)	60 (86%)
Elk (*Alces alces*)	45 (88%)	17 (89%)	52 (74%)
Grayling (*Thymallus nigrescens*)	48 (94%)	19 (100%)	67 (96%)
Red Deer (*Cervus elaphus*)	39 (76%)	16 (86%)	55 (79%)
Reindeer (*Rangifer tarandus*)	48 (94%)	19 (100%)	67 (96%)
Sable (*Martes zibellina*)	46 (90%)	17 (89%)	63 (90%)
Siberian marmot (*Marmota sibirica*)	42 (82%)	15 (79%)	57 (81%)

3.2. Knowledge of Environmental Sensitivity and Park Governance

Table 3 shows the questions asked of participants to assess perceptions on conservation and park management activities. The results reveal a high degree of environmental sensitivity among respondents with 96% of those surveyed answering that they believe environmental protection is essential for their well-being and 90% believing that human activity can irreversibly impact the environment. However, a majority of respondents were not aware of the existence of the park's governing institutions nor had any knowledge on how to report a grievance or negligent human activity to authorities—only 24% of total participants, 24% of pastoralists and 26% of settled households, were able to correctly identify the Ministry of Nature, Environment and Tourism, as the administer of park management. Furthermore, only 19% of respondents could accurately state park rules and regulations related to hunting, fishing, forestry and waste management. Trust between local residents and park management is also low with 39% of pastoralists and 42% of settled households stating an unwillingness by park authorities to redress grievances as a major source of contention. Park inhabitants also noted that outside business owners benefit the most from tourism with only 10% of total respondents believing that local residents have an equal opportunity to share in the economic benefits.

Table 3. Perceptions on conservation and park management activities.

Participant Answered ...	Demographic		Total
	Pastoralist	Settled	
Environmental protection is essential for wellbeing;	49 (96%)	18 (95%)	67 (96%)
Human disturbance can result in irreversible environmental damage;	47 (92%)	16 (84%)	63 (90%)
Is aware of park rules and regulations related to hunting, fishing, forestry and waste management;	11 (22%)	2 (11%)	13 (19%)
Can identify the governing bodies of KLNP;	12 (24%)	5 (26%)	17 (24%)
Park authorities can be trusted to redress grievances;	20 (39%)	8 (42%)	28 (40%)
Certain activities should be prohibited within the park;	35 (69%)	16 (84%)	51 (73%)
Tourism provides an equal share of economic benefits.	6 (12%)	1 (5%)	7 (10%)

In terms of associations across demographics we find little difference in opinion toward conservation and park management between settled and pastoral households. Ninety-six percent of mobile pastoralists consider the environment the most important factor in their wellbeing and consider environmental protection essential. This is also true for 95 percent of those living in settled households.

Seventy three percent of total respondents, 69% for mobile pastoralists and 84% for settled households, expressed the need to prohibit certain activities within park boundaries including hunting of birds, mining and logging of trees. Although the majority of respondents' views aligned with conservation norms a common point of contention was fishing and hunting, which, although prohibited by law, remains an important subsistence activity for many pastoral and settled households. An overwhelming majority of participants answered that they believe local households should have the right to fish and hunt although they answered unfavorably when asked if commercial fishing and hunting should be allowed to promote tourism.

Most respondents answered favorably for the development of the tourism sector; however, improved regulation and access to economic benefits were answered as necessary for local communities. Only 10% of total respondents believed that tourism provided an equal share of benefits and was often cited as a major reason for conflict and distrust with park officials and business owners.

3.3. Sources of Information

In Table 4, 94% of respondents stated they obtain some information about the environment from friends and family. Information from local authorities, such as park rangers, accounted for only 26% of information sources. School sources provided 26% of respondents with information about park management and television and radio provided 18% of participants, respectively. We account for low levels of information dissemination across television and radio due to a lack of reliable access to electricity. Twenty percent of respondents claimed to receive some information about biodiversity and conservation through ICT such as mobile phones. While access to an electricity grid is limited, many households own and operate 50-watt solar home systems that can sufficiently charge phones and other portable devices. Connection to the cellular network is also improving as telecommunication companies expand their coverage to serve popular tourist areas.

Table 4. Information sources on biodiversity, conservation and park management.

	Demographic		Total
	Pastoralist	Settled	
Personal experience	47 (92%)	19 (100%)	66 (94%)
Television, radio	9 (17%)	4 (21%)	13 (18%)
Local authorities	12 (23%)	6 (32%)	18 (26%)
School	15 (29%)	3 (16%)	18 (26%)
ICT/Internet	11 (22%)	3 (16%)	14 (20%)

Regarding ICT, 84 percent of respondents answered that they own a mobile phone, with 44% of that group using a smart phone (computing and internet capabilities) as opposed to a traditional mobile handset (basic calling features) (Table 5). Sixty-seven percent of those interviewed reported at least occasional use of SMS and 58% use their mobile phone to take photos. Sixty-nine percent of mobile pastoralists reported using SMS and 33 percent responded accessing the Internet at least occasionally with their smart phone. Common barriers to use were poor reception, limited battery life, cost and user ability.

Table 5. Mobile phone use within local communities at KLNP.

	Demographic		Total
	Pastoralist	Settled	
Own a mobile phone	42 (82%)	17 (89%)	59 (84%)
Own a smart phone	23 (45%)	8 (42%)	31 (44%)
Utilize messaging and SMS	35 (69%)	12 (63%)	47 (67%)
Occasional use of Internet	17 (33%)	4 (21%)	21 (30%)
Use of social media	11 (22%)	4 (21%)	15 (21%)
Photography and digital media	32 (62%)	9 (47%)	41 (58%)

4. Discussion

We find, in line with previous studies on indigenous values [7–10,29], that local communities at KLNP do value and have representative knowledge about their environment. We identify that poorly developed information channels and low levels of information sharing diminish trust and erode local support for conservation activities. Though a variety of information channels at KLNP exist, many are underutilized. The introduction of conservation themed curriculum in schools and increased outreach by park authorities may be two immediate actions that can be taken to enhance awareness of conservation objectives and improve information exchange between local residents and authorities in a protected area setting [30]. We find that 80 percent of respondents indicate low engagement with park officials and uneven enforcement of park policies as a source of hostility. Park authorities must make clear that they are there to work with and not against local community members. Transparency of institutional rules, regulations and activities is also crucial for achieving local participation in conservation activities [31]. Installing signs or placards within the perimeter of the park and providing households with a list of park rules are two examples that have been cited as effective for informing local citizenry on regulations. In addition, park rangers would benefit from training that helps them assess local activity as being aligned with traditional values and therefore permitted in accordance with international norms. The growing use of ICT including mobile phones and access to the Internet presents a new opportunity to engage the local community with up to date information about conservation activities and park management. Given the limited human resources of the park, ICT should be considered as a primary mechanism for improved information sharing, open dialog, transparent reporting as well as a channel to communicate opportunities for residents to access economic benefits from the tourism sector. If the local population is unlikely to see any economic benefit from tourism long-term sustainability and cooperation is unlikely. If future action plans are inclusive, coherent, and strategic and sustained communication, education and public awareness efforts are made, long-term conservation of KLNP is possible. Finally, the Ministry of Nature, Environment and Tourism should establish a clear plan, based on local and expert consultation, identifying how tourism should be developed within the park over time as well ensure a more equal share of economic benefits reach the local people. Promoting locally run tourist camps, establishing a designated market for locally produced handicrafts, requiring outside run tourist facilities hire local staff, offering entrepreneurial training programs for local residents, and setting a quota for the number of outside owned tourist facilities within park boundaries are just a few examples of efforts that park management can take to include local people in the tourism sector.

Although this study accomplished its aims several limitations would benefit from future study. First, as ICT has only recently been introduced into the local community use patterns will likely evolve as familiarity with the technology improves and more reliable access to the network is achieved. Future studies will want to explore the longitudinal effects of ICT use in more depth. Second, because of limited time and distance between households only a small number of participants were included in the survey. A more rigorous study with a greater number of participants may provide a better understanding of how local residents can actively contribute to conservation activities. Finally, future studies will also want to focus on identifying the relative messages for bridging the information gap between park management and local communities as well as which ICT channels and social media platforms can help establish transparent information sharing.

5. Conclusions

In this paper we attempt to assess the local knowledge and information sources on biodiversity, conservation and protected area governance at Khuvsgol Lake National Park in Mongolia. We find, in line with our study objectives, that a majority of participants had a high awareness of biodiversity and held positive attitudes toward nature conservation and protected areas; however, insufficient knowledge of park rules and low levels of trust between local residents and park authorities may undermine the park's conservation objectives in the long run. Limited information channels and poor communication between local residents and park authorities are also a source for low-level participation in conservation activities. The growing use of smart phones and access to the Internet presents a new opportunity to connect community members with conservation activities and provide information on park management. Our results support the accepted wisdom that conservation goals can be better achieved if local communities are engaged and included in the decision-making process and rules, regulations and activities are transparently shared, understood and agreed upon [7–9,14,15]. In the face of the expanding global protected area network, it is also necessary to anticipate how the forces of tourism may impact local economies and livelihoods in the long term. Policy must reflect local knowledge and community involvement in the management and sharing of economic benefits to ensure long-term sustainability of protected areas. Limited operating budgets make ICT, such as smart phones and the Internet, an important mechanism in improving information sharing and collaboration between park residents and conservationists.

Supplementary Materials: The following are available online at http://www.mdpi.com/2073-445X/7/4/117/s1, Table S1: Questionnaire.

Author Contributions: C.M., H.S., B.H., and E.E. conceptualized and designed the study. C.M., H.S., and E.E. carried out the investigation. Project administration was facilitated by C.M., H.S., B.H., and E.E. Visualization and map making was conducted by C.M. C.M. wrote, revised and edited the manuscript.

Funding: This study was supported by JSPS KAKENHI Grant Number JP26304045.

Acknowledgments: The authors thank S. Funakawa and Y. Okamoto (Kyoto University) for their assistance with the field studies. The author's deep appreciation is expressed to Troy Sternberg from Oxford University, Oxford, for his valuable comments on the manuscript.

Conflicts of Interest: The authors declare no conflicts of interest.

References

1. Jenkins, C.N.; Joppa, L. Expansion of the global terrestrial protected area system. *Biol. Conserv.* **2009**, *142*, 2166–2174. [CrossRef]
2. UNEP-WCMC and IUCN. *Protected Planet Report 2016*; UNEP-WCMC and IUCN: Cambridge, UK; Gland, Switzerland, 2016; Available online: https://wdpa.s3.amazonaws.com/Protected_Planet_Reports/2445%20Global%20Protected%20Planet%202016_WEB.pdf (accessed on 6 June 2018).
3. Hiss, T. Can the world really set aside half the planet for wildlife? *Smithsonian* **2014**, *45*, 66–78. Available online: https://www.smithsonianmag.com/science-nature/can-world-really-set-aside-half-planet-wildlife-180952379/ (accessed on 6 June 2018).

4. Wilson, E.O. *Half-Earth: Our Planet's Fight for Life*; Liveright Publishing Corporation: New York, NY, USA, 2016.

5. Jones, K.R.; Venter, O.; Fuller, R.A.; Allan, J.R.; Maxwell, S.L.; Negret, P.J.; Watson, J.E.M. One-third of global protected land is under intense human pressure. *Science* **2018**, *360*, 788–791. [CrossRef] [PubMed]

6. IUCN. Bio-Cultural Diversity and Indigenous Peoples Journey. In Proceedings of the 4th IUCN World Conservation Congress Forum, Barcelona, Spain, 6–9 October 2008; UNEP-WCMC and IUCN: Cambridge, UK; Gland, Switzerland, 2008.

7. Walker, K.; Rylands, A.B.; Woofter, A.; Hughes, C. (Eds.) *Indigenous Peoples and Conservation: From Rights to Resource Management*; Conservation International: Arlington, VA, USA, 2010.

8. Chatty, D.; Colchester, M. *Conservation and Mobile Indigenous Peoples: Displacement, Forced Settlement and Sustainable Conservation*; Berghahn: Oxford, UK, 2002.

9. Colchester, M. Conservation policy and indigenous peoples. *Environ. Sci. Policy* **2004**, *7*, 145–153. [CrossRef]

10. IUCN. *Guidelines for Protected Area Management Categories. Commission on National Parks and Protected Areas*; IUCN: Gland, Switzerland, 1994.

11. WWF. *Statement of Principles: Indigenous Peoples and Conservation*; World Wildlife Fund for Nature International: Gland, Switzerland, 1996.

12. WCPA. *Principles and Guidelines on Indigenous and Traditional Peoples and Protected Areas*; WCPA, IUCN, WWF (International): Gland, Switzerland, 1999.

13. Nepal, S.K. *Indigenous Communities and Protected Areas-Overview and Case Studies from Canada, China, Ethiopia, Nepal and Thailand*; Unpublished Report; WWF: Gland, Switzerland, 2000; 36p.

14. Colchester, M. Salvaging nature: Indigenous peoples and protected areas. In *Social Change and Conservation. Environmental Politics and Impacts of National Parks and Protected Areas*; Earthscan: London, UK, 1997.

15. Capacity4dev.eu. Jane Goodall: "We Have a Window of Time, But We Need a Radical Change". 2018. Available online: https://europa.eu/capacity4dev/articles/jane-goodall-we-have-window-time-we-need-radical-change (accessed on 8 June 2018).

16. USAID. Emerging Technology & Practice for Conservation Communications in Africa, June 2012. Available online: http://www.abcg.org/action/document/show?document_id=315 (accessed on 8 June 2018).

17. The Nature Conservancy. *Identifying Conservation Priorities in the Face of Future Development: Applying Development by Design in the Mongolian Gobi*; The Nature Conservancy: Arlington County, VA, USA, 2013.

18. IC4D. *Information and Communications for Development*; World Bank: Washington, DC, USA, 2009.

19. UNEP-WCMC. *Global Statistics from the World Database on Protected Areas (WDPA), August 2014*; UNEP World Conservation Monitoring Centre: Cambridge, UK, 2014.

20. BirdLife Asia. *Safeguarding Important Areas of Natural Habitat Alongside Economic Development*; Mongolia Discussion Papers; East Asia and Pacific Region Sustainable Development Department, World Bank: Washington, DC, USA, 2009.

21. Law on Special Protected Areas Act 1994, c.7. Available online: http://extwprlegs1.fao.org/docs/pdf/mon77268E.pdf (accessed on 1 June 2018).

22. Clark, E.L.; Munkhbat, J.; Dulamtseren, S.; Baillie, J.E.M.; Batsaikhan, N.; Samiya, R.; Stubbe, M. (Eds.) *Mongolian Red List of Mammals*; Regional Red List Series; Zoological Society of London: London, UK, 2006; Volume 1.

23. Goulden; Clyde, E.; Tsogtbaatar, J.; Chuluunkhuyag; Hession, W.C.; Tumurbaatar, D.; Dugarjav, C.; Cianfrani, C.; Brusilovskiy, P.; Namkgaijangtsen, G. The Mongolian LTER: Hovsgol National Park. *Korean J. Ecol.* **2000**, *23*, 135–140.

24. Keay, M. The Tsaatan Reindeer Herders of Mongolia: Forgotten lessons of human-animal systems. *Encyclopedia of Animals and Humans*. 2006. Available online: http://library.arcticportal.org/437/1/tsaatan_reindeer_herders.pdf (accessed on 6 June 2018).

25. National Statistics Office of Mongolia. *Khovsgol Aimag Statistical Yearbook, 2015*; National Statistics Office of Mongolia: Ulaanbaatar, Mongolia, 2015; Available online: http://www.khuvsgul.nso.mn/uploads/users/62/files/hun-amurhiin-undsen-uzuulelt(1).pdf (accessed on 6 June 2018).

26. National Statistics Office of Mongolia. *Аялал жуулчлал (Tourism), 2017*; National Statistics Office of Mongolia: Ulaanbaatar, Mongolia, 2017; Available online: http://www.1212.mn/stat.aspx?LIST_ID=976_L18 (accessed on 6 June 2018). (In Mongolian)

27. Mongolia: Integrated Livelihoods Improvement and Sustainable Tourism in Khuvsgol Lake National Park Project. Asian Development Bank: Mandaluyong, Philippines, 2016. Available online: https://www.adb.org/projects/documents/mon-integrated-livelihoods-improvement-sustainable-tourism-khuvsgul-lake-rrp (accessed on 6 June 2018).
28. Free, C.M.; Jensen, O.P.; Mason, S.A.; Eriksen, M.; Williamson, N.J.; Boldgiv, B. High-levels of microplastic pollution in a large, remote, mountain lake. *Mar. Pollut. Bull.* **2014**, *85*, 156–163. [CrossRef] [PubMed]
29. Nepal, S.K. Sustainable tourism, protected areas, and livelihood needs of local communities in developing countries. *Int. J. Sustain. Dev. World Ecol.* **1997**, *4*, 123–134. [CrossRef]
30. Corrigan, C.; Hay-Edie, T. *A Toolkit to Support Conservation by Indigenous Peoples and Local Communities: Building Capacity and Sharing Knowledge for Indigenous Peoples' and Community Conserved Territories and Areas (ICCAs)*; UNEP-WCMC: Cambridge, UK, 2013; Available online: http://www.iccaregistry.org/assets/ICCA%20toolkit%20final%20Version%202%20en-d28f988305a52c562d77fd2b1868a547534d5852ecb6abb05819fab8f6bae6e8.pdf (accessed on 6 June 2018).
31. Nepal, S.K. Involving Indigenous Peoples in Protected Area Management: Comparative Perspectives from Nepal, Thailand and China. *Environ. Manag.* **2002**, *30*, 748–763. [CrossRef] [PubMed]

land

MDPI

Article

Context and Opportunities for Expanding Protected Areas in Canada

Michael A. Wulder [1,*], Jeffrey A. Cardille [2], Joanne C. White [1] and Bronwyn Rayfield [3]

[1]   Canadian Forest Service (Pacific Forestry Centre), Natural Resources Canada, Victoria, BC V8Z 1M5, Canada; joanne.white@canada.ca
[2]   Department of Natural Resource Sciences and McGill School of Environment, McGill University, Ste. Anne de Bellevue, QC H9X 3V9, Canada; jeffrey.cardille@mcgill.ca
[3]   Apex Resource Management Solutions Ltd., Ottawa, ON K2A 3K2, Canada; bronwynrayfield@gmail.com
*   Correspondence: mike.wulder@canada.ca

Received: 25 September 2018; Accepted: 7 November 2018; Published: 15 November 2018

Abstract: At present, 10.5% of Canada's land base is under some form of formal protection. Recent developments indicate Canada aims to work towards a target of protecting 17% of its terrestrial and inland water area by 2020. Canada is uniquely positioned globally as one of the few nations that has the capacity to expand the area under its protection. In addition to its formally protected areas, Canada's remote regions form de facto protected areas that are relatively free from development pressure. Opportunities for expansion of formally protected areas in Canada include official delineation and designation of de facto protected areas and the identification and protection of land to improve connectivity between protected areas (PAs). Furthermore, there are collaborative opportunities for expanding PA through commitments from industry and provincial and territorial land stewards. Other collaborative opportunities include the contributions of First Nations aligning with international examples of Indigenous Protected Areas, or the incorporation and cultivation of private protection programs with documented inclusion in official PA networks. A series of incremental additions from multiple actors may increase the likelihood for achieving area-based targets, and expands stakeholder engagement and representation in Canada's PA system. Given a generational opportunity and high-level interest in expansion of protected areas in Canada and elsewhere, it is evident that as a diverse number of stakeholders and rights holders collaboratively map current and future land uses onto forest landscapes, science-based conservation targets and spatial prioritizations can inform this process.

Keywords: conservation; biodiversity; ecosystems; IUCN; land use; protected areas

1. Introduction

Given population growth and increasing pressures on land for human use [1], there is an increasing global reliance on protected areas (PAs) as cornerstones of conservation strategies. This global emphasis on PAs is evidenced by the widespread support for the Convention on Biological Diversity (CBD: an international treaty with 193 member countries) which, among other decisions, adopted the Aichi Biodiversity Target to protect 17% of the most biodiverse landscapes by 2020 [2]. The establishment of national PA systems has been the globally preferred approach to biodiversity conservation in the 20th and 21st centuries. National PA systems around the world have rapidly increased from 141 areas covering less than 1% of Earth's land area in 1911 to 130,709 areas and 13% global land coverage in 2011 [3,4] and 14.8% in 2016 [5]. Furthermore, the rate of PA expansion has been growing: protected areas have increased nearly 80% from 1990 levels. Coordination of PAs worldwide has been greatly facilitated by The World Commission on Protected Areas, administered by the International Union for Conservation of Nature (IUCN), which provides the following definition for a protected area:

"A protected area is a clearly defined geographical space, recognised, dedicated, and managed, through legal or other effective means, to achieve the long-term conservation of nature with associated ecosystem services and cultural values" [6].

This definition encompasses a variety of types of PAs, classified by management objectives [7] and most recently by governance type [8]. One of the current challenges for many developed nations is to pursue strategic expansion of their PA systems through diversification of management and governance mechanisms [9].

As of 2016, Canada has achieved protection of approximately 10.5% of its terrestrial and inland water area [5] (Table 1). More than 85% of Canada's PAs are classified as IUCN management categories I–IV [9], which prohibit industrial activities such as mining, forestry, and hydro development. Furthermore, much of the area protected exists in large PAs in excess of 3000 km^2 [9]. Canada is one of the few nations with the potential to protect large, intact landscapes, particularly in the boreal and arctic regions that are expected to be under increasing pressures based upon a number of future climate change projections [10,11]. Ultimately, a key factor in the ability of Canadian PA systems to protect their biological resources is the sustained ecological integrity of the individual PAs. Given the existing level and location of protection, combined with the fact that many additional areas are currently in de facto protection [12], Canada has an opportunity unique among developed nations for a fundamental expansion of high-value, nondegraded protected areas. Furthermore, there is a strong willingness for comprehensive conservation planning owing to recent multi-stakeholder cooperative agreements and increasing political will [13] and objectives reported in the Federal Sustainable Development Strategy to meet the 17% area target by 2020 ([14], p. 48) agreed to under the Convention on Biological Diversity and related Aichi target [2].

The goal of this research is to highlight opportunities and avenues for expanding Canada's terrestrial PA system. The areas protected by province and territory are summarized in Table 1, whereas the areas protected by federal agencies are summarized in Table 2. We motivate our work with a description of the national context for PAs in Canada and provide examples focused on Canada's nationally dominant (>550 Mha) and ecologically important boreal forests. We then provide relevant background on the science of conservation planning, the application of ecological concepts to PA selection and management, and the application of IUCN PA categories in Canada. We offer guidelines to present opportunities regarding the expansion of PA systems derived from both national and international examples. Finally, we present two hypothetical scenarios for doubling protection that build on the current mix of IUCN PA categories at national and ecozone scales.

2. Background for Protected Areas in Canada

Canada has a history of protected area designation, management, and governance with the establishment of its first national park (Banff) in 1885 and its first provincial park (Ontario's Algonquin) in 1893. The federal government established Parks Canada (then the Dominion Parks Branch) as the World's first government protected area organization, in 1911 [15]. National and provincial parks are the most common types of protected areas in Canada, yet within the boreal region alone there are more than 70 types of PAs [9] that play various cultural and ecological roles. The balance among the various PA roles has evolved over time from a recreational focus towards an ecological one. Concurrently, the source of the dominant threats to PAs has also changed from internal sources, such as heavy visitation rates, to external ones, including overdevelopment of the surrounding landscape, climate change, invasive species, and airborne pollution [16]. See Table 3 for an overview of IUCN management categories including codes, descriptions, and Canadian examples.

As Canada contemplates PA expansion, it is worth noting that relative to other nations, Canada has a disproportionately large level of PAs that are in strictly protected classes [9]. National parks (IUCN

Level II, see Table 4) comprise ~48% of Canada's PAs by area [17], and the evolution of its park system policies provide supporting evidence of the shift toward an ecological focus on protection. The shift began with the 1964 comprehensive statement of national parks policy, whose main purpose was to clarify ambiguity in the National Parks Act originally passed in 1930. The policy shift asserted that the fundamental role of national parks was to be one of protection rather than use [15]. This principle of ecological integrity was subsequently legally formalized in an amendment to the National Parks Act in 1988. The first National Park System plan was approved in 1971 to guide park expansion under a more systems-based approach. The emphasis was to increase representation of the 39 natural regions of Canada within the park system. As a result, given that roughly three quarters of the area within the national park system was designated after 1971, the majority of the park system was delineated within a value system that emphasized ecological integrity and representativeness. The mosaic of the larger set of all Canadian protected areas is more complicated, however, with management and governance carried out by diverse federal, provincial, and territorial agencies (Figure 1; Table 1) as well as First Nations and private land owners (Table 2) [5].

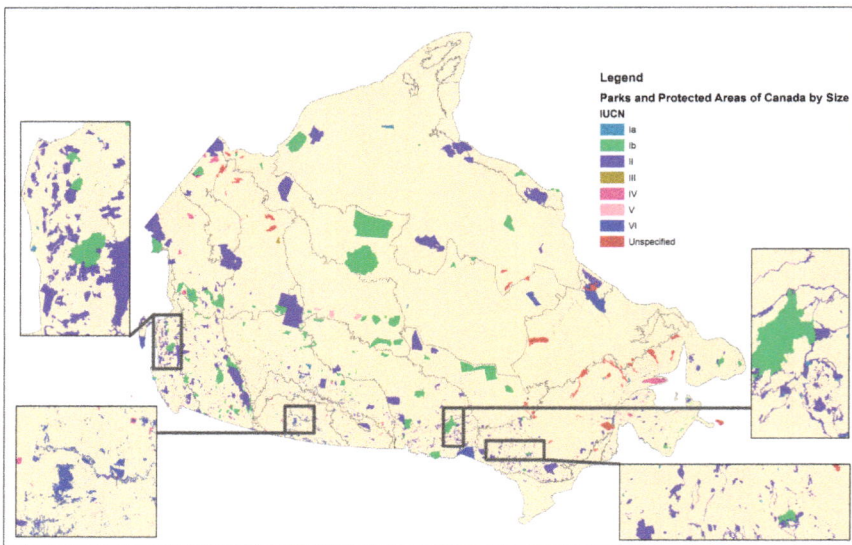

Figure 1. Map of Canadian protected areas and ecozones. Zoom views are provided of select areas to illustrate the context of smaller PAs. IUCN management categories are described in Table 3. Ecozones as described in [18] and labeled in Figure 2.

Canada's terrestrial PA system is poised for expansion for several complementary reasons: (1) to ensure representation and persistence of Canadian biodiversity; (2) to align with global protection levels and targets; and (3) to formalize the protection of de facto protected areas [12]. Recent studies have demonstrated the biases and underrepresentation of the current Canadian PA systems using a variety of biodiversity subsets and surrogates such as disturbance-sensitive mammals [19], species at risk [20], and productivity [21]. The disproportionate distribution to strictly protected IUCN categories shows an opportunity to employ some of the other IUCN classes, which would offer some flexibility in PA assignment, while remaining within international norms. As a consequence, many different jurisdictions are concurrently working to expand Canada's PA systems (e.g., [22]). Demonstrating an ongoing capacity for expansion, the total amount of protected area in Canada has more than doubled from 5.2% in 1990, to 10.5% in 2016 [5]. Many protected area management agencies have explicit expansion objectives. Nationally, Parks Canada is expanding with the objective of having at least one national park in each of Canada's 39 natural regions. There are currently 11 natural regions unrepresented in the national park system although some have interim protection [23]. Provincially, Ontario's Biodiversity Strategy sets a target to have 17% of terrestrial and aquatic ecosystems protected in line with the Aichi global target [2].

Table 1. Total percentage of terrestrial area protected by province and territory, Canada, 2016. Note that areas reported include land and freshwater, but not marine areas. Also note that not all jurisdictions in Canada report on protected areas that are privately owned.

Province or Territory	Provincial or Territorial Area (Square Kilometres)	Area Protected (Square Kilometres)	Percentage of Province or Territory Protected	Percentage of total National Area under Protection
British Columbia	944,735	144,858	15.3	13.76
Alberta	661,848	83,140	12.6	7.90
Nova Scotia	55,284	6513	11.8	0.62
Yukon	482,443	56,334	11.7	5.35
Manitoba	647,797	71,139	11	6.76
Ontario	1,076,395	114,470	10.6	10.87
Nunavut	2,093,190	211,299	10.1	20.07
Quebec	1,512,418	150,588	10	14.31
Northwest Territories	1,346,106	125,657	9.3	11.94
Saskatchewan	651,036	55,654	8.5	5.29
Newfoundland and Labrador	405,212	29,472	7.3	2.80
New Brunswick	72,908	3378	4.6	0.32
Prince Edward Island	5660	180	3.2	0.02
Totals	9,955,032	1,052,682		100.00

Table 2. Area protected by federal agency, Canada, 2016.

Jurisdiction	Terrestrial Area Protected (Square Kilometres)	Percentage of total National Area under Protection
Parks Canada	338,964	32.2
Environment and Climate Change Canada, Canadian Wildlife Service	104,854	10.0
Indigenous and Northern Affairs Canada	34,945	3.3
Fisheries and Oceans Canada	0	<0.1
National Capital Commission	462	<0.1
Correction for overlap among jurisdictions	−10,868	−1.0
Grand total	468,357	44.5

Note: Terrestrial areas include both land and freshwater. Entries represent the total area protected by each federal jurisdiction, accounting for any overlaps that may exist. This correction is made to avoid double-counting areas that benefit from more than one protection mechanism. No correction has been made for overlap between terrestrial and marine polygons resulting from variable definitions of coastlines or mapping artefacts. Areas under shared federal–provincial jurisdiction are included. Prairie Farm Rehabilitation Administration lands (Community Pastures) are being returned to provincial control and are no longer considered protected by a federal department. Source: Conservation Areas Reporting and Tracking System (CARTS) [24] (www.ccea.org/carts/).

3. Application of IUCN Protected Area Management and Governance Types in Canada

Societal values can be mapped onto the landscape as a continuum between market values and strict protection values [16]. The amount and configuration of these values on the landscape will affect the ecological, cultural, and economic potential of the landscape and will ultimately influence the future well-being of humans and of biodiversity. Different types of PAs give priority to different values; maintaining a mix of types on the landscape will thus help to achieve a balance among competing and complementary values. This will also contribute to more comprehensive and integrative conservation planning based on the principles of ecosystem management [25]. These principles emphasize the need to maintain ecosystems in the states required to achieve societal benefits while recognizing that ecosystems have some, albeit limited, resilience to natural and man-made stressors [26]. Diversified governance and management of PA types will be the key to success of planning and implementing landscapes with the right mix of values. Governance of different PA types involves decisions about PA objectives and strategies to achieve whereas management involves the actions required to achieve objectives [27].

3.1. Classifying Protected Areas Based on Management Objectives

Most PAs are established for a variety of reasons and with multiple management objectives. The IUCN management categories classify PAs based on their main management objectives (Table 3). A variety of objectives are acceptable within each category; therefore, there may even be variability amongst the management objectives of PAs within the same category. Importantly, the IUCN categories are based on the intentioned objectives. It is necessary to consider whether the area can realistically meet these objectives when assigning the area to a given IUCN category [28]. These categories facilitate international comparisons and some flexibility is required in their application [28]. Consistent application of the IUCN system by agencies that own and manage PAs in Canada is an achievable outcome, possibly guided by Canadian examples of interpretations of the IUCN system (e.g., Table 3) and the Canadian Council on Ecological Areas (CCEA) [28]. A few notable issues relevant to Canadian application of the IUCN categories are worth mentioning. Canada has large PAs in Categories I through IV, both in number and area, relative to other countries. Within these categories, some restricted human activities, which would otherwise be considered incompatible with these categories, are accommodated, such as small sections subject to extractive activities or First Nations traditional practices such as hunting and trapping. Traditionally, most areas that meet the criteria for Categories V and VI are beyond the scope of Canada's PA agencies [28]; however, this is not precluded and expected to develop as national and provincial PA systems continue to expand.

Regulations concerning the use of protected areas are not entirely consistent across international jurisdictions, nor are they consistent from province to province within Canada [29]. This presents challenges to categorizing and identifying the characteristics of PAs in Canada. For countries that are parties to the Convention on Biological Diversity, the IUCN has been instrumental in standardizing the reporting on PA systems by providing a typology of PAs consisting of six management categories [6,7]. Categories are not exclusive: many PAs have the characteristics of multiple IUCN categories, resulting in them being listed doubly or triply in existing PA databases. The CCEA produced guidelines for the application of the IUCN PA typology in Canadian PA systems [28]. Most recently, the IUCN provided a simplified typology of PAs consisting of four governance types [8]. This section provides a summary of these management and governance typologies with Canadian examples. It should be noted that this summary is not a replacement for the aforementioned publications and they remain the primary references. We also present an overview of the current representation of IUCN management categories in Canadian PAs.

Table 3. IUCN management categories for protected areas with Canadian examples.

Management Category	International Name *	Description	Canadian Example **
Ia	Strict Nature Reserve	Areas subject to strict controls on human visitation, use, and impacts for the conservation of biodiversity and geological/geomorphological features;serve as important references area for scientific research and monitoring.	The terrestrial portion of Funk Island Ecological Reserve (Newfoundland and Labrador) is primarily managed for biodiversity and habitat conservation. Access is restricted to people conducting approved scientific research.
Ib	Wilderness Area	Areas that are large and natural and unmodified (or slightly modified) protected and managed to retain their natural condition.	Tobeatic Wilderness Area (Nova Scotia) primarily manages for wilderness protection and maintenance of species/genetic diversity. Its secondary management objective is low-impact recreation and scientific research.
II	National Park	Areas that are large and natural (or near-natural) dedicated to protecting large-scale ecological processes, characteristic species and ecosystems, and offer visitor opportunities.	Wood Buffalo National Park (Alberta) has many management objectives including preservation of species and genetic diversity, recreation and tourism, scientific research, and wilderness protection.
III	Natural Monument	Areas that protect specific natural monuments (e.g., caves) or living features (e.g., ancient groves). These are often smaller areas that receive many visitors due to high historic or cultural value.	The Nisutlin River National Wildlife Area (Yukon) has management objectives that include protecting biodiversity and habitats, maintaining traditional and current use of the territory by Teslin Tlingit, and encouraging awareness through public visitation.
IV	Species/Habitat Management	Areas devoted to species or habitat conservation, often focusing on species or habitats of particular concern and requiring active management interventions.	Last Mountain Lake National Wildlife Area (Saskatchewan) manages for protection of ecosystems and biodiversity.
V	Protected Landscape/Seascape	Areas with a distinct character and significant ecological and cultural values resulting from the interaction of people and nature over time.	The National Capital Greenbelt (Ontario) is managed to protect natural and cultural resources, provide opportunities for recreation, and safeguard the working rural landscape.
VI	Protected areas with Sustainable use of Natural Resources	Areas that are large and jointly conserve ecosystems, habitats, cultural values, and traditional natural resource management systems. The majority of the area should be in a natural condition and the reminder under sustainable natural resource management.	Churn Creek Protected Area (British Columbia) is managed for the preservation of natural and cultural features, tourism and recreation, maintenance of cultural and traditional attributes, and sustainable resource use.

* International name does not necessarily correspond with Canadian usage of the same term. For example, Cypress Hills Interprovincial Park is classified as Category II National Park. Canadian examples were taken from CCEA [28] except for Category V which came from Swinnerton and Buggey [30]. Source: Adapted from IUCN [7].
** Canadian examples were taken from CCEA [28] except for Category V which came from Swinnerton and Buggey [30]. Source: Adapted from IUCN [7].

Table 4. Protected areas (PAs) in Canada, by IUCN class.

IUCN Management Category	IUCN International Name	Percentage of total National Area under Protection	Percentage of total National Area
Ia	Strict Nature Reserve	1.0	0.1
Ib	Wilderness Area	33.2	3.5
II	National Park	48.8	5.1
III	Natural Monument	0.2	<0.1
IV	Species/Habitat Management	2.0	0.2
V	Protected Landscape/Seascape	0.4	<0.1
VI	Protected Area with Sustainable Use of Natural Resources	14.4	1.5
Total		100.0	10.5

3.2. Classifying Protected Areas Based on Governance Types

Governance can be defined as "the interactions among structures, processes, and traditions that determine how power and responsibilities are exercised, how decisions are taken and how citizens or other stakeholders have their say" [31]. Governance is carried out by actors (rights holders and stakeholders) that hold power, authority, and responsibility and are accountable for meeting the protection objectives of PAs (Tables 1 and 2) The purpose of the IUCN governance classification system is to create a standardized framework in which all governance of PAs and PA systems can be assessed, compared, evaluated, and improved. In addition to the type of governance of a PA (Table 5), the quality of governance can also be assessed to the extent of how closely governance principles are followed in the process of making actual decisions. PA governance also concerns de facto decision makers that affect the conservation objectives of the PAs through informal means [8]. For example, ongoing enforcement shortcomings may result in illegal development and commercial exploitation in PAs by members of the public, such as cattle grazing and wood harvesting [32]. Parks Canada has aimed to demonstrate international leadership in establishing the principles of good governance of PAs [31]. In Table 5, we also offer Canadian examples of each category. We present these examples of governance types in Canadian PAs as a stimulus for future discussions on Canadian interpretation of the IUCN PA governance framework.

Table 5. IUCN governance categories for protected areas with Canadian examples ** (adapted from IUCN [8]).

Category	International Name	Description	Sub-types/Authority with Canadian Examples [ ]
A	Governance by government	One or more government bodies holds authority, responsibility, and management accountability. Alternately, planning and/or daily management tasks may be delegated to other actors.	• Federal or national ministry or agency [The majority of Canada's national Park System is managed by Parks Canada *] • Sub-national ministry or agency [Algonquin Park (Ontario) is managed by Ontario Parks.] • Government delegated manager [Quebec's Zones d'exploitation contrôlée (ZEC) are hunting grounds located on public lands but management and conservation is carried out by an NGO.]
B	Shared governance	Several actors formally and /or informally share authority using institutional mechanisms and processes.	• One or more sovereign State or Territories governing Transboundary PA [Canada and the United States formed Waterton–Glacier International Peace Park by the merger of the Water lakes and Glacier National Parks in 1932.] • Diverse actors and institutions working in collaborative governance [Cypress Hills Interprovincial park (Alberta, Saskatchewan)] • Pluralist board or other multi-party governing boding working in joint governance. [The Gwaii Haanas Agreement is a cooperative agreement between the Haida Nation and the Government of Canada for the governance of the Gwaii National Marine Conservation Area Reserve (British Columbia).]
C	Private governance	Land owner retains authority for managing PA subject to government legislation and restrictions. Government recognition of land owner's authority is required to ensure accountability.	• Individual land owners [Quebec's Natural Heritage Conservation Act allows the Minister of Sustainable Development, Environment, Wildfire and Parks to recognize private properties as nature reserves.] • Non-profit organizations [Gault Nature Reserve (Quebec) is owned by McGill University.] • For-profit organizations [Haliburton Forest and Wild Life Reserve Ltd. (Ontario) is a 300 km2 privately owned forest near Algonquin Provincial Park managed for conservation, tourism, and timber.]
D	Governance by indigenous people and local communities	Indigenous peoples and/or local communities hold management authority and responsibility through various forms of customary or legal institutions and rules.	• Indiginous peoples [Tla-o-qui-hat Tribal parks (British Columbia) are watersheds managed by Tla-o-qui-aht First Nations to integrate human and ecosystem well-being.] • Local communities [We are unaware of any examples of this sub-type at the present time.]

* Thirteen National Parks are governed collaboratively (Category B) through cooperative management boards between Parks Canada and the First Nations on whose territories the parks are located. ** Canadian examples are our own interpretations of the IUCN governance categories in the Canadian context with the exception of the Gwaii national Marine Conservation Area, which was presented as an example of Category B in IUCN Governance of Protected Areas [8].

3.3. Distribution of IUCN Management Categories within Canada's Protected Areas

The proportion of Canada's protected area in each IUCN category shows an uneven distribution, skewed towards categories with the strongest restrictions to human activities within them (Figure 2; Table 4). Approximately 82% of all Canada's protected area is classified as Categories Ib and II and less than 3% in Categories III and IV (Table 4). By definition, PAs in Categories Ib and II are large natural areas (e.g., national parks and wilderness reserves), and given the remote nature of much of Canada, it is perhaps not surprising that they make up a large percentage of the overall area protected. Conversely, PAs in Categories III and IV are usually small areas, because of their focus on protection of particular features (III) or requirement for active management interventions (IV). Hence, Canada's current division of protected area among these categories is in line with expectations given Canada's history, settlement patterns, and geography. However, there is little reason that Categories Ia, V, and VI should not have greater representation as they also require large areas to meet their management objectives (especially Categories Ia and VI).

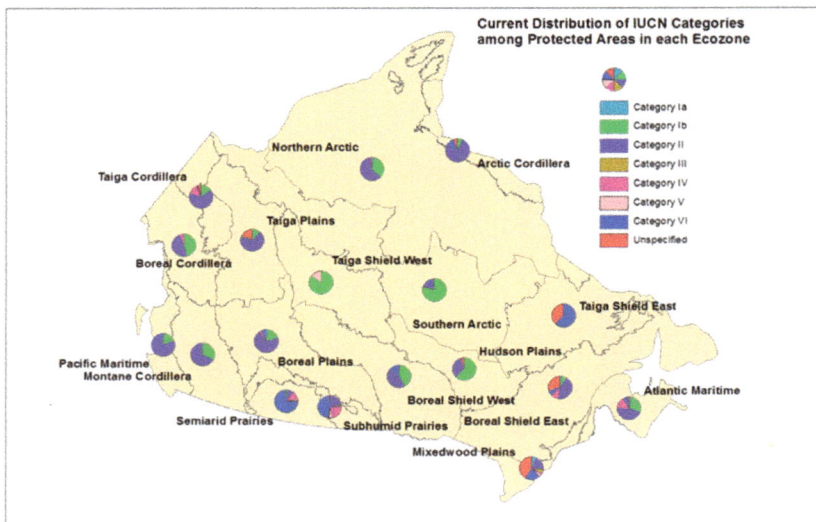

Figure 2. National map of ecozones with pie charts individually illustrating the breakdown of protected areas by ecozone into IUCN categories. Note: Ecozones differ with respect to the total amount of area protected (see Figure 3) [24].

The spatial distribution of Canada's protected areas is noteworthy for its spatial unevenness. The size of PAs in all IUCN categories generally increases from east to west and from north to south. Hence, the ecozones of Canada have considerably differing amounts, proportions, and patterns of protection, ranging from the smallest amount of area in the southern Mixedwood Plains to the highest in the Southern Arctic. Furthermore, no one category is uniformly dispersed, for example, Categories Ib and II are less abundant in the eastern ecozones (Taiga Shield East and Boreal Shield East). Only four of the eighteen ecozones have PAs in all categories and only Category II is present in all ecozones (Figure 2; Table 4), meaning that no single ecozone closely represents the national pattern. It could be expected that some PA categories would be less well-suited to some ecozones than others; for example, Category I PAs that are large wilderness areas would be difficult to establish in the populated Mixedwood Plains ecozone. Nevertheless, it would seem that managers could work at the ecozone scale to better diversify and/or rebalance PAs in light of neighboring ecozones or the national average.

4. Context for Expanding Canada's Protected Area in Forested Ecosystems

Globally, terrestrial PAs have increased from covering just 3.45% of land surfaces in 1985 [33] to 13% in 2010 [34]. This expansion has been achieved through a variety of political, legal, and other effective avenues. Given the current extent of de facto protected areas [12], Canada has an opportunity to pursue many of these avenues to improve protection in its forested ecosystems. In this section, we discuss models of PA management and governance that involve a broad range of rights holders and stakeholders. We first provide international examples of how these approaches can be useful in PA system expansion in general and follow up with IUCN categories to target for PA expansion in Canada's forested ecosystems. We acknowledge that conservation contexts will vary among forested ecozones and what may be achievable in one ecozone, or subregion, may not apply elsewhere due to factors such as industrial allocations, land ownership patterns, and the evolving status of Aboriginal and Treaty rights and land use planning exercises [35].

4.1. Recent International Precedents for Expansion of Protected Areas

Many nations have made commitments to expand terrestrial protected area systems, however progress has been limited [4]. For example, in 2004 over 80% of nations had signed the CBD, thereby committing to protect a portion of all of their ecoregions [36]. As of 2009, less than half of the terrestrial ecoregions had met the target of protecting 10% of each ecoregion by 2010 [4]. Furthermore, only 33% of terrestrial ecoregions have met the more recent target of 17% protection by 2020 [34]. At present, almost all nations have signed on to the CBD, with the exception of the USA who have signed but not ratified the treaty[1]. Some international examples of recent increases to PA systems are of particular interest, as these offer possible new opportunities or ways of understanding PA expansion. In Table 6, we share a few representative international case studies of PA expansion of relevance to Canadian conditions (Australia and Finland) and other innovative international examples that incorporate climate change adaptation and community participation (South America).

Australian researchers have been active in conservation biology and biodiversity science for the past two decades [37]. Examples of recent innovations include (1) costing the expansion of PA systems [38] and (2) optimizing the allocation of resources between protecting intact habitats versus restoring degraded habitats in order to best facilitate the persistence of threatened species under climate change [39]. Australia's terrestrial PA systems are noteworthy because of the large portion which is jointly managed with Indigenous peoples. Indigenous Protected Areas (IPAs) are defined by the Australian Government as areas:

> "of land and/or sea over which the Indigenous traditional owners or custodians have entered into a voluntary agreement with the Australia Government for the purposes of promoting biodiversity and cultural resource conservation." [40]

IPAs made up 30% of Australia's total area protected in 2012 [41]. The IPA designation was formally recognized in 1997, and has expanded rapidly such that by 2011 there were 48 declared IPAs and another 40 IPA Consultation Projects underway [42]. In theory, IPAs may qualify as any one of the seven IUCN management categories ranging from most to least natural conditions within PAs. In practice, many IPAs have nominated more than one IUCN category. The majority of IPAs are assigned to Categories V (16 IPAs) and VI (31 IPAs) incorporating human–nature interactions and sustainable resource use, respectively [42]. The remainder are assigned to Categories VI (11 IPAs), III (6 IPAs), and II (7 IPAs). IPAs have proven an effective avenue for recent and future growth of the Australian PA system [43].

Colombia, Ecuador, and Peru are three of seventeen mega-diverse countries identified by Conservation International and collectively comprise the majority of the Tropical Andes—a biodiversity

---

[1]   https://www.cbd.int/information/parties.shtml (accessed on 19 October 2018).

hotspot with an exceptionally high concentration of endemic species [44]. These three countries are contiguous and adjacent with shared borders and interconnected habitats and ecosystems. Their respective national PA systems have all undergone rapid, recent expansions within their specific political, legal, and institutional contexts. They have launched a novel initiative to coordinate a collective learning process to develop strategies for climate change adaptation across their PA systems [45].

Finland, like Australia, has been innovative in systematic conservation planning and spatial conservation prioritization [46]. Recently, Finnish scientists have developed methods to optimize the expansion of PA systems by identifying areas that, if protected, would add the most complementary conservation value to existing PAs [47]. This acknowledges the reality of most conservation planning efforts which cannot ignore existing land use patterns and therefore must account for previous limitations, land use plans and existing protected areas in spatial conservation prioritizations (see a past paper [48] for a Canadian example).

Table 6. International examples of protected area system expansion.

Country	Percentage of Protected Terrestrial Area in 2012	Increase in Percentage of Protected Area from 1990 to 2012	Lessons Learned
Australia	12.85	5.74	• Ability to subdivide private lands will greatly influence costs of PA expansion and may result in protection of some lands with lower conservation value [38]. • Habitat protection is the primary action required to maintain threatened species with reference to climate change, although restoration plays an important role for some species and locations [39]. • Comanagement of PAs with Indigenous groups, NGOs, and business organization should adapt based on ongoing evaluation of comanagement [43].
Columbia, Ecuador, Peru	21.18, 23.73, 19.06	1.89, 1.88, 14.33	• Transnational coordination of PA systems can improve the effectiveness of PA systems as they respond to climate change. • Participatory zoning can integrate local people into PA management [49]. • Comanaged public, private, and community-based PAs need management plans that define cooperation, coordination and participation frameworks for PAs and related buffer zones [50].
Finland	15.10	10.73	• Spatial conservation priorities need to account for the quality and connectivity of existing PAs [47]. • The majority of high-quality, unprotected forests (in southern Finland) are on privately owned lands [47]. • Flexible, voluntary-based conservation agreements facilitate the recognition of private PAs by government PA agencies. • Inverse spatial conservation prioritization can minimize the negative ecological effects of extractive resource use [51].

Source for all PA coverage estimates: IUCN and UNEP-WCMC [3].

4.2. Expansion through Recognition of New IUCN Management Categories and Governance Types

As illustrated by the aforementioned international case studies, diversifying management and governance types represented in the PA systems can expedite and facilitate expansion [31]. There many ways that management and governance diversification can benefit PA system expansion [8]:

- meeting expansion area targets faster by incorporating into official PA systems de facto protected areas and areas governed by indigenous peoples, the private sector, or shared governance;
- building connectivity in PA networks by adding stepping stones and corridors that are not publicly owned lands;
- improving the representativeness and adequacy of PA systems particularly in countries with a large proportion of private ownership; and
- increasing the number of people involved in conservation which may lead to social acceptance and sustainability of the system.

The success of a diversified PA system will hinge on ensuring that all types of PAs prioritize the long-term conservation of biodiversity based on the ecological principles outlined earlier.

Combinations of IUCN management and governance categories can be expanded into an IUCN Protected Area Matrix (see Table 7). Such a matrix helps to visualize and classify the diversity of PA types that can currently occur in a PA system [8]. It can also be used to consider options for expanding the PA system and for communicating these options to PA agencies and stakeholders. For example, this management–governance matrix could assist in the development of a management framework by filling in all applicable rows of a single column as though they represent nested management arrangements within a single PA or PA system. Nested within a national park, it is possible to define a species management area for a vulnerable species [23] or allocate land for sustainable resource use [49]. There is also a current IUCN adaptation possibility, akin to that of the USA, whereby US National Forests are classified as IUCN Category VI protected areas (Managed Resource Protected Area[2]). This allows for some forest harvesting, but not comprehensive industrial activity.

Based upon IUCN literature [8] and our understanding of management and governance types, shaded cells in the matrix highlight potential categories that could contribute to the expansion of Canada's PA system. These are presented as options for consideration, but not necessarily as exhaustive. The "Governance by indigenous peoples" column is highlighted as an ongoing avenue for expansion. Parks Canada is already recognizing the important governance role of aboriginal communities in a variety of PA types either autonomously or through shared governance. The one national park created in 2012–2013 was Nááts'ihch'oh National Park Reserve in the Northwest Territories [23]. Aboriginal communities have developed comprehensive land-use plans that balance cultural and ecological values with opportunities for long-term economic development such as the Tla-o-qui-hat Tribal Parks (coastal British Columbia).

Another key opportunity for expansion highlighted in Table 7 is "Private governance", informed by the South American and Finnish examples noted earlier. Indeed many more countries have identified private lands as priority areas for PA expansion including the United States, Paraguay, Costa Rica, Bolivia, South Africa, and New Zealand summarized in IUCN [8] and in Jackson and Gaston [52]. They are thought to have the potential to improve the representativeness and connectivity of PA systems [53,54]. However, in Great Britain where PAs are not constrained to public lands (more than half of all PAs are on private lands), there is still biased representation of biodiversity surrogates within the PA system [52]. Biodiversity conservation on private lands may be improved by diversifying landowner engagement strategies beyond the two most common conservation tools: conservation easements and direct payment programs [55]. Current research is focussed on identifying when different conservation tools are most likely to be effective across diverse private lands (e.g., based on ecological dynamics and interactions between individual landholders [56]). More research is needed into the circumstances under which private PAs should be incorporated into national PA systems [8,57].

---

2   http://www.iucn.org/about/work/programmes/gpap_home/gpap_quality/gpap_pacategories/gpap_category6.

Table 7. IUCN Protected Area Matrix. Light grey cells indicate key opportunities for expansion of Canada's PA systems. Source: Adapted from IUCN [8].

		Governance Category										
		A. Governance by Government			B. Shared Governance			C. Private Governance			D. Governance by Indigenous People and Local Communities	
		Federal or national ministry	Sub-national ministry or agency	Government-delegated manager	One or more sovereign states or territories governing transboundary	Diverse actors and institutions working collaboratively	Pluralist board or other multi-party governing body working in joint governance	Individual land owners	Non-profit organizations	For-profit organizations	Indigenous peoples	Local communities
Management Category	Ia			▓	▓	▓	▓				▓	
	Ib			▓							▓	
	II			▓	▓	▓	▓				▓	▓
	III			▓	▓	▓	▓		▓		▓	▓
	IV			▓	▓	▓	▓				▓	▓
	V				▓	▓	▓	▓	▓	▓	▓	▓
	VI				▓	▓	▓	▓		▓	▓	▓

4.3. Spatial Conservation Planning Opportunities to Enhance Canadian PA Systems

As a large nation with extensive areas with little access or development [12], Canada has globally unique opportunities for using and implementing spatial conservation planning [9]. Powers et al. [48] reviewed scenarios for integrating accessibility and intactness into conservation planning in Canada, using the boreal forest region to illustrate how biological elements, costs, and size considerations can provide analytical criteria for implementation of spatial conservation planning scenarios and to inform prioritization. Beyond criteria-based spatial conservation planning, Powers et al. [58] conducted analyses to incorporate climate change projections and related impact upon vegetation into a Canadian boreal conservation assessment. Similarly, remote sensing and model-derived biodiversity indicators [59] were utilized by Holmes et al. [60] to show that climate change will alter the future vegetation characteristics expected in British Columbia, impacting the nature of protection offer by PA. Spatial data sets, especially those from remote sensing, offer unique opportunities for understanding not only landscape status and protection status but understanding what spatial differences in physical environments relate to biodiversity [61]. In general, these studies show that there is available information and methodological options to inform spatial conservation planning in Canada.

Building upon the above insights and practical considerations, there are a number of strategies/pathways to increase representation and effectiveness goals for PA systems in Canada. We outline here some opportunities with Canadian examples where possible.

Privately conserved areas are increasingly being recognized as a legitimate and effective governance type. They can play a role in increasing the representativeness and connectivity of PA systems, while at the same time engaging actors other than government, which increases conservation awareness and engagement, and ensures the long-term sustainability of the PA system. Examples of this PA type already exist in Canada. The province of Quebec is an example for incorporating private sector actors and NGOs in PA management. In Ontario, Oak Ridges Moraine land-use planning process has protected long, wide corridors on private lands through negotiations between government, NGOs, and the private sector [62].

De facto protected areas [12] could be brought into formal PA systems. To meet IUCN constraints, these areas would have to be primarily managed for nature conservation with up to 25% managed for other compatible reasons, such as sustainable resource use [6]. Given that constraint, not all de facto protected areas could qualify as PAs, but could be considered as "other effective area-based conservation measures" [8]. For example, where conservation is an indirect outcome of other management goals, the area would not meet the IUCN PA definition. More northerly regions may readily meet IUCN PA definitional requirements, with more southerly areas subject to sustainable forest management requiring additional investigation. Of note, Canada has more than 167 Mha of managed forest lands under third party forest certification [63], with ecological considerations, via provision of ecosystem services, included as an element of certification.

Zoning within and around PAs could enhance rather than expand the current PA system by improving its ecological integrity. Zoning within PAs could increase representation of different IUCN categories by creating strong protection zones within lower-level PAs and vice versa. For example, Algonquin Provincial Park has strong protection zoning for 12.6% of the total park area in the form of Natural Environment, Nature Reserve, Wilderness, or Historical zones. Another 65.3% of the total park area is zoned as Recreation/Utilization and includes a sustainable forest industry [64]. Another example of PA's with multiple levels of protection is the UNESCO Biosphere reserve designation, of which there are currently 13 in Canada. Zoning around PAs provides a means to balance conservation and development across broader spatial scales and among diverse stakeholders and rights holders [65]. Creating buffer or multiple-use transition zones can help maintain the ecological integrity of PAs [13]. Zoning for ecological corridors through developed areas will also be important to allow species to move among PAs to meet their resource requirements or to track suitable habitat as the climate changes.

5. Expansion Scenarios for Canada's Protected Area Systems: An Example in the Boreal

The protected area targets suggested specifically for Canada range from 12% [66,67], to 20% [68], to 50% [69]. Canada's PAs are neither balanced in terms of total amount protected, proportion protected, nor balance among IUCN types. For the boreal region specifically, the balance among IUCN types differs substantially across the ecozones (Figure 3b). Two notably large and underprotected ecozones are Boreal Shield East and Taiga Shield East, which currently protect 7% and 3% of their large areas respectively. Expansion to the current 17% federal target [14] in these ecozones is made possible by the de facto opportunities for protection in the boreal relative to the more heavily populated areas further south [12,48], which may have more restrictive pre-existing land-use commitments [9]. The 17% scenario currently under consideration represents more than a doubling of the 8% of the boreal currently under protection (Figure 3a).

Due to political, economic, and social factors that are a necessary part of conservation in practice, there are circumstances in which a government may be limited in its ability to employ systematic, quantitative methods for identifying areas for protection. In those cases, governments may consider other approaches to guide the expansion of PA systems. Two potential approaches (among many) towards expanding Canada's protected areas would involve (i) expanding the area of existing IUCN categories or (ii) expanding the range of IUCN categories within the PA system (while also increasing the area under protection). The first approach would maintain the same types of IUCN categories that are currently found within each ecozone, and focus on expanding the areas represented by each of these categories. For the boreal example, this would result in more IUCN category Ib (wilderness areas) and II (national parks), as these are currently the most prevalent IUCN categories currently found in the boreal. An alternative approach would be to broaden the range of IUCN categories that are currently represented in each ecozone, thereby introducing new models of protection that are tailored to regional conditions.

(a)

(b)

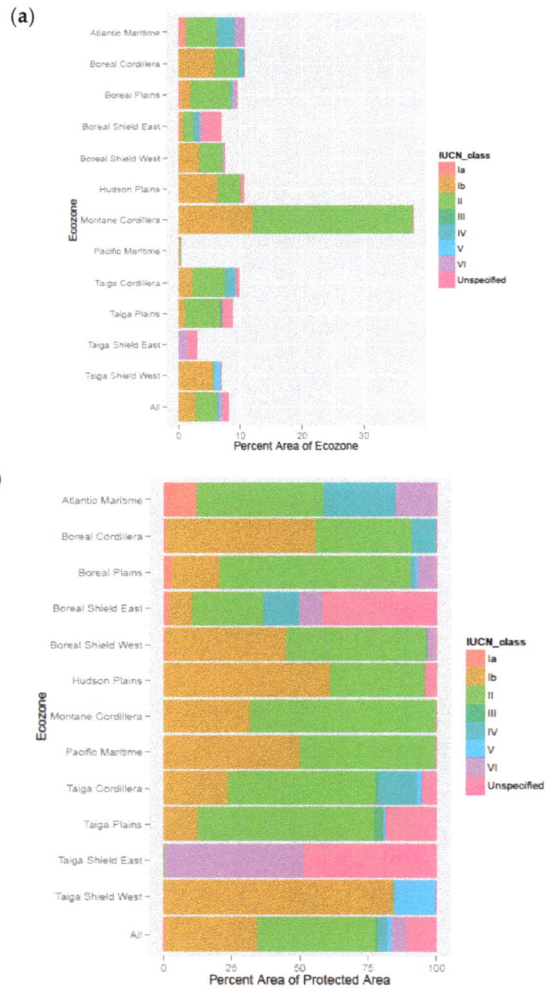

Figure 3. IUCN management categories for PAs in boreal ecozones, aiming to show balance among IUCN types. The stacked bar for each ecozone shows (**a**) the proportion by total amount of protected area found in each boreal ecozone ([9] and (**b**) the proportion by amount of protected area in each ecozone.

6. Opportunities and Conclusions

Currently, PA systems in Canada are dominated by relatively large PAs with strict protection in comparison with other countries. This is especially true for protection in Canada's forested ecosystems where Categories Ib and II make up 80% of all PAs and 34 PAs are larger than 3000 km^2 [9]. Nonetheless, only a small fraction of PAs in Canada meet the size requirements to maintain minimum viable populations of large, wide-ranging mammals or to protect landscape dynamics and natural disturbance regimes [9]. A resilient PA network consists of multiple types of PAs including a system of representative core areas, corridors, stepping stones, and buffers. In this research, we posit that strategic expansion of protection in Canada could benefit from diversification of PA types to make greater use of less restrictive PA types (IUCN Categories V and VI). This would extend biodiversity conservation values across the landscape while strengthening connections among existing PAs. We encourage an ongoing dialogue about PA system diversification by providing Canadian interpretations of IUCN management and governance categories for PAs along with Canadian and international examples of their application. Towards this goal, our synthesis of the literature and current context in Canada indicated the following opportunities.

1. **Canadian interpretation and implementation of IPA**

International examples of IPA offer insights for implementation in Canada. National circumstances inform the nature of governance possible from IPA. For instance, agreements between Indigenous traditional owners or custodians in Australia have resulted in voluntary stewardship agreements that now result in 30% of national PA. The IPA designation is relatively new, with formal recognition in 1997, leaving opportunities for definition and implementation in a Canadian context. IPA can be created over a number of IUCN categories, although most frequently those (e.g., Categories V, VI) that allow for some human–nature interaction or sustainable resource use.

2. **Identify and augment private PAs**

An important next step to inform PA system diversification will be to identify opportunities for private PAs. We propose mapping privately owned lands within de facto PAs [12] to determine candidate private PAs. As a second step, the map of candidate PAs should be assessed for their complementarity to the existing PA portfolio [47], particularly within its ecozone. This assessment would be based on the degree to which candidate PAs contribute to the representativeness and effectiveness of the PA system and their degree of irreplaceability. Private lands offer opportunities for PA expansion through provision of habitat and harbors for endangered or threatened species [70]. Knowledge of the ecological contribution of private lands can be combined with cultural, economic, and sociopolitical considerations to designate new PAs.

3. **Clarify IUCN categories of PAs**

Reporting IUCN categories for PAs is particularly challenging because the data is dispersed across the different governing agencies, noting that there are guidebooks to help standardize the assignment of IUCN categories [6,28]. However, despite best efforts, the present assignment of protected areas to IUCN categories globally does not correspond to the expected gradient of naturalness [29]. The management and governance categories discussed herein can help to enhance biodiversity representation and persistence by identifying balanced solutions when working with diverse rights holders and stakeholders, as well as considering formalizing protection of large, de facto PAs. Future work will also require assigning PAs to governing categories that may prove even more challenging since PA governance may be subject to change over time.

4. **Value connectivity in PA assessments**

Maintaining and enhancing the connectivity of protected landscapes will be critical as climatic change redistributes species habitats [71] and alters the frequency, extent, and intensity of disturbances [72]. This may be especially (though not exclusively) relevant in boreal regions, where disturbances can be large and contiguous [73]. For species to respond to climate-induced changes and maintain viable populations, PA systems must be conscious of the evolutionary continuum [74] and designed to allow for broad-scale traversability of the landscape. Improving the connectivity among PAs is a way to increase the persistence of biodiversity across the network of PAs. Recent work across large areas, including with Zonation [47,75] and circuit theory [76–78] point toward spatial data informed decision support tools that explicitly consider a variety of measures of connectivity.

5. **Consider tradeoffs**

Systematic conservation planning [79] will require the a sophisticated understanding of a wide number of tradeoffs among multiple actors. As planners contemplate PA expansion, they can consider some of the following tradeoffs: biodiversity representativeness versus ease of implementing protection across different ownerships; the economic costs of implementing protection on private lands and public perception of payments; the tradeoff between connecting existing PAs versus protecting isolated but important habitats; tradeoffs between megafauna-heavy conservation areas and other important ecological criteria; tradeoffs between different aspects of biodiversity; and tradeoffs between known critical needs of today versus the less certain estimated needs of tomorrow.

6. PA data

Any future research extending the scope of this research should begin by revisiting the PA data used for this analysis. Conservation Areas Reporting and Tracking System [24] is the primary source for information on Canada's PAs. It is unclear, however, to what extent this source is consistent with other sources such as Environment Canada's Environmental Indicators on Protected Areas and the World Database on Protected Areas. Many countries are currently in the process of putting in place formalized reporting structures for PAs. Canada is relatively advanced in this regard although we suspect that the reporting of private PAs is less consistent than that of public PAs.

Protection of natural spaces for habitat, wildlife, and provision of ecosystem services are an important component of land and forest management. Given the stated national target, that "by 2020, at least 17% of terrestrial areas and inland water are conserved through networks of protected areas and other effective area-based conservation measures" [5] (p. 45); this is a critical period in the history of Canadian land/water management. Ongoing technological advances in remote sensing, GIS, and modeling have made it newly possible to analyze the state of Canada's many landscapes, model the impacts of management decisions, and plan for the future. As a number of different stakeholders and rights-holders collaboratively map current and future land uses onto forest landscapes, science-based conservation targets and spatial prioritizations must be ready and accessible to inform this process. To this end, we have provided an overview of the theoretical framework of spatial conservation prioritization rooted in ecological concepts. By synthesizing the relevant literature in the Canadian context, our objective was to highlight the opportunities available to expand protected areas in Canada, while also broadening engagement and recognizing that conservation is not one-size-fits-all. This work offers a source of options and opportunities to inform conservation agencies with ideas, options, and approaches to expand Canada's PA systems. Multiple actors and agencies, each working within their own stewardship constraints and opportunities, can combine to respond to the stated national priority of protecting more of Canada's natural area.

Author Contributions: This paper was a collaborative effort of the authorial team, with specific, but not limited, roles. Conceptualization, M.A.W., J.C.W.; Methodology, M.A.W., J.C.W., J.A.C., and B.R.; Formal Analysis, B.R., J.C.W., and J.A.C.; Investigation, B.R.; Writing, M.A.W, J.A.C., J.C.W., and B.R.; Funding Acquisition, M.A.W.

Funding: This research received no funding external to government. It was undertaken as part of the "Earth Observation to Inform Canada's Climate Change Agenda (EO3C)" project jointly funded by the Canadian Space Agency (CSA), Government Related Initiatives Program (GRIP), and the Canadian Forest Service (CFS) of Natural Resources Canada (NRCan).

Acknowledgments: Geordie Hobart of the Canadian Forest Service is thanked for assistance with preparation of the manuscript for submission.

Conflicts of Interest: The authors declare no conflicts of interest.

References

1.   Small, C.; Sousa, D. Humans on Earth: Global extents of anthropogenic land cover from remote sensing. *Anthropocene* **2016**, *14*, 1–33. [CrossRef]

2.   CoP10. Convention on Biological Diversity. In Proceedings of the Conference of the Parties to the Convention on Biological Diversity: Updating and Revision of the Strategic Plan for the Post-2010 Period, Nagoya, Japan, 18–29 October 2010.

3.   International Union for Conservation of Nature (IUCN) and United Nations Environment World Conservation Monitoring Centre (UNEP-WCMC). *The World Database on Protected Areas (WDPA) October 2012*; UNEP-WCMC: Cambridge, UK, 2012.

4.   Jenkins, C.N.; Joppa, L. Expansion of the global terrestrial protected area system. *Biol. Conserv.* **2009**, *142*, 2166–2174. [CrossRef]

5.   ECCC (Environment and Climate Change Canada). Canadian Environmental Sustainability Indicators: Canada's Protected Areas. Available online: www.ec.gc.ca/indicateurs-indicators/default.asp?lang=en&n= 478A1D3D-1 (accessed on 20 March 2018).

6.   Dudley, N. *Guidelines for Applying Protected Area Management Categories*; IUCN: Gland, Switzerland, 2008.

7.   International Union for Conservation of Nature (IUCN). *Guidelines for Protected Area Management Categories. CNPPA with the Assistance of WCMC*; IUCN: Cambridge, UK, 1994.

8.   International Union for Conservation of Nature (IUCN). *Governance of Protected Areas: From Understanding to Action*; IUCN: Cambridge, UK, 2014.

9.   Andrew, M.E.; Wulder, M.A.; Cardille, J.A. Protected areas in boreal Canada: A baseline and considerations for the continued development of a representative and effective reserve network. *Environ. Rev.* **2014**, *22*. [CrossRef]

10. Field, C.B.; Barros, V.R.; Dokken, D.; Mach, K.; Mastrandrea, M.; Bilir, T.; Chatterjee, M.; Ebi, K.; Estrada, Y.; Genova, R. *IPCC, 2014: Climate Change 2014: Impacts, Adaptation, and Vulnerability. Part A: Global and Sectoral Aspects*; Contribution of Working Group II to the Fifth Assessment Report of the Intergovernmental Panel on Climate Change; Cambridge University Press: Cambridge, UK; New York, NY, USA, 2014.

11. Price, D.T.; Alfaro, R.; Brown, K.; Flannigan, M.; Fleming, R.; Hogg, E.; Girardin, M.; Lakusta, T.; Johnston, M.; McKenney, D. Anticipating the consequences of climate change for Canada's boreal forest ecosystems. *Environ. Rev.* **2013**, *21*, 322–365. [CrossRef]

12. Andrew, M.E.; Wulder, M.A.; Coops, N.C. Identification of *de facto* protected areas in boreal Canada. *Biol. Conserv.* **2012**, *146*, 97–107. [CrossRef]

13. Leroux, S.J.; Kerr, J.T. Land-use development in and around protected areas at the wilderness frontier. *Conserv. Biol.* **2013**, *27*, 166–176. [CrossRef] [PubMed]

14. ECCC (Environment and Climate Change Canada). Achieving a Sustainable Future: A Federal Sustainable Development Strategy for Canada, 2016–2019. Available online: http://fsds-sfdd.ca/downloads/FSDS_ 2016-2019_Final.pdf (accessed on 20 March 2018).

15. McNamee, K. From Wild Places to Endangered Spaces: A History of Canada's National Parks. In *Parks and Protected Areas in Canada: Planning and Management*; Dearden, P., Rollins, R., Eds.; Oxford University Press: Don Mills, ON, Canada, 2009; pp. 24–54.

16. Dearden, P.; Rollins, R. Parks and Protected Areas in Canada. In *Parks and Protected Areas in Canada: Planning and Management*; Dearden, P., Rollins, R., Eds.; Oxford University Press: Don Mills, ON, Canada, 2009; pp. 3–23.

17. Environment Canada. Environmental Indicators. Protected Areas. Available online: https://www.ec.gc.ca/ indicateurs-indicators/default.asp?lang=en&n=8390800A-1 (accessed on 26 March 2013).

18. Ecological Stratification Working Group. *A National Ecological Framework for Canada*; Environment Conservation Service, Environment Canada: Ottawa, ON, Canada, 1996.

19. Wiersma, Y.F.; Nudds, T.D. Efficiency and effectiveness in representative reserve design in Canada: The contribution of existing protected areas. *Biol. Conserv.* **2009**, *142*, 1639–1646. [CrossRef]

20. Deguise, I.E.; Kerr, J.T. Protected areas and prospects for endangered species conservation in Canada. *Conserv. Biol.* **2006**, *20*, 48–55. [CrossRef] [PubMed]

21. Andrew, M.E.; Wulder, M.A.; Coops, N.C. Patterns of protection and threats along productivity gradients in Canada. *Biol. Conserv.* **2011**, *144*, 2891–2901. [CrossRef]

22. Ontario Biodiversity Council. *Renewing Our Commitment to Protecting What Sustains Us*; ntario Biodiversity Council: Peterborough, ON, Canada, 2011.

23. Parks Canada. Parks Canada Agency 2012–13 Departmental Performance Report. Available online: http://www.pc.gc.ca/eng/docs/pc/rpts/rmr-dpr/index.aspx (accessed on 6 November 2018).

24. CARTS. *Conservation Areas Reporting and Tracking System. Canadian Council on Ecological Areas*; CARTS: Ottawa, ON, Canada, 2011.

25. Slocombe, D.S.; Dearden, P. Protected Areas and Ecosystem-based Management. In *Parks and Protected Areas in Canada: Planning and Management*; Dearden, P., Rollins, R., Eds.; Oxford University Press: Don Mills, ON, Canada, 2009; pp. 342–370.

26. Szaro, R.C.; Sexton, W.T.; Malone, C.R. The emergence of ecosystem management as a tool for meeting people's needs and sustaining ecosystems. *Landsc. Urban Plan.* **1998**, *40*, 1–7. [CrossRef]

27. Borrini-Feyerabend, G. Governance of protected areas: Innovations in the air. *Policy Matters* **2003**, *12*, 92–101.

28. Canadian Council on Ecological Areas (CCEA). *Canadian Guidebook for the Application of IUCN Protected Area Categories 2008*; CCEA Occasional Paper No. 18. Canadian Council on Ecological Areas Secretariat; CCEA: Ottawa, ON, Canada, 2008; p. 66.

29. Leroux, S.J.; Krawchuk, M.A.; Schmiegelow, F.; Cumming, S.G.; Lisgo, K.; Anderson, L.G.; Petkova, M. Global protected areas and IUCN designations: Do the categories match the conditions? *Biol. Conserv.* **2010**, *143*, 609–616. [CrossRef]

30. Swinnerton, G.S.; Buggey, S. Protected Landscapes in Canada: Current Practice and Future Significance. In *The George Wright Forum*; George Wright Society: Hancock, MI, USA, 2004.

31. Graham, J.; Amos, B.; Plumptree, T. *Governance Principles for Protected Areas in the 21st Century*; Institute on Governance in Collaboration with Parks Canada and CIDA: Ottawa, ON, Canada, 2003.

32. Office of the Auditor General of Canada. *Chapter 4—Ecosystems—Federal Protected Areas for Wildlife, 2008 March Status Report of the Commissioner of the Environment and Sustainable Development*; Office of the Auditor General of Canada: Ottawa, ON, Canada, 2008.

33. Zimmerer, K.S.; Galt, R.E.; Buck, M.V. Globalization and multi-spatial trends in the coverage of protected-area conservation (1980–2000). *Ambio* **2004**, *33*, 520–529. [CrossRef] [PubMed]

34. Bertzky, B.; Corrigan, C.; Kemsey, J.; Kenney, S.; Ravilious, C.; Besançon, C.; Burgess, N. *Protected Planet Report 2012: Tracking Progress Towards Global Targets for Protected Areas*; IUCN: Gland, Switzerland; UNEP-WCMC: Cambridge, UK, 2012.

35. Boreal Leadership Council. *Free, Prior, and Informed Consent in Canada: A Summary of Key Issues, Lessons, and Case Studies towards Practical Guidance for Developers and Aboriginal Communities*; Boreal Leadership Council: Ottawa, ON, Canada, 2012.

36. CoP7. Convention on Biological Diversity. In Proceedings of the Conference of the Parties to the Convention on Biological Diversity, Kuala Lumpur, Malaysia, 9–20 February 2004.

37. Liu, X.; Zhang, L.; Hong, S. Global biodiversity research during 1900–2009: A bibliometric analysis. *Biodivers. Conserv.* **2011**, *20*, 807–826. [CrossRef]

38. Adams, V.M.; Segan, D.B.; Pressey, R.L. How much does it cost to expand a protected area system? Some critical determining factors and ranges of costs for Queensland. *PLoS ONE* **2011**, *6*, e25447. [CrossRef]

39. Maggini, R.; Kujala, H.; Taylor, M.F.J.; Lee, J.R.; Possingham, H.P.; Wintle, B.A.; Fuller, R.A. *Protecting and Restoring Habitat to Help Australia's Threatened Species Adapt to Climate Change*; National Climate Change Adaptation Research Facility: Gold Coast, Australia, 2013.

40. Australian Government Department of Environment, Water, Heritage and the Arts (DEWHA). The Indigenous Protected Area Program: Background Information and Advice to Applicants. Available online: http://www.environment.gov.au/indigenous/pubs/ipa/ipa-advice.pdf (accessed on 30 March 2014).

41. Government of Australia. Collaborative Australian Protected Area Database. Available online: http://www.environment.gov.au/topics/land/nrs/science-maps-and-data/capad (accessed on 30 March 2014).

42. Hill, R.; Walsh, F.; Davies, J.; Sandford, M. Our Country Our Way: Guidelines for Australian Indigenous Protected Area Management Plans. In Proceedings of the Queensland Coastal Conference, Cairns, Australia, 19–21 October 2011; CSIRO Ecosystem Sciences and Australian Government Department of Sustainability, Water, Environment, Population and Communities: Cairns, Australia, 2011.

43. Ross, H.; Grant, C.; Robinson, C.J.; Izurieta, A.; Smyth, D.; Rist, P. Co-management and Indigenous protected areas in Australia: Achievements and ways forward. *Aust. J. Environ. Manag.* **2009**, *16*, 242–252. [CrossRef]

44. Myers, N. Biodiversity hotspots for conservation priorities. *Nature* **2000**, *403*, 853–858. [CrossRef] [PubMed]

45. Deutsche Gesellschaft für Internationale Zusammenarbeit (GIZ). Trinational Regional Project Initiative: Strengthening National Protected Area Systems in Colombia, Ecuador and Peru, 2012–2015. Available online: http://www.giz.de/en/worldwide/12717.html (accessed on 30 March 2014).

46. Kukkala, A.S.; Moilanen, A. Core concepts of spatial prioritisation in systematic conservation planning. *Biol. Rev. Camb. Philos. Soc.* **2013**, *88*, 443–464. [CrossRef] [PubMed]

47. Lehtomäki, J.; Tomppo, E.; Kuokkanen, P.; Hanski, I.; Moilanen, A. Applying spatial conservation prioritization software and high-resolution GIS data to a national-scale study in forest conservation. *For. Ecol. Manag.* **2009**, *258*, 2439–2449. [CrossRef]

48. Powers, R.P.; Coops, N.C.; Nelson, T.; Wulder, M.A.; Drever, C.R. Integrating accessibility and intactness into large-area conservation planning in the Canadian boreal forest. *Biol. Conserv.* **2013**, *167*, 371–379. [CrossRef]

49. Naughton-Treves, L.; Alvarez-Berríos, N.; Brandon, K.; Bruner, A.; Buck Holland, M.; Ponce, C.; Saenz, M.; Suarez, L.; Treves, A. Expanding protected areas and incorporating human resource use: A study of 15 forest parks in Ecuador and Peru. *Sustain. Sci. Pract. Policy* **2006**, *2*, 32–44. [CrossRef]

50. Solano, P. *Legal Framework for Protected Areas: Peru*; IUCN: Gland, Switzerland, 2010.

51. Kareksela, S.; Moilanen, A.; Tuominen, S.; Kotiaho, J.S. Use of inverse spatial conservation prioritization to avoid biological diversity loss outside protected areas. *Conserv. Biol.* **2013**, *27*, 1294–1303. [CrossRef] [PubMed]

52. Jackson, S.F.; Gaston, K.J. Incorporating private lands in conservation planning: Protected areas in Britain. *Ecol. Appl.* **2008**, *18*, 1050–1060. [CrossRef] [PubMed]

53. Keeley, A.T.; Basson, G.; Cameron, D.R.; Heller, N.E.; Huber, P.R.; Schloss, C.A.; Thorne, J.H.; Merenlender, A.M. Making habitat connectivity a reality. *Conserv. Biol.* **2018**. [CrossRef] [PubMed]

54. Mönkkönen, M.; YLISIRNIÖ, A.L.; Hämäläinen, T. Ecological efficiency of voluntary conservation of boreal-forest biodiversity. *Conserv. Biol.* **2009**, *23*, 339–347. [CrossRef] [PubMed]

55. Bennett, D.E.; Pejchar, L.; Romero, B.; Knight, R.; Berger, J. Using practitioner knowledge to expand the toolbox for private lands conservation. *Biol. Conserv.* **2018**, *227*, 152–159. [CrossRef]

56. McDonald, J.A.; Helmstedt, K.J.; Bode, M.; Coutts, S.; McDonald-Madden, E.; Possingham, H.P. Improving private land conservation with outcome-based biodiversity payments. *J. Appl. Ecol.* **2018**, *55*, 1476–1485. [CrossRef]

57. Di Minin, E.; Macmillan, D.C.; Goodman, P.S.; Escott, B.; Slotow, R.; Moilanen, A. Conservation businesses and conservation planning in a biological diversity hotspot. *Conserv. Biol.* **2013**, *27*, 808–820. [CrossRef] [PubMed]

58. Powers, R.P.; Coops, N.C.; Tulloch, V.J.; Gergel, S.E.; Nelson, T.A.; Wulder, M.A. A conservation assessment of Canada's boreal forest incorporating alternate climate change scenarios. *Remote Sens. Ecol. Conserv.* **2017**, *3*, 202–216. [CrossRef]

59. Nelson, T.A.; Coops, N.C.; Wulder, M.A.; Perez, L.; Fitterer, J.; Powers, R.; Fontana, F. Predicting climate change impacts to the Canadian boreal forest. *Diversity* **2014**, *6*, 133–157. [CrossRef]

60. Holmes, K.R.; Nelson, T.A.; Coops, N.C.; Wulder, M.A. Biodiversity indicators show climate change will alter vegetation in parks and protected areas. *Diversity* **2013**, *5*, 352–373. [CrossRef]

61. Powers, R.P.; Coops, N.C.; Morgan, J.L.; Wulder, M.A.; Nelson, T.A.; Drever, C.R.; Cumming, S.G. A remote sensing approach to biodiversity assessment and regionalization of the Canadian boreal forest. *Prog. Phys. Geogr.* **2013**, *37*, 36–62. [CrossRef]

62. Whitelaw, G.S.; Eagles, P.F.J. Planning for long, wide conservation corridors on private lands in the Oak Ridges Moraine, Ontario, Canada. *Conserv. Biol.* **2007**, *21*, 675–683. [CrossRef] [PubMed]

63. Natural Resources Canada. The State of Canada's Forests: Annual Report 2017. Available online: http://cfs.nrcan.gc.ca/stateoftheforests (11 September 2018).

64. Ontario Parks. *Algonquin Park Management Plan Amendment. Ministry of Natural Resources*; Ontario Parks: Peterborough, ON, Canada, 2013.

65. Naughton-Treves, L.; Holland, M.B.; Brandon, K. The role of protected areas in conserving biodiversity and sustaining local livelihoods. *Ann. Rev. Environ. Resour.* **2005**, *30*, 219–252. [CrossRef]

66. Hummel, M. *Protecting Canada's Endangered Spaces*; Key Porter Books: Bolton, ON, Canada, 1995.

67. Environment Canada. *Canadian Protected Areas Status Report 2000–2005*; Environment Canada: Ottawa, ON, Canada, 2006.

68. Canadian Boreal Initiative. *The Boreal Forest at Risk: A Progress Report*; Canadian Boreal Initiative: Ottawa, ON, Canada, 2003.

69. Canadian Boreal Initiative. *The Boreal in the Balance: Securing the Future of Canada's Boreal Region, a Status Report*; Canadian Boreal Initiative: Ottawa, ON, Canada, 2005.

70. Groves, C.R.; Kutner, L.S.; Stoms, D.M.; Murray, M.P.; Scott, J.M.; Schafale, M.; Weakley, A.S.; Pressey, R.L. Owning up to our responsibilities: Who owns lands important to biodiversity. In *Precious Heritage: The Status*

of Biodiversity in the United States; Stein, B.A., Kutner, L.S., Adams, J.S., Eds.; Oxford University Press: New York, NY, USA, 2000; pp. 275–300.

71. Loarie, S.R.; Duffy, P.B.; Hamilton, H.; Asner, G.P.; Field, C.B.; Ackerly, D.D. The velocity of climate change. *Nature* **2009**, *462*, 1052. [CrossRef] [PubMed]

72. Turner, M.G. Disturbance and landscape dynamics in a changing world1. *Ecology* **2010**, *91*, 2833–2849. [CrossRef] [PubMed]

73. Frazier, R.J.; Coops, N.C.; Wulder, M.A. Boreal Shield forest disturbance and recovery trends using Landsat time series. *Remote Sens. Environ.* **2015**, *170*, 317–327. [CrossRef]

74. Carvalho, S.B.; Velo-Antón, G.; Tarroso, P.; Portela, A.P.; Barata, M.; Carranza, S.; Moritz, C.; Possingham, H.P. Spatial conservation prioritization of biodiversity spanning the evolutionary continuum. *Nat. Ecol. Evol.* **2017**, *1*, 0151. [CrossRef] [PubMed]

75. Lehtomäki, J.; Moilanen, A. Methods and workflow for spatial conservation prioritization using Zonation. *Environ. Model. Softw.* **2013**, *47*, 128–137. [CrossRef]

76. Leonard, P.B.; Duffy, E.B.; Baldwin, R.F.; McRae, B.H.; Shah, V.B.; Mohapatra, T.K. gflow: Software for modelling circuit theory-based connectivity at any scale. *Methods Ecol. Evol.* **2017**, *8*, 519–526. [CrossRef]

77. Pelletier, D.; Clark, M.; Anderson, M.G.; Rayfield, B.; Wulder, M.A.; Cardille, J.A. Applying circuit theory for corridor expansion and management at regional scales: Tiling, pinch points, and omnidirectional connectivity. *PLoS ONE* **2014**, *9*, e84135. [CrossRef] [PubMed]

78. Pelletier, D.; Lapointe, M.-É.; Wulder, M.A.; White, J.C.; Cardille, J.A. Forest connectivity regions of Canada using circuit theory and image analysis. *PLoS ONE* **2017**, *12*, e0169428. [CrossRef] [PubMed]

79. Margules, C.R.; Pressey, R.L. Systematic conservation planning. *Nature* **2000**, *405*, 243–253. [CrossRef] [PubMed]

Perspective

Globalization and Biodiversity Conservation Problems: Polycentric REDD+ Solutions

Mwangi Githiru [1,2,]* and Josephine W. Njambuya [3]

[1] Wildlife Works, P.O. Box 310, Voi 80300, Nairobi 00100, Kenya
[2] Department of Zoology, National Museums of Kenya, P.O. Box 40658, Nairobi 00100, Kenya
[3] School of Natural Resources and Environmental Studies, Karatina University, P.O. Box 1957, Karatina 10101, Kenya; jnjambuya@gmail.com
* Correspondence: mwangi@wildlifeworks.com; Tel.: +254-720-433455

Received: 21 November 2018; Accepted: 15 December 2018; Published: 19 February 2019

Abstract: Protected areas are considered the cornerstone of biodiversity conservation, but face multiple problems in delivering this core objective. The growing trend of framing biodiversity and protected area values in terms of ecosystem services and human well-being may not always lead to biodiversity conservation. Although globalization is often spoken about in terms of its adverse effects to the environment and biodiversity, it also heralds unprecedented and previously inaccessible opportunities linked to ecosystem services. Biodiversity and related ecosystem services are amongst the common goods hardest hit by globalization. Yet, interconnectedness between people, institutions, and governments offers a great chance for globalization to play a role in ameliorating some of the negative impacts. Employing a polycentric governance approach to overcome the free-rider problem of unsustainable use of common goods, we argue here that REDD+, the United Nations Framework Convention on Climate Change (UNFCCC) climate change mitigation scheme, could be harnessed to boost biodiversity conservation in the face of increasing globalization, both within classic and novel protected areas. We believe this offers a timely example of how an increasingly globalized world connects hitherto isolated peoples, with the ability to channel feelings and forces for biodiversity conservation. Through the global voluntary carbon market, REDD+ can enable and empower, on the one hand, rural communities in developing countries contribute to mitigation of a global problem, and on the other, individuals or societies in the West to help save species they may never see, yet feel emotionally connected to.

Keywords: carbon finance; global commons; jurisdictional; nested approaches; public goods

---

1. Globalization, Biodiversity, and Protected Areas

Globalization can be a confusing term because it typically represents several different processes occurring simultaneously across the world. Perhaps its most universal notion is that of allowing goods, ideas, capital and to some extent people to move more freely, but not necessarily equally, between countries [1,2]. The process of integration and interaction among people, organizations, institutions, and governments of different nations, largely driven by cheaper or better coverage and penetration of Internet connectivity and increasing accessibility and affordability of modern transport is unprecedented. This is also supported by other enablers like technology and advancements in international policy. It is even reflected in research, where funding bodies increasingly have calls for consortia involving partners from multiple countries, often prizing collaborations between institutions in the North and South [3].

While globalization can bring much needed innovation, technology, jobs, and other resources to areas where there is scarcity, for the environment and natural resources, the ever-increasing connectedness comes with an attendant risk of externalizing costs, e.g., related to resource extraction and waste disposal [4]. Positive bends in national environmental Kuznets curves that suggest that as a country grows wealthier it reduces its resource-use intensity, may in fact only reflect externalization to other regions [5,6]. For instance, millions of hectares of agricultural land in developing countries, particularly in sub-Saharan Africa and Latin America, have been sold or leased to help meet the rapidly growing global demand for food and other bio-resources [7,8], with potential ramifications for local communities and the environment [9].

Consequently, globalization has mixed effects on the environment, culture and livelihoods of human and non-human societies around the world [10], and remains a deeply polemic topic. To some, it allows poor countries and/or people to grow economically and raise their standards of living, but others claim that it largely benefits multinational corporations at the expense of local people, cultures, enterprises, and environments due to a lack of adequate structure and controls; for instance, exports of agricultural products have been found to be correlated with forest loss with local consequences [11]. Indeed, globalization is considered as a major driver or enabler of major environmental and biodiversity damage due to increased consumption, production, movement of goods and associated greenhouse gas (GHG) emissions [2,6,12]. To exploit the opportunities presented by globalization for biodiversity conservation, there is a need to understand how it works and how the problems come about, before we can consider where potential solutions may lie.

Protected Areas (PAs)—defined as geographical spaces, recognized, dedicated, and managed, through legal or other effective means, to achieve the long-term conservation of nature with associated ecosystem services and cultural values [13]—are considered important tools for the conservation of biological diversity, and cornerstones of sustainable development strategies [14]. PAs assume several forms and diverse governance systems (either government/public, private/NGO, communities/indigenous groups or mixture of these), with two categories on the IUCN classification (V and VI) explicitly building in a human element to the protection of biodiversity, including culture, aesthetics and sustainable use [13]. PAs increased by more than 50% between 1990 and 2010 [15], but without a concomitant increase in public funding for their operations [16]. With the realization that PAs are consequently not as effective as they could be, novel means for funding them have been suggested (e.g., [17]). Besides funding, various forms of incentives are required to both directly or indirectly support the biodiversity conservation goals of PAs. This can be directly by improving operations within the PAs themselves, or indirectly by improving connectivity and reducing threats emanating from the landscapes surrounding the PAs.

2. Polycentric Solutions to Common Resource Problems

For globalization to avert some of these associated biodiversity problems, appropriate safeguards are required [12]. Such safeguards can range from secure property rights, better transparency and accountability, effective anti-corruption measures, and participation through free, prior, and informed consent (FPIC) [18,19]; sustainability certification is also growing in popularity, but impacts remains unclear or controversial [20,21]. Nonetheless, implementing these safeguards in some form is becoming an increasingly required and normal way of conducting business across many sectors (e.g., see [22]).

Such safeguards notwithstanding, globalization presents an additional unique problem for biodiversity conservation. The tragedy of the commons principle [23] suggests that common pool resources will almost always be overused in situations where any benefits are gained individually while the costs are shared communally. Natural resources like biodiversity generally fit snugly to this original definition of common pool resources. Overexploitation of fisheries provides a poignant example where scientific work points to clear problems and recommended solutions, yet human institutions seem incapable of implementing such solutions [24]. Amongst other issues, this can be attributed to the tension between individual (often short-term) benefits and group (often long-term) costs [25]. Likewise, when examining a historical microcosm of today's global dilemmas, Jared Diamond mused what the Easter Islanders said as they cut down the last trees, dooming themselves to a canoeless future and consequent extinction: did they say "we'll wait until the others stop" [26]?

In his book *The Logic of Collective Action*, Olson explained how beyond a given group size or number of people involved in a system—from tax payment to natural resource conservation—social links that ensure people work for the common good or fulfil their responsibilities start breaking down [27]. This is an enduring problem for environmental and biodiversity conservation, one that is accentuated by globalization. In his valedictory speech, Robert May [28] showed how cooperative systems—from marmots' sentry to humans' vaccinations—are vulnerable to 'cheating', whereby individuals enjoy the group benefit without incurring the costs. Such 'cheats' prosper in evolutionary terms, which makes it difficult for the cooperative benefits or system to be maintained. Only when such free riders (analogous to social and environmental externalities) are fully accounted for or excluded, will the net benefits of globalization be more uniformly and equitably felt and distributed across group members.

Elinor Ostrom in *Governing the Commons*, methodically laid out an eight-point plan on how to overcome some of these common-pool resource (CPR) problems, based on an extensive analysis of diverse systems [29]. Her eighth point on the eight-point plan specifically addressed CPRs that were part of larger systems where she argued that appropriation, monitoring, enforcement, conflict resolution, and governance activities would need to be organized in multiple layers of nested enterprises. In other words, there is need to design a system that builds responsibility for governing the common resource in nested tiers from the lowest level up to the entire interconnected system.

For climate change, she argued that single policies adopted only at a global scale were unlikely to generate sufficient trust among citizens, institutions and governments, so that collective action can take place in the inclusive and transparent manner necessary to reduce global warming [30]. She suggested that polycentric governance, founded upon the principle of subsidiarity, provided the potentially viable way to overcome this challenge [31]. It resolves the scale problem since it involves having multiple foci that are consistent with the various levels that decisions are made. It is worth noting here that polycentric governance does not allude to levels of government simply carrying out orders from those at higher levels; it requires a certain level of independence, as well as interdependence between governance institutions and organizations at various levels [32].

3. The Globalization, Biodiversity and REDD+ Nexus

Ecosystem goods and services range from the most tangible ones, like food and fiber, to least tangible and abstract ones, like options and existence values [33]. Because they also occur in a continuum from private to public based on ability to exclude other users, the requirements for their provision and conservation varies significantly, which reflects in the values placed on them [34]. Real or perceived scarcity can be used to partly explain the paradox of valuation, e.g., whereby water is essential but cheap and diamonds non-essential but expensive [35]. In essence, whether someone pays or is willing to pay for a given good or service, is based upon how they believe they benefit from paying for it, and whether they can get away with enjoying the benefit without paying the price [34].

Ecosystem services related to global warming, public forests and biodiversity (both tangible and intangible values), remain very susceptible to overuse. Yet, they also have the greatest potential for showing how polycentric governance might work towards resolving globalization problems associated with free-riding. Towards this, global warming could be considered superior to biodiversity because it more directly completes both sides of the user-provider equation:

- User: I know that I directly contribute to global warming through GHG emissions, e.g., by driving to work every day. For the large part, this is unlike biodiversity conservation whereby most people's contribution to the global biodiversity problem is more indirect, e.g., through externalities across supply chains, and not directly by, say, poaching.
- Provider: Notwithstanding what my neighbour opts to do, because of my direct contribution to the problem, I know that my doing something about it, like driving less or offsetting, also contributes directly to solving a small part of the problem.

Nonetheless, even this does not fully assuage the fear that, because I do not know what my neighbour shall do, assuming they are gaming the system and not reducing their emissions, then there is no point in me doing anything because we will be equally (adversely) affected. This is where the polycentric approach comes in, with multiple scales of active oversight to ensure wide-ranging compliance [30]. One of the avenues by which such a polycentric approach can be implemented to solve some of the globalization-driven biodiversity problems is through the proposed nested REDD+ process (e.g., [36]).

3.1. REDD+ and Biodiversity Conservation

Reduced Emissions from Deforestation and Degradation, Improved Forest Management and Afforestation, Reforestation and Revegetation (collectively referred to as REDD+), is a climate change mitigation scheme under the United Nations Framework Convention on Climate Change (UNFCCC) to help stem destruction of the world's tropical forests, by offering financial incentives to forest-right holders not to cut down trees [37]. It is a quintessential product of globalization, fully relying on actions at the global or regional levels, whether through carbon markets or other forms of global climate financing, to drive solutions at the global, national, and local levels [38]. Besides simply reducing CO_2 emissions, REDD+ also introduces an opportunity for climate financing to enhance biodiversity conservation, secure ecosystem services, and improve rural livelihoods [39]. Various international standards exist to ensure that biodiversity and social impacts are accorded the appropriate importance during implementation of REDD+ projects, at par with the carbon emissions reduction goal [40]. As a caveat, the appropriate regulatory framework is needed to ensure that multiple goals are considered, especially when areas of high biodiversity richness do not align with potential for reducing emissions [41].

Depending on the nature and source of threats, REDD+ operates across the four PA governance typologies mentioned earlier—namely public/government, private/NGO, community/indigenous peoples, and a mix of these. It operates on discrete and defined land units, which may be part of classic PAs (falling in any one of the six IUCN management categories), or other lands which may not fit snugly into these but come under some form of protection and conservation through REDD+ activities. While this can directly enhance biodiversity conservation in designated or gazetted PAs, its major strength perhaps is through indirectly propping these PAs by helping conserve adjacent and surrounding landscapes, which enhances the PA conservation goals through improved movement and connectivity, reduced threats, and expanded habitat for wide-ranging species not easily confined into PAs.

Ensconced in the expansive Tsavo Conservation Area, the Kasigau Corridor REDD+ Project (KCRP) in SE Kenya offers a good example. The KCRP comprises privately-owned land units classified as group ranches straddled by the Tsavo East and Tsavo West National Parks [42,43]. These group ranches were part of a contiguous ecosystem with the two Tsavo National Parks and hence, although not formally designated as wildlife protection areas, they still remain vital habitat and dispersal areas (including corridors) especially for wide-ranging species such as the African Elephant *Loxodonta africana* and African Wild Dog *Lycaon pictus* [44,45]. In this case, the role of the REDD+ project in protecting critical wildlife habitat and corridor is evident, which also improves protection for other species [46].

Global efforts to develop national REDD+ programmes indicate that various countries are designing their programs to deliver positive biodiversity impacts from REDD+. Yet, because most these national programs are still in their infancy, their biodiversity goals, planned conservation actions, and monitoring plans remain relatively unclear. Only a few countries such as the DRC and Costa Rica explicitly indicate their biodiversity goals and actions in their National REDD+ Strategy documents [47]. These countries further build into their national strategies detailed guidance on how their strategies intend to generate positive biodiversity impacts. This is largely through voluntary safeguard frameworks, such as the REDD+ Social and Environmental Standards and multiple-benefit standards designed for forest carbon projects.

3.2. The Polycentric Approach Extends this Nexus

Sub-national or national approaches to REDD+ have been extolled over project-level REDD+, mainly due to the potential for broader accounting of leakage [48]. Yet, project-level approaches confer two key advantages: besides enabling early involvement and experimentation, they allow for wide participation by a cross-section of stakeholders including the private sector. This aligns well with the stated advantages of polycentric approaches to conservation [32]. Fittingly, a nested (polycentric) scheme that combines both national and project level approaches has been deemed the most flexible mechanism towards delivering core REDD+ goals [49,50] (Figure 1).

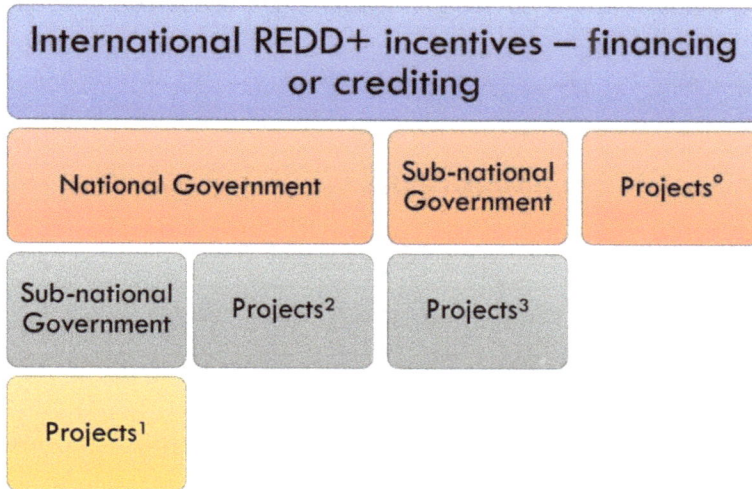

Figure 1. Alternative nesting strategies for capturing and distributing international REDD+ incentives. In the nested approach, rather than going directly to projects ("Projects°"), REDD+ financing passes through either national or sub-national government levels before cascading to projects (Projects [1,2,3]).

Such systems must carefully combine elements of top-down and bottom-up approaches [51]: top-down to ensure no double-counting and bottom-up to ensure that there is enough incentive to undertake the REDD+ actions required on-the-ground [52]. Nested or polycentric REDD+ helps sidestep potential free-rider problems at both the local and higher levels. First, having independent projects (through any of the four options in Figure 1) at the lowest scales ensures that drivers of deforestation are addressed at source, where it is relatively easy to apply the social pressures and economic incentives needed to ensure comprehensive compliance. Second, when this is linked to a higher jurisdictional level that builds up to the national level (project options 1, 2 & 3, Figure 1), it ensures that jurisdictions and governments also contribute to the process, by formulating and implementing the complementary policy and regulatory instruments needed to stem forest loss and retain the system's overall integrity [53].

4. Conclusions

While commitments to reduce emissions at the local or national levels will be necessary for coping with the climate change problem, most climate policy is still focused at the global level towards clinching global agreements. Given that international agreements are both difficult to clinch and remain susceptible to free-rider problems, a polycentric approach at various levels with active oversight of local, regional, and national stakeholders could hold greatest promise for success. Nested or polycentric

REDD+ schemes help avert potential free-rider problems by engendering small- to medium-scale action, as well as monitoring and reporting at all levels.

Further, REDD+ has the potential to harness resources for PA management and biodiversity conservation by using the appeal of GHG emissions as a more amenable global commodity. Using carbon to build polycentric policy frameworks and infrastructure could facilitate future development of a similar system for biodiversity. The power of globalization enables a rural farmer in Kenya to play a role in global climate change mitigation, while a social worker in downtown New York can help conserve the elephants in Africa. These persons need never meet, need never directly experience the impacts of climate change or the delight of seeing an elephant in the wild, but their worlds remain intimately connected. Now, thanks to these global schemes, they can begin positively influencing one another for the betterment of all.

Author Contributions: Conceptualization, M.G. and J.W.N.; Writing—Original Draft Preparation, M.G.; Writing—Review & Editing, M.G. and J.W.N.

Funding: This research received no external funding.

Acknowledgments: We would like to thank Wildlife Works Sanctuary Ltd. for the opportunity to work and experience the Kasigau Corridor REDD+ Project in Kenya. We are also grateful to colleagues at Wildlife Works and the REDD+ fraternity for stimulating conversations that drove the desire to write this paper. Comments and suggestions from three anonymous reviewers greatly enriched this manuscript.

Conflicts of Interest: The authors declare no conflict of interest.

References

1. Gruen, D.; O'Brien, T. Introduction. In *Globalisation, Living Standards and Inequality: Recent Progress and Continuing Challenges*; Gruen, D., O'Brien, T., Lawson, J., Eds.; J.S. McMillan Printing Group: Sydney, Australia, 2002; p. 274.

2. Ehrenfeld, D. Globalisation: Effects on Biodiversity, Environment and Society. *Conserv. Soc.* **2003**, *1*, 99–111.

3. Habel, J.C.; Eggermont, H.; Günter, S.; Mulwa, R.K.; Rieckmann, M.; Koh, L.P.; Niassy, S.; Ferguson, J.W.H.; Gebremichael, G.; Githiru, M.; et al. Towards more equal footing in north–south biodiversity research: European and sub-Saharan viewpoints. *Biodivers. Conserv.* **2014**, *23*, 3143–3148. [CrossRef]

4. Hoff, H. *Understanding the Nexus. Background Paper for the Bonn2011 Conference: The Water, Energy and Food Security Nexus*; SEI: Stockholm, Sweden, 2011.

5. Nepstad, D.C.; Stickler, C.M.; Almeida, O.T. Globalization of the Amazon soy and beef industries: Opportunities for conservation. *Conserv. Biol.* **2006**, *20*, 1595–1603. [CrossRef] [PubMed]

6. Lenzen, M.; Moran, D.; Kanemoto, K.; Foran, B.; Lobefaro, L.; Geschke, A. International trade drives biodiversity threats in developing nations. *Nature* **2012**, *486*, 109–112. [CrossRef] [PubMed]

7. Cotula, L.; Vermeulen, S.; Leonard, R.; Keeley, J. *Land Grab or Development Opportunity? Agricultural Investment and International Land Deals in Africa*; Iied: London, UK; Rome, Italy, 2009.

8. Deininger, K.; Byerlee, D. *Rising Global Interest in Farmland: Can it Yield Sustainable and Equitable Benefits?* World Bank Group: Washington, DC, USA, 2011.

9. Aabø, E.; Kring, T. *The Political Economy of Large-Scale Agricultural Land Acquisitions: Implications for Food Security and Livelihoods/Employment Creation in Rural Mozambique*; WP 2012; UNDP: Maputo, Mozambique, 2012.

10. Cowen, T.; Barber, B. Globalization and Culture. *Cato Policy Rep.* **2003**, *25*, 8–10.

11. DeFries, R.S.; Rudel, T.; Uriarte, M.; Hansen, M. Deforestation driven by urban population growth and agricultural trade in the twenty-first century. *Nat. Geosci.* **2010**, *3*, 178–181. [CrossRef]

12. Huwart, J.-Y.; Verdier, L. *Economic Globalisation: Origins and Consequences*; OECD Publishing: Paris, France, 2013.

13. IUCN. *Governance of Protected Areas: From Understanding to Action*; IUCN: Cambridge, UK, 2014.

14. CBD. *Protected Areas in Today's World: Their Values and Benefits for the Welfare of the Planet*; Technical Series No. 36; CBD: Montreal, QC, Canada, 2008.

15. Bertzky, B.; Corrigan, C.; Kemsey, J.; Kenney, S.; Ravilious, C.; Besançon, C.; Burgess, N. *Protected Planet Report 2012: Tracking Progress towards Global Targets for Protected Areas*; IUCN: Gland, Switzerland; Cambridge, UK, 2012.

16. Emerton, L.; Bishop, J.; Thomas, L. *Sustainable Financing of PAs: A Global Review of Challenges and Options*; IUCN: Gland, Switzerland; Cambridge, UK, 2006.

17. Githiru, M.; King, M.W.; Bauche, P.; Simon, C.; Boles, J.; Rindt, C.; Victurine, R. Should biodiversity offsets help finance underfunded Protected Areas? *Biol. Conserv.* **2015**, *191*. [CrossRef]

18. IFC. *IFC Performance Standards on Environmental and Social Sustainability*; IFC: Washington, DC, USA, 2012.

19. TEP. *The Equator Principles*; TEP: Geneva, Switzerland, 2013.

20. Edwards, D.P.; Fisher, B.; Wilcove, D.S. High conservation value or high confusion value? Sustainable agriculture and biodiversity conservation in the tropics. *Conserv. Lett.* **2012**, *5*, 20–27. [CrossRef]

21. Blackman, A.; Rivera, J. Producer-level benefits of sustainability certification. *Conserv. Biol.* **2011**, *26*, 1176–1185. [CrossRef]

22. Gupta, A. Transparency in Global Environmental Governance: A Coming of Age? *Glob. Environ. Polit.* **2010**, *10*, 1–9. [CrossRef]

23. Hardin, G. The tragedy of the commons. *Science* **1968**, *162*, 1243–1248.

24. Pauly, D.; Christensen, V.; Guenette, S.; Pitcher, T.J.; Sumaila, U.R.; Walters, C.J.; Watson, R.; Zeller, D. Towards sustainability in world fisheries. *Nature* **2002**, *418*, 689–695. [CrossRef] [PubMed]

25. Lawton, J.H. Ecology, politics and policy. *J. Appl. Ecol.* **2007**, *44*, 465–474. [CrossRef]

26. Diamond, J.M. *Collapse: How Societies Choose to Fail or Survive*; Allen Labe; Viking Press: New York, NY, USA, 2005.

27. Olson, M. *The Logic of Collective Action: Public Goods and the Theory of Groups*; Harvard University Press: Cambridge, MA, USA, 1965.

28. May, R.M. Threats to Tomorrow's World. Available online: http://www.royalsoc.ac.uk/downloaddoc.asp?id=2414 (accessed on 28 November 2018).

29. Ostrom, E. *Governing the Commons: The Evolution of Institutions for Collective Action*; Cambridge University Press: Cambridge, UK, 1990.

30. Ostrom, E. *A Polycentric Approach for Coping with Climate Change*; Policy Research Working Paper; Annals of Economics and Finance: Washington, DC, USA, 2009.

31. Ostrom, V.; Tiebout, C.M.; Warren, R. The organization of government in metropolitan areas: A theoretical inquiry. *Am. Polit. Sci. Rev.* **1961**, *55*, 831–842. [CrossRef]

32. Cole, D.H. Advantages of a polycentric approach to climate change policy. *Nat. Clim. Chang.* **2015**, *5*, 114–118. [CrossRef]

33. Jantzen, J. *The Economic Value of Natural and Environmental Resources*; TME, Institute for Applied Environmental Economics: Voorschoten, The Netherlands, 2006.

34. Krutilla, J.V. Conservation reconsidered. *Am. Econ. Rev.* **1967**, *57*, 777–786.

35. Simpson, R.D. *The "Ecosystem Service Framework": A Critical Assessment*; Ecosystem Services Economics (ESE); UNEP: Nairobi, Kenya, 2011.

36. Chagas, T.; Streck, C.; O'Sullivan, R.; Olander, J.; Seifert-Granzin, J. *Nested Approaches to REDD+ an Overview of Issues and Options*; Gobierno Regional Puno: Washington, DC, USA, 2011.

37. Ebeling, J.; Yasué, M. Generating carbon finance through avoided deforestation and its potential to create climatic, conservation and human development benefits. *Philos. Trans. R. Soc. Lond. B Biol. Sci.* **2008**, *363*, 1917–1924. [CrossRef] [PubMed]

38. Biedenkopf, K.; Müller, P.; Slominski, P.; Wettestad, J. A Global Turn to Greenhouse Gas Emissions Trading? Experiments, Actors, and Diffusion. *Glob. Environ. Polit.* **2017**, *17*, 1–11. [CrossRef]

39. Harvey, C.A.; Dickson, B.; Kormos, C. Opportunities for achieving biodiversity conservation through REDD. *Conserv. Lett.* **2010**, *3*, 53–61. [CrossRef]

40. CCBA. *Climate, Community and Biodiversity Standards*, 3rd ed.; CCBA: Arlington, VA, USA, 2013.

41. Paoli, G.D.; Wells, P.L.; Meijaard, E.; Struebig, M.J.; Marshall, A.J.; Obidzinski, K.; Tan, A.; Rafiastanto, A.; Yaap, B.; Slik, J.W.F.; et al. Biodiversity Conservation in the REDD. *Carbon Balance Manag.* **2010**, *5*, 7. [CrossRef] [PubMed]

42. WWC. *The Kasigau Corridor REDD Project Phase II—The Community Ranches*; Wildlife Works: Mill Valley, CA, USA, 2011.

43. Githiru, M. Implementing REDD+: Issues, opportunities and challenges with an example from the Kasigau Corridor REDD+ project, SE Kenya. In *XIV World Forestry Congress*; FAO (Food and Agriculture Organization of the United Nations: Durban, South Africa, 2015; pp. 1–10.

44. Williams, H.F.; Bartholomew, D.C.; Amakobe, B.; Githiru, M. Environmental factors affecting the distribution of African elephants in the Kasigau wildlife corridor, SE Kenya. *Afr. J. Ecol.* **2018**, *56*. [CrossRef]

45. Githiru, M.; Kasaine, S.; Mdamu, D.M.; Amakobe, B. Recent records and conservation issues affecting the African wild dog Lycaon pictus in the Kasigau Corridor, SE Kenya. *Canid Biol. Conserv.* **2014**, *17*, 1–8.

46. Becker, C.D.; Ginsberg, J.R. Mother-infant behaviour of wild Grevy's zebra: Adaptations for survival in semidesert East Africa. *Anim. Behav.* **1990**, *40*, 1111–1118. [CrossRef]

47. Panfil, S.N.; Harvey, C.A. *REDD+ and Biodiversity Conservation: Approaches, Experiences and Opportunities for Improved Outcomes*; USAID: Washington, DC, USA, 2014.

48. Karsenty, A.; Tulyasuwan, N.; de Blas, D.E. *Financing Options to Support REDD+ Activities*; European Commission: Brussels, Belgium, 2012.

49. Angelsen, A. *Moving Ahead with REDD: Issues, Options and Implications*; Center for International Forestry Research (CIFOR): Bogor, Indonesia, 2008.

50. Pedroni, L.; Dutschke, M.; Streck, C.; Porrua, M.E. Creating incentives for avoiding further deforestation: The nested approach. *Clim. Policy* **2009**, *9*, 207–220. [CrossRef]

51. Manolache, S.; Nita, A.; Ciocanea, C.M.; Popescu, V.D.; Rozylowicz, L. Power, influence and structure in Natura 2000 governance networks. A comparative analysis of two protected areas in Romania. *J. Environ. Manag.* **2018**, *212*, 54–64. [CrossRef] [PubMed]

52. VCS. *Jurisdictional and Nested REDD+ (JNR) Requirements*; VCS: Washington, DC, USA, 2014.

53. Dimitrov, R.S. The Paris Agreement on Climate Change: Behind Closed Doors. *Glob. Environ. Polit.* **2016**, *16*, 1–11. [CrossRef]

Article

Indigenous Protected and Conserved Areas (IPCAs), Aichi Target 11 and Canada's Pathway to Target 1: Focusing Conservation on Reconciliation

Melanie Zurba [1,2,*], Karen F. Beazley [1], Emilie English [1] and Johanna Buchmann-Duck [1]

[1] School for Resource and Environmental Studies, Dalhousie University, P.O. BOX 15000, Halifax, NS B3H 4R2, Canada; Karen.Beazley@Dal.ca (K.F.B.); Emilie.English@Dal.ca (E.E.); jbuchmannduck@dal.ca (J.B.-D.)
[2] College of Sustainability, Dalhousie University, P.O. BOX 15000, Halifax, NS B3H 4R2, Canada
* Correspondence: Melanie.Zurba@Dal.ca; Tel.: +1-902-494-2966

Received: 30 November 2018; Accepted: 4 January 2019; Published: 7 January 2019

Abstract: This article provides analysis of the issues relating to movement towards new models for Indigenous-led conservation in light of Canada's initiatives for greater protected areas representation through Target 1. We provide a background on Canada's Pathway to Target 1, which is based on Target 11 from the Aichi Biodiversity Targets set forth by the Convention on Biological Diversity (CBD). We contemplate the past, present and future of colonization and reconciliation in Canada, and consider the influence of international declarations, programs and initiatives on the potential for the formation of Indigenous Protected and Conserved Areas (IPCAs). We then provide an analysis of "wicked problems" that Indigenous communities, governments, and other stakeholders in protected areas will need to navigate towards implementing the IPCA approach in Canada. We outline the different types of Indigenous involvement in protected areas and how they potentially fit within the principles for the development of IPCAs. We then turn our discussion to the need to refocus conservation on reconciliation by restoring nation-to-nation relationships and relationships between the land and peoples. The lessons we draw have potential parallels for other nation states, particularly those signatory to the CBD and with a colonial history, aiming for biodiversity conservation and reconciliation with Indigenous peoples through IPCAs.

Keywords: biodiversity; conservation targets; protected areas; Indigenous peoples; IPCAs; reconciliation; Aichi Biodiversity Targets

1. Introduction

Events over the last two decades have significantly changed the landscape for conservation, protected areas, biodiversity protection, as well as the individual and collective rights of Indigenous peoples. One of the markers of this significant shift in the conservation paradigm is the Durban Accord, which was adopted by the International Union for the Conservation (IUCN) at the Vth IUCN World Parks Congress in Durban South Africa in 2005. The IUCN is the world's largest conservation organization, bringing together 1300 member organizations, including governments, non-governmental organizations (NGOs) and communities from across the globe to address the most pressing conservation issues [1]. The Durban Accord represents a historical moment in conservation and the human rights movement and identifies a "new protected areas paradigm", which not only rejects many long-held assumptions, policies and practices, but perhaps more importantly proposes an entirely different method for establishing, governing, and managing national parks and protected areas [2]. This new paradigm directly confronts and works to decolonize colonial conservation [2]:

> In this changing world, we need a fresh and innovative approach to protected areas and their role in broader conservation and development agendas. This approach demands the

maintenance and enhancement of our core conservation goals, equitably integrating them with the interests of all affected people. In this way, the synergy between conservation, the maintenance of life support systems and sustainable development is forged. We see protected areas as vital means to achieve this synergy efficiently and cost-effectively. We see protected areas as providers of benefits beyond boundaries—beyond their boundaries on a map, beyond the boundaries of nation states, across societies, genders and generations. [3] (p. 220)

Coinciding with the Durban Accord, new approaches to governance of protected areas have been emerging globally and have brought greater equity into governance systems with mixed outcomes. Such approaches have been vexed by a variety of issues relating, but not limited to, power sharing, capacity building, and navigating legal frameworks and jurisdictions [4]. Concurrent with the challenges of bringing greater equity into protected areas' governance systems are the challenges of developing new strategies for protecting the planet's unprecedented losses of biodiversity [5]. In 2010, the Tenth meeting of the Conference of the Parties to the Convention on Biological Diversity (CBD) adopted a revised *Strategic Plan for Biodiversity*, which included the launching of the Aichi Biodiversity Targets [6]. The Aichi Targets, as they are commonly known, include 20 individual targets, which are meant to be achieved by the parties of the CBD by 2020. In a delayed response to the Aichi Targets, Canada established a series of national goals known as "The 2020 Biodiversity Goals and Targets for Canada" [7]. Aichi Target 11[1] is partially reflected within Canada Target 1, which states:

By 2020, at least 17% of terrestrial areas and inland water, and 10% of marine and coastal areas of Canada are conserved through networks of protected areas and other effective area-based measures. [7] (p. 23)

In addition to attempting to advance Target 1, the Government of Canada is also attempting to move forward with reconciliation efforts and putting into practice the Truth and Reconciliation Commission's Calls to Action (described in greater detail in Section 3.1) as well as the United Nations Declaration on the Rights of Indigenous Peoples (UNDRIP). In 2017, the Indigenous Circle of Experts (ICE) was invited, created and their collective work commemorated by national Indigenous organizations and federal, provincial and territorial jurisdictions engaged in the Pathway to Target 1 [8,9]. The members of the ICE were tasked with examining how Canada Target 1 and consequently Canada's global commitment to the CBD could be met in an equitable manner, including the development of Indigenous-led conservation, which ICE would come to call "Indigenous Protected and Conserved Areas (IPCAs)" [8] (p. iii). In the Spring of 2018, the ICE released a report entitled *We Rise Together: Achieving Pathway to Canada Target 1 through the creation of Indigenous Protected and Conserved Areas in the spirit and practice of reconciliation* [9].

Our paper reflects on the *We Rise Together* report as a monumental shift in Canada's guiding frameworks for protected areas. We explore how the proposed IPCAs relate to international biodiversity conservation targets, Canada's Target 1, the context of complex histories and relationships between Indigenous peoples and the Canadian government, and national and international guidance on protected areas. We provide an analysis of the "wicked problems" that Indigenous communities, governments, and other stakeholders in protected areas will need to navigate towards implementing the Indigenous-led approach in Canada. These issues are of relevance to other nations grappling with implementing Indigenous-led protected areas in the context of international biodiversity conservation targets, particularly those with a colonial history of dispossession of Indigenous territories and displacement of Indigenous peoples for colonial purposes, including protected areas.

---

[1] Aichi Target 11 reads, "By 2020, at least 17 per cent of terrestrial and inland water areas and 10 per cent of coastal and marine areas, especially areas of particular importance for biodiversity and ecosystem services, are conserved through effectively and equitably managed, ecologically representative and well-connected systems of protected areas and other effective area-based conservation measures, and integrated into the wider landscape and seascape" [6] (p. 9).

2. Protected Areas in Canada, Aichi Biodiversity Targets and Canada's Pathway to Target 1

The Government of Canada has adopted the IUCN definition of a protected area as: "a clearly defined geographical space, recognized, dedicated and managed, through legal or other effective means, to achieve the long-term conservation of nature with associated ecosystem services and cultural values" [7,10]. Once an area has been identified as a protected area, one of the IUCN's seven protected area management categories may be applied to further define the parameters of protection and use for the area [10]. The wording "legal or other effective means" is important for the establishment of IPCAs in Canada because it provides a mechanism through which IPCAs can be recognized and reported without being co-opted by traditional colonial models for protected areas, which is an important aspect of self-determination [11]. The international policy discourse, put forth by the IUCN, clarifies that "other effective means" of conservation can refer to "recognized traditional rules" [10].

Language exists to describe what "other effective area-based conservation measures" (OECMs) entail in the context of Aichi Target 11 and thus of relevance to Canada target 1, and draft guidelines have been released by the IUCN-World Commission on Protected Areas (WCPA) [12]. The inclusion of OECMs in Aichi Target 11 acknowledges that " ... areas outside the recognized protected area networks also contribute to the effective in-situ conservation of biodiversity .... [including] territories and areas governed by ... Indigenous Peoples ... , and shared governance" [12] (p. 11). Whereas protected areas have conservation as a primary objective, an OECM "should deliver the effective in-situ conservation of biodiversity, *regardless of its objectives*" [12] (p. 14, emphasis added). Protected area practitioners and scholars have expressed concern about the lack of a precise definition for OECMs potentially leading to inconsistent designations, which fall short of conservation objectives [13–17]. For example, Lemieux et al. [15] demonstrated how this concern became a reality in Canada when the Department of Fisheries and Oceans designated marine refuges as OECMs even though these areas are left exposed to external industrial pressures. Accordingly, newly drafted international guidelines aim to provide sufficient information for interested parties to apply the OECM concept and to recognize these areas in reporting to the CBD [12]. Such guidance should support good practice in establishing IPCAs in Canada, regardless of their primary objectives, and in recognizing Indigenous peoples, territories and governments and their roles, responsibilities and contributions to effective in-situ conservation of biodiversity.

The prominence of Aichi Target 11 has resulted in Canada's establishment of Target 1, which initiates the race to meet area-based quantitative targets by 2020, with a risk of insufficient consideration of qualitative aspects [14,15], including those relevant to IPCAs. MacKinnon et al. (2015) [14] pointed out that, in *Canada's Fifth Report to the CBD*, the qualitative aspects of Target 11 are not evaluated, and thus there is a sole focus on the quantitative aspects. While establishing greater area of protection is important, emphasis on an area-based focus has resulted in a proliferation of parks that lack demonstrable conservation impact and leave many questions around how Indigenous peoples might play meaningful roles in the future of protected areas [15–17]. Canada's focus on Aichi Target 11, has also led to the neglect of other Aichi Targets that could serve as important guidelines for structural changes leading to meaningful Indigenous involvement and leadership in protected areas. Of the 20 Aichi Targets, Aichi Targets 14 and 18 speak the most directly to Indigenous peoples. Target 14 states:

> By 2020, ecosystems that provide essential services, including services related to water, and contribute to health, livelihoods and well-being, are restored and safeguarded, taking into account the needs of women, indigenous and local communities, and the poor and vulnerable. [6] (p. 9)

Target 18 also has significant implications for Indigenous peoples and reconciliation in Canada:

> By 2020, the traditional knowledge, innovations and practices of indigenous and local communities relevant for the conservation and sustainable use of biodiversity, and their customary use of biological resources, are respected, subject to national legislation and relevant international obligations, and fully integrated and reflected in the implementation

of the Convention with the full and effective participation of indigenous and local communities, at all relevant levels. [6] (p. 9)

Collectively Targets 14 and 18 have significant implications for Indigenous peoples and reconciliation in Canada. Both international guidance [18,19] and the ICE report clearly state that IPCAs should be Indigenous-led, and as such IPCAs should allow for the integration of traditional knowledge in defining conservation and sustainable use for those communities. Thus, it should be called into question whether the "equitable" management goal within Target 11 could be achieved in IPCAs without the inclusion of Targets 14 and 18. To support the achievement of multiple targets, it will be important that indicators for each be developed and equally considered. In their evaluation of Aichi Target progress, Tittensor et al. [20] noted that there are no identifiable indicators for Target 18, suggesting an urgent need for the development and integration of indicators across targets.

3. Past, Present and Future: Canada's Colonial History and Guiding Frameworks for Indigenous Protected and Conserved Areas (IPCAs)

The increasing prominence of Indigenous-led protected areas, combined with heightened discourse and action on reconciliation efforts in Canada, presents an opportunity to achieve conservation and reconciliation concurrently. This explicit connection between the status of Indigenous peoples' culture and nature is clearly acknowledged in *We Rise Together*. As Eli Enns, Co-Chair of the ICE, states, "Whenever you find intact ecological biodiversity, you find intact, thriving, cultural holistic diversity" [9] (p. 73). Similarly, Niigaan Sinclair, Anishinaabe professor of Native Studies at the University of Manitoba, observes, "You can't have a bear clan with no bears. Indigenous nations know this. It's called diversity. Canada's built on it. We just need to remember who gave us it" [21] (n.p.).

Modern frameworks for IPCAs demonstrate due consideration of important facets of reconciliation (such as Indigenous-led), which, if applied correctly, may serve as a pathway to reparation and decolonization of peoples and nature. As recognized by the Truth and Reconciliation Commission of Canada (TRC) [22], acknowledging the truth is an important precursor to reconciliation, and thus examining Canada's colonial past and present is a critical component of moving forward with IPCAs.

3.1. Canada's Protected Areas: The Colonial Past and Present

Protected areas, as embodied in Canadian law and imagination, have been developed through a paradigm which holds paramount the preservation of a people-free "wilderness". National parks in Canada were early tools that deliberately perpetuated colonial injustices and continue to operate as such today despite some recent improvements [9,23]. Many well-established and world-renowned national parks in Canada, such as Banff, Jasper, and Riding Mountain National Parks, as well as provincial parks such as Quetico in Ontario, were founded on and continue to struggle with a legacy of colonial dispossession of Indigenous peoples [23–26]. The creation of these defined regions, which continue to act as conservation enclosures, not only intentionally removed Indigenous peoples from their lands and dispossessed them of their territories, but also perpetuated the illusion of "wilderness" as being pure and devoid of human life and influence [27]. This "wilderness" model of national parks concurrently advanced capitalist enterprises such as sport hunting, recreation, and tourism, while excluding Indigenous peoples as beneficiaries in any capacity or form [9,23]. It became an exercise of "primitive accumulation", entailing both the dispossession of land and the enclosure of a commons in favor of specific interests [28]. The new concept for IPCAs, conversely, is holistic and based on conservation that includes people and culture.

Many Indigenous peoples distrust the concept and nature of protected areas in Canada. Marylyn Baptiste (former Chief of the Xeni Gwet'in) stated that "since my dad was chief, as far as I've always learned, governments in Canada have reserved parks for their own benefit for later use", and cites Bill F4 in BC which proposed opening parks to mining exploration [29] (p. 20). Other objectives, such as species protection, can prompt conservation efforts from the government that perpetuate the exclusionary colonial model of conservation and wilderness. For example, a caribou range plan proposed by the Government of Alberta was not well received by Indigenous peoples. They believed that the plan was spurred by the province's desire to turn their traditional territory into parkland and suppress Indigenous land rights, rather than efforts to protect the caribou herds [30]. However, alternative perspectives amongst Indigenous leaders exist in favor of protected areas. As John Amagoalik, Inuit Elder and Statesman, stated, "The Government of Canada needs to fulfill its overdue promise to create a national marine conservation area in the Landcaster Sound. Inuit rely on the abundance of Tallurutiup Tariunga and expect it to be protected" [31]. In December 2018, the negotiation of the Tallurutiup Imanga Inuit Impact and Benefit Agreement (IIBA) was announced, and once approved by the Canadian government, this will complete "the largest unified land and water protected area in Canada" [32] (n.p.).

While the colonial model of protected areas persists, there is an increasing acknowledgement within Canadian parks institutions (federal, provincial, and territorial) of the "history of exclusion" [9], and in some cases attempts at reparations are being made [23]. As stated by a participant in Exploring Empowerment for Indigenous Protected and Conserved Areas in B.C., a workshop held by the David Suzuki Foundation, "reconciliation is also restitution—returning things taken" [33] (p. 17). Furthermore, Canadian governments have a fiduciary responsibility to consult and accommodate Indigenous peoples when establishing a new protected area, according to Section 35 of the Constitution Act, 1982.[2] Significant policy and legislative changes occurring over the past few decades are beginning to affirm the special relationship Indigenous peoples have with lands, including those within national parks [36]. Nevertheless, Indigenous and legal scholars agree that many restitutions and reparations for communities and structural changes to protected areas institutions will be essential if protected areas are to become inclusive of Indigenous values, aspirations and knowledge systems [36–39]. Furthermore, fundamental considerations relating to territory and sovereignty are due when considering parks and protected areas. It is imperative for Canada, as well as provincial and territorial jurisdictions, to acknowledge the relationship and continuing ownership that Indigenous peoples have with their traditional territories if protected-area systems are to truly move to a stage of decolonization [40]. Connecting to this imperative are state- and community-driven processes of reconciliation. Both processes include acknowledging the need for reparations for historical wrongdoings, as well as the development of new pathways for relationships rooted in truth, justice, and healing [22]. However, several critiques of state-driven processes exist, including that formal reconciliation processes have been used to divert attention away from legal restitutions and disputes over lands and resources [37]. Despite such critiques, it is important to understand the context of formal reconciliation processes in Canada because they will continue to shape the development of policy affecting Indigenous peoples and their lands.

As a first step towards reconciliation, the Government of Canada established the Truth and Reconciliation Commission of Canada in 2008 as a part of the Indian Residential Schools Settlement Agreement.[3] Following the end of the hearings involving testimonials from Residential School survivors in 2015, the TRC developed a set of "Calls to Action" that would serve as a guide to the processes of reconciliation for the Government of Canada and all Canadians [22]. The need to acknowledge Indigenous lands is referred to several times in the TRC's Calls to Action. In addition to

---

[2] Canada's supreme law is written in the Constitution Acts of 1867 and 1982. The Constitution Acts outline Canada's system of government (as a federation), as well as the civil rights of all Canadians and those in Canada. According to section 91.24 in the Constitution Act of 1867, the federal government has jurisdiction over "Indians, and Lands reserved for Indians". Through section 35 of the Constitution Act 1982, "the existing Aboriginal and Treaty rights of Aboriginal people in Canada are hereby affirmed"; the Act also clarifies that "Aboriginal Peoples of Canada include the Indian, Inuit, and Métis Peoples of Canada" [34,35].

[3] Indian residential schools were first established in Canada in the 1880s, with the last one closing in 1996. The government's intention with this system was to "remove the Indian from the child" by forcibly removing them from their families and communities, and forbidding the use of Indigenous languages and any expressions of Indigenous cultures [41].

the Calls to Action, the TRC outlines "principles of reconciliation", which include recognizing it as a process of " … healing relationships through truth sharing, … and redress [for] past harms", as well as " … constructive action on addressing the ongoing legacy of colonialism … " [42] (p. 3). The principles of reconciliation can be interpreted in many different ways, according to each (Indigenous) nation's understanding of what it should encompass [22]. Some factors that may pose challenges to reconciliation are deeply entrenched in the structural roots of the nation-state known as Canada, and will require long-term political will if they are to be gradually overcome. Encouragingly, some solutions and models for reconciliation can be found by looking to the roots of Indigenous-settler relations, as in the case of the nation-to-nation spirit of many treaties.[4]

While broader discussion about sovereignty, treaties, UNDRIP, Indigenous law, Aboriginal law, and Canadian law are beyond the scope of this paper, it is important to acknowledge the weight of these issues as they are at the root of the set of problems and opportunities that reconciliation is attempting to address. In *We Rise Together*, the ICE states that reconciliation "means identifying the appropriate healing process for restoring relationships: first, between Crown and Indigenous Peoples, recognizing what has not worked in the past so it is corrected moving forward in the spirit of peace and friendship; and second, between all people (Indigenous and non-Indigenous) and the lands" [9] (p. 7). Creating an appropriate forum for honest discussion around targets and IPCAs is important, and will likely need to happen within the context of "ethical space" as indicated in *We Rise Together* [9]. The concept of ethical space includes characteristics such as the creation "of a place for knowledge systems to interact with mutual respect, kindness, generosity, and other basic values and principles" and is "a space for collaboration and achieving common ground" [9] (p. 17).

3.2. Guiding Frameworks for Indigenous Protected and Conserved Areas (IPCAs)

The IUCN, as the leading global conservation organization, brings governments, non-governmental organizations (NGOs), the United Nations (UN), and communities together "to forge and implement solutions to environmental challenges," including those related to protected areas governance through various programs and initiatives [1] (para. 3). Since the Durban Accord, there has been increasing participation of Indigenous peoples and Indigenous Peoples' Organizations (IPOs) in developing global protected area (and other forms of environmental) governance frameworks through IUCN programs and initiatives. The IUCN Programme on Protected Areas administers standards for global protected areas such as the Green List of Protected and Conserved Areas, which sets "the new global standard for protected areas in the 21st Century" [44] (para. 2). Within the guiding policy document, Green List of Protected and Conserved Areas: Standards, Version 1.1, under criterion for "Good governance" of protected areas, is the indicator that states: "The site's local governance structures and mechanisms recognise the legitimate rights of Indigenous Peoples and local communities" [45] (p. 13). There is also reference in the standards to the need to consider Indigenous Community Conserved Areas and OECMs under "Sound Design and Planning" of protected area systems [45]. Several other programs and initiatives focus on developing frameworks for Indigenous participation in protected areas, as well as specialized working groups. In particular, the Indigenous and Community Conserved Areas (ICCA) Consortium [18] has been highly influential in advancing frameworks for Indigenous rights with regards to protected areas.

The ICCA Consortium is a "membership-based civil society organization" that was formed in 2010 with the mission "to promote the appropriate recognition of, and support to, Indigenous peoples' and community conserved areas and territories (ICCAs) at local, national and international levels" [46] (para. 1). At an international level, the ICCA Consortium outlines three defining characteristics of

---

[4] Treaties, as defined by the Government of Canada [43], are "agreements made between the Government of Canada, Indigenous groups, and often provinces and territories that define ongoing rights and obligations on all sides". Both historical (70 treaties between 1701 and 1923) and modern-day treaties (25 treaties since 1971) are relevant in the context of protected areas.

ICCAs, which are echoed in the IUCN Protected Area Guidelines. These are that the community has a close and meaningful relationship to the area; the community is the primary decision maker for site management and has the power to develop and enforce regulations; and management by the community leads to conservation regardless of motivation [46,47]. Similarly, the ICE (2018) states three defining elements of IPCAs for the Canadian context: "they are Indigenous-led; they represent a long-term commitment to conservation; and they elevate Indigenous rights and responsibilities" [9] (p. 5). The concept of "Indigenous-led" forms one of three essential elements of an IPCA in both the ICE [9] and the ICCA Consortium [18] characterizations.

Given the history of colonization in Canada, Indigenous-led is a critical element to ensuring reconciliation is a part of the IPCA process. The ICE defines Indigenous-led as where "Indigenous governments have the primary role in determining the objectives, boundaries, management plans and governance structures for IPCAs as part of their exercise of self-determination" [9] (p. 36). The concept is also being used in environmental assessment and has been defined as being on the terms of Indigenous people, with their approval, and where Indigenous peoples "are involved in the scoping, data collection, assessment, management planning, and decision-making about a project" [48] (p. 10). These definitions and characteristics of Indigenous-led are important to keep in mind in discussions of how IPCAs relate to Canada's conservation targets. Importantly, an IPCA should not be evaluated outside of the Indigenous community's concerns or by the expectations of external actors [47]. Steven Nitah, negotiator and former Chief of the Lutsel K'e Dene First Nation, described this imperative for self-determination and effective partnerships:

> Fundamentally, the starting place for an (Indigenous Protected Area) IPA must be self-determination by the [I]ndigenous peoples themselves, but once declared, IPAs become the basis for building effective partnerships between [I]ndigenous and public governments and other entities, including NGOs, research institutions, and the philanthropic community. [49] (p. 4)

The consideration of Indigenous-led is critical in light of the Aichi Targets in Canada, as well as elsewhere. These targets have created a potentially problematic set of standards, by which IPCAs may be evaluated by criteria that have been defined outside of the Indigenous community, and the impetus for IPCA establishment and national/international reporting may be imposed by the nation state, rather than "led" by the Indigenous community.

In Canada, management arrangements with the federal, provincial, or territorial governments that, while highlighted as Indigenous-led, may be worth reexamining as part of the reconciliation process. Partnerships should be considered Indigenous-led where (and only where) the choice to partner with the government comes from the Indigenous community or Indigenous government itself and not from the federal, provincial, or territorial government [29]. An example of this is Dınàgà Wek'èhodì Candidate Area in the Northwest Territories where, after proposing that an area of ecological and cultural importance be recognized, the Tłıchǫ Government (an Indigenous self-government in the Northwest Territories) put forward a proposal to protect the area using territorial legislation, which they are developing in partnership with the Government of the Northwest Territories [50,51]. Recently, the Dehcho First Nations and the Government of Canada announced their collaborative effort to establish an IPCA in traditional Dehcho territory, called Edéhezíe Protected Area [52,53]. For Canada, this is the first officially announced IPCA [53].

The Government of Canada finalized an agreement with the Dehcho First Nations in association with the designation of Edéhezíe Protected Area as a National Wildlife Area by 2020 [52,54]. Dahti Tsetso, Resource Management Coordinator for the Dehcho First Nation, stated that the agreement " . . . will give us some capacity to start addressing the goals of our communities and approaching protection in ways that make sense to them, that helps our communities approach stewardship in a meaningful way" [55] (para. 10). Under this arrangement, Dehcho guardians would clearly be important in monitoring and management, but it may be unclear how much control they have had in establishment of the area. Designating an IPCA under a pre-existing federal protected area category may present issues and risk an imbalance of power with more resting in the federal government's hands. Up until the designation process, however, this initiative appears to have been Indigenous-led: the Dehcho First Nations and Tłı̨chǫ government requested in June of 2010 that Edéhezíe be designated as a National Wildlife Area [52]; and, in 2018, the Dehcho First Nations (DFN) Annual Assembly,

> ... resolved that the Assembly: 1. approves and enacts the Dehcho Protected Area law (2018); 2. authorizes the DFN Grand Chief to enter into the Edéhezíe Establishment Agreement with Canada on behalf of the Dehcho First Nations; and 3. authorizes the Dehcho First Nations to finalize the Establishment agreement with Canada and to do such other things as may be necessary to permanently protect Edéhezíe ... [56] (p. 2)

The Dehcho First Nations community has been pushing for protection of their cultural lands for years considering the threat of mining and other resource development in the area [55]. Indigenous leadership in this case is clear; however, it is also worth questioning if this request was made because of inadequacy of the legislative tools that are available to protect the region from industrial development, and what this could mean for further interference from the federal government. Although the Dehcho are finalizing an establishment agreement with Canada, it remains an open question as to whether, over time, the power imbalance of a colonial designation will or will not be redressed, even though the process has been Indigenous-led, and with an agreement in place for Indigenous management.

4. Potential Types of Indigenous Protected and Conserved Areas (IPCAs) in Canada

IPCAs have the potential to fit into Target 1 as either a protected area or an OECM depending on whether it meets the definition and criteria, including among others its objectives for conservation. However, despite such categorizations, Indigenous peoples have been managers of their lands and resources since time immemorial [13]. While the decision to establish an IPCA must come from Indigenous leadership, a range of partnerships with government or other outside parties may be appropriate [9]. Through their processes of engagement, the ICE [9] concluded that the type of partnerships and the degree to which decision-making authority is shared should be determined by the Indigenous group based on the objectives and needs of their nation or government. Within Canada, Indigenous peoples have been involved in co-management and joint management agreements for several decades, and have also created independent governance models to protect their lands and culture from harms arising from industrial and governmental pressures [9,57]. The different types of Indigenous involvement in protected areas are outlined in Table 1, which also provides examples and information on the traditional territory and jurisdiction(s). Outlining the different potential types and examples of IPCAs provides insights into how IPCAs might be categorized and defined according to Canada's Pathway to Target 1. Another important consideration is whether the Indigenous community wants to have their involvement in protected areas counted at all. In the future, such processes should be initiated only where the Indigenous government has indicated a distinct interest.

Table 1. Indigenous involvement in protected areas management in Canada. Models adapted from *We Rise Together* by the ICE [9].

Type of Indigenous Involvement in Protected Area	Example	Traditional Territory/Province/Territory	Jurisdiction/Decision-Making Authority
Advisory board	Fundy National Park	Mi'kmaq/Wolastoqiyik/New Brunswick [58]	Federal authority
Joint or cooperative management	Torngat Mountains National Park	Inuit from Nunavik/Inuit from Nunatisavut/Newfoundland and Labrador [59]	Federal authority
Conservancy	Bear Island Conservancy	Nat'oot'en Nation/British Columbia [60]	Shared governance (with the province)
Tribal park	Tla-o-qui-aht Tribal Parks	Tla-o-qui-aht First Nation/British Columbia [61]	Provincial authority
Indigenous management	Wehexlaxodiale	Tłı̨chǫ First Nation/Northwest Territories [62]	Indigenous governance
Indigenous governance	Edéhezíe Protected Area	Dehcho First Nation/Northwest Territories [53]	Indigenous governance

With the ICE's defining qualities of IPCAs in mind (i.e., the process is Indigenous-led; there is a long-term commitment to conservation; and they are in support of Indigenous rights and responsibilities), the designations in Table 1 that would qualify as IPCAs are those that are "conservancies", "Tribal Parks", "Indigenous management", and "Indigenous governance". Unlike Tribal Parks and Indigenous management, conservancies are recognized and defined under provincial law. This means that all conservancies in British Columbia will meet the same criteria: they meet the IUCN definition of a protected area as they are geographically defined, legally protected under provincial law, and are set aside for the protection of biodiversity as well as cultural values of Indigenous peoples [60]. Conservancies allow for the development of natural resources if it is consistent with the protection of biodiversity, uses by Indigenous peoples, and recreational values [60]. However, conservancies cannot be considered "Indigenous-led", as the province still maintains the power to approve decisions. While joint authority might support "Indigenous rights and responsibilities" and show a "long-term commitment to conservation", it cannot be assumed that the process is adequately Indigenous-led. Such adequacy can only be determined by the Indigenous group that is part of the agreement and involved in decision-making.

Tribal Parks are not officially recognized by the Government of Canada, which has the potential to conflict with the need to support Indigenous rights and responsibilities. Tla-o-qui-aht Tribal Parks is a well-established model of a Tribal Park located in British Columbia. These parks are geographically defined, and while Tla-o-qui-aht law may not be officially recognized in Canada [9], the Tribal Park can still meet the protected area definition of "other effective means" noted earlier in this paper. These Tribal Parks are intended to focus on the integration of human wellbeing with healthy ecosystems [61]. Tribal Parks are currently not considered under Canada Target 1 and are not fully recognized as a protected area by the provincial or federal governments [9]. It is important to note that, should government recognition and support come in future, it would need to be the Tla-o-qui-aht leadership that would determine if it ought to be counted as an IPCA under Canada Target 1 as a protected area, OECM designation, or at all.

Indigenous management involves area management that is established and governed by Indigenous peoples without co-management or another type of partnership arrangement, and is a third type of Indigenous management that could fit with the IPCA model. Wehexlaxodıale is an example of Indigenous management, which involves a protected zone established and managed by the Tłįchǫ Government [62]. It is geographically defined and managed under the Tłįchǫ Land Use Plan [62]. Wehexlaxodıale is fully protected for the cultural and heritage value it provides to the Tłįchǫ people. To the Tłįchǫ people, the protection of nature, culture and heritage are intertwined [62]. This differs from typical western notions of conservation that often view nature, culture and heritage as separate aspects of environment and society. Despite this complexity, the Government of the Northwest Territories has reported Wehexlaxodıale as a protected area under "Indigenous Government Administration" on behalf of the Tłįchǫ Government and with their approval. This reporting is to the Conservation Areas Reporting and Tracking System (CARTS) database, which contains all federal, provincial and territorial data on protected areas [63], and has been used to date by Canada to report progress towards national and international commitments to biodiversity conservation.

The newly established Edéhezíe Protected Area involves a new type of arrangement that is more representative of Indigenous governance. Thus far, of all the types of protected areas involving Indigenous communities, Edéhezíe Protected Area shares the most in common with the ICE's definition for IPCAs. It is yet to be determined if this will give support to the formal establishment of an IPCA category in Canada.

5. "Wicked Problems"

Canada has a complex historical, legal, political, and social landscape with several factors that can confound those grappling with whether to recognize IPCAs under Canada's Target 1. These factors can be understood as "wicked problems" [64,65], which are "incomprehensible and resistant to solution" [66]. While innumerous wicked problems could be discussed, we have identified six major types of wicked problems that will need to be addressed before IPCAs can achieve both conservation and reconciliation objectives.

5.1. Exclusionary "Wilderness" Paradigm for Protected Areas

As discussed above with relation to the colonial past and present, the "wilderness" paradigm for protected areas, which is based on the nature/culture dichotomy, continues to be pervasive. Concepts of pristine nature and wilderness are often considered central to many conservation approaches, including protected areas designation and management. This paradigm has been underpinned by a static and linear view of nature, the concept of wilderness, and the equation of human presence with ecosystem degradation [27]. Until relatively recently, the underlying assumptions and philosophies guiding the designation of protected areas were not adequately acknowledged or explored [9,67]. Despite the development of new concepts and more inclusionary paradigms, such as IPCAs, which link Indigenous peoples with conservation lands, the dominant exclusionary paradigm has so far proven difficult to supplant, especially when moving from theory to practice [27]. Thus, a profound shift in how the Canadian state, its conservation organizations and its public think about conservation is required to support and align with the new paradigm for protected areas. As is articulated by Steven Nitah,

> I will say that [I]ndigenous contributions have largely gone unrecognized in Canada, in a system that still recognizes only federally, provincially, and territorially legislated protected areas as valid and ignores the fact that for tens of thousands of years our peoples managed the land so well that you thought it was empty. We need to move past those misconceptions and embrace the fact that long before Canadians even knew what a national park was, our peoples were successfully protecting and managing our special places under our own laws and using our own knowledge. [49] (p. 3)

5.2. Siloed Colonial Governance

The spaces for interpretation and misinterpretation of existing types of Indigenous-led management discussed in the last section highlight one of the "wicked problems" that can emerge when the dominant Western colonial government aims to categorize Indigenous governance and stewardship. The traditional resource management style of the Canadian government has been to separate natural resources into different management categories, and disconnect them from human wellbeing and cultural continuity [29]. This structure was imposed to "make sense" of a geographically large nation with diverse social, political, economic, ecological, and geological regions [68]. The separation of marine and terrestrial environments, for example, has resulted in different jurisdictions and conservation targets, with significant implications for protected area establishment, reporting and biodiversity outcomes [15], including those important to Indigenous peoples and their interrelated traditional-territorial lands, waters and lifeways. While structurally dividing the systems responsible for nature and for people's wellbeing reflected the colonial law and approach to governance [69], it has significant incompatibilities with Indigenous governance systems, which tend to be more holistic [70]. The example of Wehexlaxodiale presents a perfect representation of this issue. The intention of the region is to protect cultural values, but in protecting cultural values the Tłı̨chǫ are also protecting nature as these are inseparable concepts.

5.3. Variation in Crown–Indigenous Treaties and Land-Claim Agreements

Over a long history of settler–Indigenous relations, a diversity of treaties and land agreements have been entered into by a variety of Indigenous and colonial governments. While much of Canada is covered by historic treaties, most of British Columbia, for example, does not have signed treaties [71]. Several court cases such as *Delgamuukw v. British Columbia 1997 3 S.C.R. 1010* have resulted in decisions that have enabled a modern treaty process that facilitates better Indigenous involvement in decision making [71]. In eastern Canada, many are Treaties of Peace and Friendship in which traditional lands and territories have never been ceded. The ICE in *We Rise Together* notes the diverse nature of these legal agreements and the ability of those communities who did not cede title under treaty to use the court systems to obtain title [9]. Notably, most examples of IPCAs in Canada can be found in BC or the North. Jones et al. [71] argued that in some contexts the reasoning behind this may be due to a lack of certainty around Indigenous land ownership, which has motivated collaboration. Lloyd-Smith [72] stated that the modern land claim agreements in the Arctic present more opportunity for Indigenous peoples to lead conservation and land-use planning initiatives. In contrast, those regions under historic treaty may have limited federal and provincial incentive to enter into truly collaborative models [71]. Even in situations where there may be an appearance of "true" collaboration there may remain a lack of willingness within governments to interfere with Ministerial authority. New models to address this situation are needed, such as the *Northwest Territories Wildlife Act*, which was drafted collaboratively and goes significantly further than other pieces of legislation with respect to collaborative management [73]. For example, Part 2, Section 8 of the *Northwest Territories Wildlife Act* lays out the aim "to promote cooperative and collaborative working relationships for effective wildlife management at the local, regional and territorial levels" [73] (p. 22).

5.4. The Non-Devolution of Power by the Government of Canada

In order for IPCAs to be Indigenous-led, there will need to be significant divestments of power made by the Government of Canada. However, power remains highly centralized and the Government of Canada may be reluctant to devolve power to Indigenous governments. Nicol [74] pointed to the dichotomy between the image of a Canadian government that seeks to protect Indigenous rights while at the same time asserting its sovereignty and thereby appropriating Indigenous sovereignty. While power devolution has occurred in the territories to a degree, the rigid bounds of constitutional power in the provinces makes the territorial model of devolution inapplicable [75]. In the recent past, the Canadian government has understood aspects of UNDRIP such as free, prior, and informed consent (FPIC) as a way of undermining state power over decision making [76]. This stance is problematic and demonstrates the unwillingness of the federal government to devolve or share power. Furthermore, if power is fully vested in the current government, then existing governance systems may be vulnerable to changes in government (i.e., through elections or other political means), which may lead to changes in land-based policies [77]. Nicol [76] demonstrated that the Canadian government only values Indigenous rights within the bounds of treaties and land claims, therefore Indigenous rights are constrained to Indigenous lands in the eyes of the state.

5.5. The Diverse and Federalist Nature of Canadian Law and Law Making

The constitutional division of powers in Canada between the federal government and provincial and territorial governments adds complexity to the processes by which protected areas become formalized. Significantly, Indigenous law and Indigenous constitutional orders are not formally recognized under Canadian law. The Canadian constitution was created in a colonial context and in a time when economic gain was the primary objective of resource management, even within parks [78,79]. The division of power between the federal, provincial and territorial governments, as outlined in the *Constitution Act 1867*, further complicates Indigenous–Crown relations [76]. While the creation of Section 35 of the *Constitution Act 1982* and subsequent court cases have begun to transfer some power held by federal, provincial and territorial governments back to Indigenous peoples, these colonial legal structures remain and define environmental decision-making in Canada. For example, Canadian law is structured to recognize written law as superior to oral law, which creates a disadvantage for Indigenous peoples [80]. This relates to issues surrounding sovereignty and the potential for IPCAs to be Indigenous-led, as discussed earlier. This issue also connects to the wicked problem outlined above (Section 5.2), such that a siloed approach to law is also problematic. Curran [69] argued that colonial law operates in silos, with ecosystem and health regarded as separate domains, for example, and, consequently, responsibility is divided by jurisdiction and department. In contrast, responsibilities for the land, health and wellbeing are inseparable in Indigenous laws and lifeways.

5.6. Reporting

Monitoring and reporting on implementation and management is a fundamental component of conservation. Conservation reporting under Canada Target 1 thus far has been complex and inconsistent. Each provincial, territorial, or federal jurisdiction self-reports to the Canadian Council on Ecological Areas (CCEA) who tracks the data in their CARTS database [14]. While the CCEA has guidelines for reporting, there is no mandatory requirement for jurisdictions to follow this guidance, and thus, each jurisdiction reports based on its own process and understanding [15]. For example, the CCEA CARTS report from December 2017 shows that Saskatchewan had reported privately protected area while Ontario had not. Furthermore, the Northwest Territories and the Yukon are the only jurisdictions that have reported "Indigenous Government Administration" and these same territories along with Manitoba are the only jurisdictions to report and distinguish government-Indigenous partnership areas [63]. These jurisdictional data are not currently audited for consistency across jurisdictions and are accepted as reported [15].

Ideally, where Indigenous leadership has indicated an interest in reporting IPCAs under Target 1, the community leaders would carry out the reporting processes. However, in Canada, as a result of imbalanced power relationships with other government bodies, many Indigenous governments face capacity issues such as a lack of funding [81]. While reporting is an important part of ensuring conservation objectives are met, it is also critical that the process work to further self-determination. There are currently many structural barriers that stand in the way of Indigenous reporting, including challenges in standardization of reporting, poor recognition of Indigenous governments, and capacity issues. Consistent with international guidance, MacKinnon et al. [14] recommended standardized reporting to avoid inconsistencies across jurisdictions in Canada. This may present a challenge for IPCAs, however, if an Indigenous group must report and demonstrate a certain level of biodiversity protection through an imposed language and understanding of conservation and external methods of practice [13]. Thus far, reporting on protected areas in Canada has only been done through provinces, territories, and federal governments. Ideally, Indigenous governments would be able to control if and how they report their own IPCAs.

5.7. Tension and Uncertainty between Scientific and Traditional Knowledge in Protected Area Management

The application of Indigenous knowledge (IK) to protected area management is becoming increasingly prominent in Canada and represents the beginning of a broader acceptance and acknowledgement of various ways of knowing/knowledge systems and worldviews. For example, the Northwest Territories Government has developed a "Traditional Knowledge Policy" to formally recognize the value of these knowledge systems [82]. The ICE [9] represents IPCAs as a manifestation of Indigenous knowledge systems and states that their promotion in Canada should enhance respect for these knowledge systems. Indigenous knowledge has been used to enhance and complement western science with mixed results [71,83], and pre-conceived notions of what IK is and how it should be incorporated by non-Indigenous partners can present a significant barrier to its application and use [84]. Aichi Target 18 indicates the importance of IK in biodiversity conservation; however, as discussed around reporting, it can also be challenging to apply IK in regions that require external methods of reporting. Houde [85] outlined several challenges such as a lack of confidence in IK by non-Indigenous people, issues around how IK is shared and how the control and ownership of IK might be determined by Indigenous communities, and the complexity of incorporating opposing values into a single framework. Communities face a difficult balance between sharing their knowledge towards asserting self-determination and having their knowledge appropriated by colonial application and resource management initiatives [86]. Elder Albert Marshall articulated the need to redress this dichotomy:

> We need the English language and mainstream knowledge, but at the same time, we want the prerogative of determining for ourselves what we want from the mainstream. We already have the best of who we are from the Ancestors. Let's decide how to put these together in a two-eyed seeing approach. [87] (p. 41)

6. Refocusing Conservation for Reconciliation

While frameworks, targets, and broader governmental initiatives may provide the impetus for establishing IPCAs in Canada, it is important to consider that, primarily, these initiatives will require concerted efforts at fundamentally changing the nature of existing relationships. Primarily, restoring nation-to-nation relationships between Indigenous peoples, settlers, and Canadian states is a precondition to just, respectful, and equitable relations. Perhaps less tangibly, but equally important, is creating the space for restoration of the relationship between land and peoples. Whether these are achieved through the establishment of IPCAs or some other means is secondary to the broader objective of using conservation as a tool for reconciliation.

6.1. Restoring Nation-to-Nation Relationships

Restoring or (re)building a nation-to-nation relationship is fundamental to reconciliation, and the original treaties may serve as guides on this journey [9,38]. Fundamentally, treaties serve as a moral basis of alliance between Indigenous peoples and settlers, are relationship-based, are intended to be living documents, and secondarily serve as legal documents defining and detailing that relationship [39]. However, it is important to note that treaties were often recorded and interpreted differently by Indigenous nations and subjects of the Crown. The TRC's Call to Action 45 recommends that "the government of Canada jointly develop with Aboriginal peoples a Royal Proclamation of Reconciliation to be issued by the Crown. The proclamation would be built on the Royal Proclamation of 1763 and the Treaty of Niagara 1764, and reaffirm the nation-to-nation relationship between Aboriginal peoples and the Crown" [88] (p. 4). This would be enabled by revisiting the original spirit and intent of the treaties and through enacting a repudiation of concepts (such as the Doctrine of Discovery and terra nullius) used to justify European sovereignty over Indigenous peoples. Other equally important commitments on the part of the Crown are the full implementation of UNDRIP, and commitments to establishing "treaty relationships based on mutual recognition, mutual respect, and shared responsibility for maintaining those relationships in the future" [88] (p. 5). The importance of genuine partnerships underpinned by strong relationships cannot be underestimated. Another important step in rebuilding a nation-to-nation relationship will be to "reconcile Aboriginal and Crown constitutional and legal orders to ensure that Aboriginal peoples are full partners in Confederation, including the recognition and integrations of Indigenous laws and legal traditions in negotiation and implementation processes involving Treaties, lands claims, and other constructive agreements" [88] (p. 5). It can be reasonably asserted that reconciliation efforts that are connected to state-driven conservation cannot be fully actualized until nation-to-nation relationships grounded in the treaties and important documents like UNDRIP are fully implemented and affirmed.

6.2. Restoring Relationships with the Land

Just as it is important to restore a nation-to-nation relationship, institutionally as well as socially, it is equally important to restore, or create the conditions that allow for re-connection between people and the land; for many Indigenous peoples, caring for land is an important part of respecting their cultural responsibilities [89]. As previously discussed, through the creation of protected areas and many other provincial/territorial and federal policies and developments, Indigenous peoples were systematically and forcibly displaced from their territories, consequently loosing access to these sites and their resources, as well as suffering other intergenerational impacts (social, cultural, economic, and spiritual) [9,27]. *We Rise Together* states that the disconnection between Indigenous peoples and their territories "prevented the full functionality of Indigenous legal orders ... weakened the necessary linkages to inter-generational knowledge transmission and sustainable use ... " and may have led to the loss of other important cultural and spiritual practices (such as songs, ceremonies, dances, stories, etc.) that were intrinsically tied to the land [9] (p. 28). Indigenous-led conservation efforts and IPCAs enable Indigenous peoples to determine the future of their lands and peoples [89]. A large part of reconciliation through conservation, as well as structural and social reconciliation, is making space for diverse Indigenous knowledge systems, worldviews, laws, etc. This is highlighted in UNDRIP, through which the UN General Assembly recognizes the "urgent need to respect and promote the inherent rights of Indigenous peoples which derive from their political, economic and social structures and from their cultures, spiritual traditions, histories and philosophies, especially their rights to their lands, territories, and resources" [90] (p. 2), and that "respect for Indigenous knowledge, cultures, and traditional practice contributes to the sustainable and equitable development and proper management of the environment" [90] (p. 2). Such considerations will be essential for promoting self-determination, which is integral to reconciliation in the development of meaningful pathways to IPCAs.

7. Conclusions

Recent global and national trends towards recognition of Indigenous peoples' leadership in protected areas presents hope for the future of conservation and reconciliation. The spirit and intent of IPCAs reflect the new paradigm and has the potential for positive cultural and social outcomes in addition to the conservation of biodiversity [10]. Countries responding to the new paradigm for biodiversity conservation are now grappling with their colonial histories and modern-day approaches to conservation and are exploring the implementation of protected areas frameworks that are supportive of Indigenous leadership. Each country has their own set of "wicked problems" that they must navigate. Australia has become a model for the shift towards Indigenous leadership in protected areas, boasting health, education, employment and social cohesion outcomes through their Indigenous Protected Area (IPA) program, which is part of the Australian National Reserve System [91]. Australia currently has 75 IPAs over both land and sea jurisdictions, which contribute more than 45% of the National Reserve System's total area [91]. Despite the impressive enhanced representation of protection of biodiversity through the National Reserve System, IPAs continue to be vexed by numerous issues, such as those relating to who has ultimate decision-making authority [92].

IPCAs have the potential to play a major role in meeting many signatory nations' commitments to biodiversity conservation targets under the CBD, along with their associated national targets. Canada's commitments to Aichi Target 11 as reflected in Canada's Target 1 is no exception. Indeed, Canada is in a good position to follow Australia's lead in modeling shifts in IPCA practice both at home and in the global arena. It is important, however, to be mindful of the political motivations behind state support for IPCAs. If the primary objective of government support for IPCAs is to achieve national obligations to meet international conservation commitments and the 2020 target deadline, then this could undermine the meaningfulness of IPCAs and the IPCA process for Indigenous communities [9,13]. In Canada, the *We Rise Together* report states that the federal government must be willing to support IPCAs regardless of their relationship to conservation targets, as they have a purpose beyond meeting quantitative national objectives [9]. With less than two years left to meet the Aichi Targets, pushing for area-based targets could lead to rushed collaboration and risk inappropriate designations that do not support the objectives of reconciliation or conservation.

Given the "wicked problems" associated with forwarding IPCAs within Canada's Pathway to Target 1, a process that allows for reconciliation through adaptability, capacity building and the strengthening of relationships will be necessary. For example, Jonas et al. [13] argued that, similar to recognition, reporting requires the free prior and informed consent of the Indigenous peoples involved. The ICE [9] identifies some key features necessary for Indigenous peoples to support the shift in paradigm they envision through IPCAs. These include: dedicating sufficient time and resources to explore Indigenous-led conservation and engagement with Indigenous governments; supporting innovative funding models; identifying new partnerships, as well as allies and champions; and creating resources that would support Indigenous governments in their work on IPCAs. It is important to move forward with caution, however, to avoid a situation in which IPCAs become another, hopefully unintended, colonization of Indigenous peoples, their lands and biodiversity to meet Canada's Target 1 for protected areas. This potentiality calls into question the appropriateness of Canada's recognition and reporting of IPCAs if entered into quickly and primarily to aid Canada in meeting its quantitative commitments under the CBD.

Ideally, Indigenous communities would have the ability to define IPCAs according to their particular contexts, thus contributing Indigenous leadership in the process of protected area and OECM recognition. To achieve this, the government could support the "three essential elements" of IPCAs put forward by the ICE (i.e., the process is Indigenous-led; there is a long-term commitment to conservation; and they are in support of Indigenous rights and responsibilities). In so doing, the government would also uphold Indigenous rights and support the self-determination of each community to define their IPCAs. This would also be consistent with IUCN governance guidelines, which note that nations need to be careful not to force top-down management of IPCAs, resulting in the displacement of existing governance structures [47]. As articulated by the ICE [9], close attention will need to be paid to jurisdiction, financial solutions and capacity development, as well as addressing disconnections between—and the dominance of Canadian over—Indigenous worldviews, ontologies, and epistemologies. These factors among others must be considered and overcome in efforts aimed at developing IPCAs as living examples of reconciliation, as well as potential contributions to biodiversity conservation through Pathway to Canada Target 1 and Aichi Target 11.

Considerations around Aichi Targets other than Target 11 are also important. IPCAs inherently offer a commitment to conservation, however they might be better suited to contribute to targets other than Target 11. For example, Aichi Target 18 has a focus on the full and appropriate recognition of Indigenous knowledge systems in conservation, and Target 14 aims to protect ecosystem services and the needs of Indigenous peoples [6]. Accordingly, IPCAs that embed Indigenous knowledge systems and serve to protect ecosystem values of importance to Indigenous communities could contribute to Targets 14 and 18, regardless of whether they fit within the parameters of Aichi Target 11. In these and other cases, such as areas established for cultural regeneration, IPCAs may meet the definition of an OECM and be counted toward Aichi Target 11 or Canada Target 1. Regardless, an Indigenous government's decision about whether they want their IPCA to be counted must be respected by the provincial, territorial, or federal government so as to honor reconciliation processes and FPIC under UNDRIP. Alternatively, an IPCA may have a purpose that does not meet the requirements of a protected area or an OECM under Target 1. In this case, it should still be recognized as a governance model to be supported and protected, as it may serve an important role in reconciliation and conservation processes. It will be critical, however, that colonial governments maintain support for IPCAs, other forms of Indigenous-led conservation and reconciliation without imposing control or regulation.

Even though Canada and other nation states are on a tight timeline to achieve national targets and international biodiversity conservation commitments by 2020, there should also be an understanding that land- and sea-based reconciliation will require due process so that past injustices to Indigenous peoples related to protected areas can be addressed [9]. Inadequately addressing past injustices could create further wrongdoings and reveal Canada's and other countries' interests in establishing IPCAs as opportunistic, rather than being based in genuine efforts to move forward on reconciliation with regards to protected areas. Taking the time to listen to communities and address structural injustices (associated with the wicked problems above) will be important if nation states, including the Government of Canada, are to truly move forward in good faith in a way that acknowledges the intrinsic and holistic value of IPCAs beyond the race to meet targets. As discussed by the ICE [9],

IPCAs can be places of cultural regeneration, learning, restoration, and reconciliation, while actively contributing to the protection of biodiversity.

Author Contributions: Conceptualization, all authors; methodology, all authors; validation, all authors; formal analysis, all authors; investigation, all authors; resources, E.E. and J.B.-D.; data curation, all authors; writing—original draft preparation, all authors; writing—review and editing, all authors; and supervision, M.Z. and K.F.B.

Funding: This research received no external funding.

Acknowledgments: We would like to acknowledge Claudia Haas and Lillith Brook of the Government of Northwest Territories, and the reviewers for their helpful feedback.

Conflicts of Interest: The authors declare no conflict of interest.

References

1. IUCN. About. 2018. Available online: https://www.iucn.org/about (accessed on 29 November 2018).
2. Stevens, S. A new protected area paradigm. In *Indigenous Peoples, National Parks, and Protected Areas*, 1st ed.; Stevens, S., Ed.; University of Arizona Press: Tuscon, AZ, USA, 2014; pp. 47–83.
3. The Durban Accord: Our Global Commitment for People and Earth's Protected Areas. 2005. Available online: https://cmsdata.iucn.org/downloads/durbanaccorden.pdf (accessed on 18 December 2018).
4. Garnett, S.T.; Burgess, N.D.; Fa, J.E.; Fernaández-Llamazares, Á.; Molnaár, Z.; Robinson, C.; Watson, J.E.M.; Zander, K.K.; Austin, B.; Brondizio, E.S.; et al. A spatial overview of the global importance of Indigenous lands for conservation. *Nat. Sustain.* **2018**, *1*, 369–374. [CrossRef]
5. Ceballos, G.; Ehrlick, P.; Barnosky, A.; Garcia, A.; Pringle, R.; Palmer, T. Accelerated modern human-induced species losses: Entering the sixth mass extinction. *Sci. Adv.* **2015**, *1*, e1400253. [CrossRef] [PubMed]
6. United Nations Environment Programme (UNEP) Conference of the Parties (COP) to the UN Convention on Biological Diversity (CBD). Decision Adopted by the Conference of the Parties to the Convention on Biological Diversity at its Tenth Meeting. 2010. Available online: https://www.cbd.int/doc/decisions/cop-10/cop-10-dec-02-en.doc (accessed on 29 November 2018).
7. Environment and Climate Change Canada. 2020 Biodiversity Goals & Targets for Canada. 2016. Available online: http://www.biodivcanada.ca/ (accessed on 29 November 2018).
8. Pathway to Canada Target 1. Introduction. Available online: http://www.conservation2020canada.ca/the-pathway/ (accessed on 29 November 2018).
9. ICE. We Rise Together. 2018. Available online: https://static1.squarespace.com/static/57e007452e69cf9a7af0a033/t/5ab94aca6d2a7338ecb1d05e/1522092766605/PA234-ICE_Report_2018_Mar_22_web.pdf (accessed on 29 November 2018).
10. Dudley, N.; Jonas, H.; Nelson, F.; Parrish, J.; Pyhälä, A.; Stolton, S.; Watson, J.E.M. The essential role of other effective area-based conservation measures in achieving big bold conservation targets. *Glob. Ecol. Conserv.* **2018**, *15*, e00424. [CrossRef]
11. Stevens, S. Implementing the UN Declaration on the Rights of Indigenous Peoples and International Human Rights Law Through Recognition of ICCAs. *Policy Matters* **2010**, *17*, 181–194.

12. IUCN WCPA. (Draft) Guidelines for Recognising and Reporting Other Effective Area-Based Conservation, IUCN, Switzerland. Version 1. 10 January 2018. Available online: https://www.iucn.org/sites/dev/files/content/documents/guidelines_for_recognising_and_reporting_oecms_-_january_2018.pdf (accessed on 16 December 2018).

13. Jonas, H.D.; Barbuto, V.; Jonas, H.C.; Kothari, A.; Nelson, F. New Steps of Change: Looking Beyond Protected Areas to Consider Other Effective Area-Based Conservation Measures. *Parks* **2014**, *20*, 111–128. [CrossRef]

14. MacKinnon, D.; Lemieux, C.J.; Beazley, K.; Woodley, S.; Helie, R.; Perron, J.; Elliott, J.; Haas, C.; Langlois, J.; Lazaruk, H.; et al. Canada and Aichi Biodiversity Target 11: Understanding 'other effective area-based conservation measures' in the context of the broader target. *Biodivers. Conserv.* **2015**, *24*, 3559–3581. [CrossRef]

15. Lemieux, C.J.; Gray, P.A.; Devillers, R.; Wright, P.A.; Dearden, P.; Halpenny, E.A.; Groulx, G.; Beechey, T.J.; Beazley, K. How the race to achieve Aichi Target 11 could jeopardize the effective conservation of biodiversity in Canada and beyond. *Mar. Policy* **2019**, *99*, 312–323. [CrossRef]

16. Watson, J.E.M.; Venter, O.; Lee, J.; Jones, K.R.; Robinson, J.G.; Possingham, H.P.; Allan, J.R. Protect the last of the wild. *Nature* **2018**, *563*, 27–30. [CrossRef]

17. Barnes, M.D.; Glew, L.; Wyborn, C.; Craigie, I.D. Prevent perverse outcomes from global protected area policy. *Nat. Ecol. Evolut.* **2018**, *2*, 759–762. [CrossRef]

18. ICCA Consortium. Discover: Territories and Areas Conserved by Indigenous Peoples and Local Communities. Available online: https://www.iccaconsortium.org/index.php/discover/ (accessed on 29 November 2018).

19. Stolton, S.; Shadie, P.; Dudley, N. *IUCN WCPA Best Practice Guidance on Recognising Protected Areas and Assigning Management Categories and Governance Types*; Best Practice Protected Area Guidelines Series No. 21; IUCN: Gland, Switzerland, 2013.

20. Tittensor, D.P.; Walpole, M.; Hill, S.L.L.; Boyce, D.G.; Britten, G.L.; Burgess, N.D.; Butchart, S.H.M.; Leadley, P.W.; Regan, E.C.; Alkemade, R.; et al. A mid-term analysis of progress toward international biodiversity targets. *Science* **2014**, *346*, 241–245. [CrossRef]

21. Sinclair, N. Diversity a gift from Indigenous peoples to Canada. *Winnipeg Free Press*, 19 November 2018. Available online: https://www.winnipegfreepress.com/local/diversity-a-gift-from-indigenous-peoples-to-canada-500870831.html (accessed on 16 December 2018).

22. Truth and Reconciliation Commission. *Honouring the Truth, Reconciling for the Future: Summary Report of the Truth and Reconciliation Commission of Canada*; Truth and Reconciliation Commission of Canada: Winnipeg, MB, Canada, 2015.

23. Youdelis, M. "They could take you out for coffee and call it consultation!": The colonial antipolitics of Indigenous consultation in Jasper National Park. *Environ. Plan.* **2016**, *48*, 1374–1392. [CrossRef]

24. Sandlos, J. Not wanted in the boundary: The expulsion of the Keeseekoowenin Ojibway Band from Riding Mountain National Park. *Can. Hist. Rev.* **2008**, *89*, 189–221. [CrossRef]

25. Binnema, T.; Niemi, M. "Let the line be drawn now": Wilderness, conservation, and the exclusion of Aboriginal People from Banff National Park in Canada. *Environ. Hist.* **2006**, *11*, 724–750. [CrossRef]

26. Moola, F.; Roth, R. Moving beyond colonial conservation models: Indigenous Protected and Conserved Areas offer hope for biodiversity and advancing reconciliation in the Canadian boreal forest. *Environ. Rev.* **2018**, *7*. [CrossRef]

27. Shultis, J.; Heffner, S. Hegemonic and emerging concepts of conservation: A critical examination of barriers to incorporating Indigenous perspectives in protected area conservation policies and practice. *J. Sustain. Tour.* **2016**, *24*, 1227–1242. [CrossRef]

28. Kelly, A.B. Conservation practice as primitive accumulation. *J. Peasant Stud.* **2011**, *38*, 683–701. [CrossRef]

29. David Suzuki Foundation. Tribal Parks and Indigenous Protected and Conserved Areas: Lessons learned from BC Examples. 2018. Available online: https://davidsuzuki.org/wp-content/uploads/2018/08/tribal-parks-indigenous-protected-conserved-areas-lessons-b-c-examples.pdf (accessed on 16 December 2018).

30. Troian, M. Questions about Indigenous land rights in Alberta surround provincial plan to support caribou herds. *APTN*, 9 March 2018. Available online: https://aptnnews.ca/2018/03/09/questions-indigenous-land-rights-alberta-surround-provincial-plan-support-caribou-herds/ (accessed on 16 December 2018).

31. Oceans North Canada. The Arctic Heritage and Beauty of Tallurutiup Imanga (Lancaster Sound). *The Pew Charitable Trusts*, 6 November 2015. Available online: https://youtu.be/Z5AnTPFtMYc (accessed on 17 December 2018).

32. Ganey, S.; Inuit Stand to Benefit from New Canadian Marine Park. The Pew Charitable Trusts. 5 December 2018. Available online: https://www.pewtrusts.org/en/research-and-analysis/articles/2018/12/05/inuit-stand-to-benefit-from-new-canadian-marine-park?utm_campaign=environment&utm_source=twitter&utm_medium=social&utm_content=lancastersound_goal (accessed on 17 December 2018).

33. Gardner, J.; Consulting, D. "Let Us Teach You"—Exploring Empowerment for Indigenous Protected and Conserved Areas in B.C. Report of the 27 September 2018 IPCA Workshop. Available online: https://davidsuzuki.org/wp-content/uploads/2018/11/let-us-teach-you-exploring-empowerment-for-indigenous-protected-and-conserved-areas-in-b-c.pdf (accessed on 16 December 2018).

34. The Constitution Act, 1867, 30 & 31 Vict, c 3. Available online: http://canlii.ca/t/ldsw (accessed on 16 December 2018).

35. The Constitution Act, 1982, Schedule B to the Canada Act 1982 (UK), 1982, c 11. Available online: http://canlii.ca/t/ldsx (accessed on 16 December 2018).

36. Ruru, J.A. Settling Indigenous place: Reconciling legal fictions governing Canada and Aotearoa New Zealand's National Parks. Doctoral Thesis, University of Victoria, Victoria, BC, Canada, 2012.

37. Coulthard, G. *Red Skin, White Masks*; University of Minnesota Press: Minneapolis, MN, USA, 2014; pp. 105–109. ISBN 978-0-8166-7965-5.

38. Borrows, J.; Coyle, M. *The Right Relationships*; University of Toronto Press: Toronto, ON, Canada, 2017; p. 332.

39. Poezler, G.; Coates, K.S. *From Treaty Peoples to Treaty Nation*; UBC Press: Vancouver, BC, Canada, 2015; pp. 272–273.

40. Stevens, S. Introduction. In *Indigenous Peoples, National Parks, and Protected Areas*, 1st ed.; Stevens, S., Ed.; University of Arizona Press: Tuscon, AZ, USA, 2014; pp. 3–14.

41. Indigenous Foundations. The Residential School System. 2009. Available online: https://indigenousfoundations.arts.ubc.ca/the_residential_school_system/ (accessed on 16 December 2018).

42. Truth and Reconciliation Commission. *What We Have Learned: Principles of Truth and Reconciliation*; Truth and Reconciliation Commission of Canada: Winnipeg, MB, Canada, 2015.

43. Government of Canada. Treaties and Agreements. Available online: https://www.rcaanc-cirnac.gc.ca/eng/1100100028574/1529354437231#chp3 (accessed on 16 December 2018).

44. IUCN. Green List. 2018. Available online: http://www.iucn.org/theme/protected-areas/our-work/green-list (accessed on 29 November 2018).

45. IUCN. IUCN Green List of Protected and Conserved Areas: Standard, Version 1.1. 2017. Available online: https://www.iucn.org/sites/dev/files/iucn_green_list_standard_version_1.1_nov_2017_3.pdf (accessed on 29 November 2018).

46. ICCA Consortium. Mission. Available online: https://www.iccaconsortium.org/index.php/movement/mission/ (accessed on 29 November 2018).

47. Borrini-Feyerabend, G.; Dudley, N.; Jaeger, T.; Lassen, B.; Pathak Broome, N.; Phillips, A.; Sandwith, T. *Governance of Protected Areas: From understanding to action. Best Practice Protected Area Guidelines Series No. 20*; IUCN: Gland, Switzerland, 2013; Available online: http://cmsdata.iucn.org/downloads/governance_of_protected_areas___from_understanding_to_action.pdf (accessed on 29 November 2018).

48. Gibson, G.; Hoogeveen, D.; MacDonald, A.; The Firelight Group. *Impact Assessment in the Arctic: Emerging Practices of Indigenous-Led Review*; Prepared for the Gwich'in Council International. 2018. Available online: https://gwichincouncil.com/sites/default/files/Firelight%20Gwich%27in%20Indigenous%20led%20review_FINAL_web_0.pdf (accessed on 29 November 2018).

49. Canada. House of Commons. Standing Committee on Environment and Sustainable Development. Evidence. Meeting 25, September 27. 42nd Parliament, 1st Session. 2016. Available online: https://www.ourcommons.ca/Content/Committee/421/ENVI/Evidence/EV8446035/ENVIEV25-E.PDF (accessed on 16 December 2018).

50. GNWT. Dìnàgà Wek'èhodì. Available online: https://www.enr.gov.nt.ca/en/services/conservation-network-planning/dinaga-wekehodi (accessed on 29 November 2018).

51. Tłıcho Government. Dìnàgà Wek'èhodì. Available online: https://tlicho.ca/government/departments/culture-lands-protection/lands-protection/dinaga-wekehodi (accessed on 29 November 2018).

52. Government of Canada. First New Indigenous Protected Area in Canada: Edéhzhíe Protected Area. Available online: https://www.canada.ca/en/environment-climate-change/news/2018/10/first-new-indigenous-protected-area-in-canada-edehzhie-protected-area.html (accessed on 29 November 2018).

53. Lavoie, J. Canada's New Indigenous Protected Area heralds New Era of Conservation. *The Narwhal*, 12 October 2018. Available online: https://thenarwhal.ca/canadas-new-indigenous-protected-area-heralds-new-era-of-conservation/(accessed on 29 November 2018).

54. Campbell, J. Dene celebrate protection of Edéhzhíe. *CKLB Radio*, 12 October 2018. Available online: https://cklbradio.com/dene-celebrate-protection-of-edehzhie/(accessed on 18 December 2018).

55. Galloway, G. Vast region of Northwest Territories declared an Indigenous Protected Area. *The Globe and Mail*, 11 October 2018. Available online: https://www.theglobeandmail.com/politics/article-vast-region-of-northwest-territories-declared-an-indigenous-protected/(accessed on 29 November 2018).

56. Dehcho First Nations. Dehcho First Nations Annual Assembly, Resolution # 01—Dehcho Protected Area Law 24–26 July 2018, Wrigley, NT. Available online: https://dehcho.org/docs/REGARDINGEdehzhieDehchoAnnualAssembly2018.pdf (accessed on 29 November 2018).

57. Murray, G.; King, L. First Nations Values in Protected Area Governance: Tla-o-qui-aht Tribal Parks and Pacific Rim National Park Reserve. *Hum. Ecol.* **2012**, *40*, 385–395. [CrossRef]

58. Parks Canada. Fundy National Park of Canada: Management Plan. Available online: http://parkscanadahistory.com/publications/fundy/mgt-plan-e-2011.pdf (accessed on 29 November 2018).

59. Parks Canada. Draft 2018: Torngat Mountains National Park of Canada. Available online: https://www.pc.gc.ca/en/pn-np/nl/torngats/info/index (accessed on 29 November 2018).

60. BC Parks. Summary of the Parks and Protected Areas System. Available online: http://www.env.gov.bc.ca/bcparks/about/park-designations.html#Conservancy (accessed on 29 November 2018).

61. Tla-o-qui-aht Tribal Parks and The Wilderness Committee. *Welcome to Tla-o-qui-aht: Tribal Parks!* Wilderness Committee: Vancouver, BC, Canada, 2017; Volume 32, Available online: https://www.wildernesscommittee.org/sites/all/files/publications/2013_tla-o-qui-aht_Paper-Web-2.pdf?_ga=2.158268986.1454244689.1542313948-902515130.1542313948 (accessed on 29 November 2018).

62. Tłı̨chǫ Government. Tłı̨chǫ Wenek'e—Tłı̨chǫ Land Use Plan. 2013. Available online: https://www.tlicho.ca/sites/default/files/105-LandUsePlan_FINAL%20VERSION%5B%5D.pdf (accessed on 16 December 2018).

63. CARTS. Report of Protected Area in Canada. 2017. Available online: http://ccea.org/CARTS/CARTS%202017/CARTS2017ReportEN.pdf (accessed on 29 November 2018).

64. Churchman, C. Guest Editorial: Wicked Problems. *Manag. Sci.* **1967**, *14*, B141–B142.

65. Rittel, H.W.J.; Webber, M.M. Dilemmas in a general theory of planning. *Policy Sci.* **1973**, *4*, 155–169. [CrossRef]

66. Head, B.W.; Alford, J. Wicked Problems: Implications for Public Policy and Management. *Adm. Soc.* **2015**, *47*, 711–739. [CrossRef]

67. Kalamandeen, M.; Gillson, L. Demything "wilderness": Implications for protected areas designation and management. *Biodivers. Conserv.* **2007**, *16*, 165–182. [CrossRef]

68. Bakker, K.; Cook, C. Water Governance in Canada: Innovation and Fragmentation. *Water Resour. Dev.* **2011**, *27*, 275–289. [CrossRef]

69. Curran, D. "Legalizing" the Great Bear Rainforest Agreements: Colonial Adaptations Toward Reconciliation and Conservation. *McGill Law J.* **2017**, *62*, 813–860. [CrossRef]

70. Black, K.; McBean, E. Increased Indigenous participation in environmental decision-making: A policy analysis for the improvement of Indigenous health. *Int. Indig. Policy J.* **2016**, *7*, 5. [CrossRef]

71. Jones, R.; Rigg, C.; Lee, L. Haida marine planning: First Nations as a partner in marine conservation. *Ecol. Soc.* **2010**, *15*, 12. [CrossRef]

72. Lloyd-Smith, G. Indigenous protected areas gaining momentum—But are they recognized by law? *Environ. Law Alert Blog*, 15 August 2017. Available online: https://www.wcel.org/blog/indigenous-protected-areas-gaining-momentum-are-they-recognized-law (accessed on 16 December 2018).

73. Government of the Northwest Territories. Wildlife Act, S.N.W.T. 2017, c.19. Available online: https://www.justice.gov.nt.ca/en/files/legislation/wildlife/wildlife.a.pdf (accessed on 14 November 2018).

74. Nicol, H.N. From Territory to Rights: New Foundations for Conceptualising Indigenous Sovereignty. *Geopolitics* **2017**, *22*, 794–814. [CrossRef]

75. Sabin, J. *A Federation within a Federation: Devolution and Indigenous Government in the Northwest Territories*; Institute for Research on Public Study 66; Institute for Research on Public Policy: Montreal, QC, Canada, 2017; ISBN 1920-9436.

76. Wilson, P.; McDermott, L.; Johnston, N.; Hamilton, M. An Analysis of International Law, National Legislation, Judgements, and Institutions as they Interrelate with Territories and Areas Conserved by Indigenous Peoples and Local Communities: No. 8 Canada. 2012. Available online: https://www.iccaconsortium.org/wp-content/uploads/2015/08/legal-review-8-canada-2012-en.pdf (accessed on 29 November 2018).
77. Zurba, M.; Diduck, A.P.; Sinclair, A.J. First Nations and industry collaboration for forest governance in northwestern Ontario, Canada. *For. Policy Econ.* **2016**, *69*, 1–10. [CrossRef]
78. Woo, G. *Ghost Dancing with Colonialism Decolonization and Indigenous Rights at the Supreme Court of Canada*; University of British Columbia Press: Vancouver, BC, Canada, 2011; pp. 15–25. ISBN 978-0-7748-1887-2.
79. McNamee, K. From Wild Places to Endangered Spaces: A History of Canada's National Parks. In *Parks and Protected Areas in Canada: Planning and Management*, 3rd ed.; Dearden, P., Rollins, R., Eds.; Oxford University Press: Don Mills, ON, Canada, 2009; pp. 24–55. ISBN 978-0-19-542734-9.
80. Miller, B. *Oral History on Trial Recognizing Aboriginal Narratives in the Courts*; UBC Press: Vancouver, BC, Canada, 2011; pp. 1–21. ISBN 978-0-7748-2070-7.
81. Singleton, S. Native people and planning for marine protected areas: How "stakeholder" processes fail to address conflicts in complex, real-world environments. *Coast. Manag.* **2009**, *37*, 421–440. [CrossRef]
82. GNWT. Traditional Knowledge. Available online: https://www.enr.gov.nt.ca/en/services/traditional-knowledge (accessed on 29 November 2018).
83. Mistry, J.; Berardi, A. Bridging indigenous and scientific knowledge. *Science* **2016**, *352*, 1274–1275. [CrossRef] [PubMed]
84. Reo, N.; Whyte, K.; McGregor, D.; Smith, M.; Jenkins, J. Factors that support Indigenous involvement in multi-actor environmental stewardship. *AlterNative* **2017**, *13*, 58–68. [CrossRef]
85. Houde, N. The six faces of traditional ecological knowledge: Challenges and opportunities for Canadian co-management arrangements. *Ecol. Soc.* **2007**, *12*, 34. [CrossRef]
86. Ens, E.J.; Pert, P.; Clarke, P.A.; Budden, M.; Clubb, L.; Doran, B.; Douras, C.; Gaikwad, J.; Gott, B.; Leonard, S.; et al. Indigenous biocultural knowledge in ecosystem science and management: Review and insight from Australia. *Biol. Conserv.* **2015**, *181*, 122–149. [CrossRef]
87. ICE. Pathway to Canada Target 1: Indigenous Circle of Experts (ICE) Regional Gatherings Report. 2018. Available online: https://static1.squarespace.com/static/57e007452e69cf9a7af0a033/t/5ab9504c0e2e7246a9551a5a/1522094157137/Regional+Gathering+Reports+EN.pdf (accessed on 18 December 2018).
88. Truth and Reconciliation Commission. *Calls to Action*; Truth and Reconciliation Commission of Canada: Winnipeg, MB, Canada, 2015. Available online: https://www.documentcloud.org/documents/2091412-trc-calls-to-action.html (accessed on 8 November 2018).
89. Courtois, V.; Nitah, S. Indigenous-led conservation offers a path to global leadership and reconciliation. *The Star*, 23 January 2018. Available online: https://www.thestar.com/opinion/contributors/2018/01/23/indigenous-led-conservation-offers-a-path-to-global-leadership-and-reconciliation.html (accessed on 16 December 2018).
90. United Nations. United Nations Declaration on the Rights of Indigenous Peoples. 2008. Available online: https://www.un.org/esa/socdev/unpfii/documents/DRIPS_en.pdf (accessed on 17 December 2018).
91. Australian Government. Indigenous Protected Areas. n.d. Available online: http://www.environment.gov.au/land/indigenous-protected-areas (accessed on 16 December 2018).
92. Dussart, F.; Poirier, S. (Eds.) *Entangled Territorialities: Negotiating Indigenous Lands in Australia and Canada*; University of Toronto Press: Toronto, ON, Canada, 2017; ISBN 9781487521592.

Review

Conservation through Biocultural Heritage—Examples from Sub-Saharan Africa

Anneli Ekblom [1,2,*], **Anna Shoemaker** [1], **Lindsey Gillson** [3], **Paul Lane** [1,4,5]
and Karl-Johan Lindholm [1]

[1]   Department of Archaeology and Ancient History, African and Comparative Archaeology,
     Uppsala University, Box 626, SE-751 26 Uppsala, Sweden; anna.shoemaker@arkeologi.uu.se (A.S.);
     paul.lane@arkeologi.uu.se (P.L.); karl-johan.lindholm@arkeologi.uu.se (K.-J.L.)
[2]   Natural Resources and Sustainable Development, Department of Earth Sciences, Uppsala University,
     Villavägen 16, 75236 Uppsala, Sweden
[3]   Plant Conservation Unit, Botany Department, University of Cape Town, Private Bag X3, Rondebosch 7701,
     South Africa; lindsey.gillson@uct.ac.za
[4]   Department of Archaeology, University of Cambridge, Downing Street, Cambridge CB2 3DZ, UK
[5]   School of Geography, Archaeology and Environmental Studies, University of the Witwatersrand,
     Johannesburg 2000, South Africa
*   Correspondence: anneli.ekblom@arkeologi.uu.se

Received: 30 November 2018; Accepted: 23 December 2018; Published: 2 January 2019

Abstract: In this paper, we review the potential of biocultural heritage in biodiversity protection
and agricultural innovation in sub-Saharan Africa. We begin by defining the concept of biocultural
heritage into four interlinked elements that are revealed through integrated landscape analysis.
This concerns the transdisciplinary methods whereby biocultural heritage must be explored, and here
we emphasise that reconstructing landscape histories and documenting local heritage values needs
to be an integral part of the process. Ecosystem memories relate to the structuring of landscape
heterogeneity through such activities as agroforestry and fire management. The positive linkages
between living practices, biodiversity and soil nutrients examined here are demonstrative of the
concept of ecosystem memories. Landscape memories refer to built or enhanced landscapes linked
to specific land-use systems and property rights. Place memories signify practices of protection or
use related to a specific place. Customary protection of burial sites and/or abandoned settlements,
for example, is a common occurrence across Africa with beneficial outcomes for biodiversity and forest
protection. Finally, we discuss stewardship and change. Building on local traditions, inclusivity
and equity are essential to promoting the continuation and innovation of practices crucial for local
sustainability and biodiversity protection, and also offer new avenues for collaboration in landscape
management and conservation.

Keywords: biocultural heritage; sub-Saharan Africa; traditional ecological knowledge; hotspots;
sacred forests; conservation

---

1. Introduction

Globally, a high proportion of biodiversity resides outside of protected areas. Incentives for
biodiversity protection, therefore, must be built and fostered amongst diverse stakeholders, in areas
where biodiversity and communities co-exist [1,2]. In keeping with this principle, biocultural heritage
is an emerging concept drawing on local knowledge, land-use practices and heritage values to define
sustainability and resilience from the perspective of local inhabitants [3–10]. The concept is particularly
relevant in African contexts, as many landscapes can be defined as *continuing* cultural landscapes
following the International Union for Conservation of Nature (IUCN)'s definition of its category V

landscapes as those "where the interaction of people and nature over time has produced an area of distinct character with significant ecological, biological, cultural and scenic value" [11]. In rural areas of Africa, the most common forms of agriculture entail low-intensity land-use practices, often based on various customary systems of access and ownership rights. Globally, there has been increasing realisation that the discontinuation of small scale, low intensity agricultural practices contributes to the recent reductions in biodiversity [12–16]. However, in many African settings environmental debates are still centred on the assumption that local practices of fire management, cultivation and/or grazing cause degradation (for summaries and critiques of such arguments, see, e.g., [17–24]). As will be exemplified here, local low intensity and customary practices may hold the key to strengthening, adapting and re-innovating forms of land-use that accommodate biodiversity and cultural heritage and promote adaptive management and resilience [25]. Building on and reinvigorating such local practices is important given that the effects of climate change are accelerating and climatic insecurity and its effects on food production and security are increasingly pertinent issues [26–28]. At the same time, ongoing competition for land from industrial agriculture, biofuel production, carbon off-setting projects and conservation initiatives make local communities increasingly vulnerable to both climate change and socio-economic transformations that are detrimental to particular livelihood traditions [29–33].

2. Background

As an emerging field, biocultural heritage has been explored from different disciplinary perspectives ranging from those focused on socio-cultural practices explored using ethnographic methods, to those rooted in understanding and modelling biological systems on a grand scale [5,6,9,34,35]. The origins of the concept can be traced back to the emerging interest in community-based resource management and traditional ecological knowledge in the 1980s [36–39], the adoption of the Convention on Biological Diversity (CBD), and the aftermath of the 2003 IUCN World Parks Conference in Durban, South Africa [4]. The United Nations Educational, Scientific and Cultural Organization (UNESCO) uses the term 'biological cultural heritage' to refer to ecosystems (including habitats and species) originating or developing from human practices [40]. More broadly, biocultural heritage is considered to encompass the natural–cultural components of human–environment interactions including knowledge, practices and innovation. Practices related to biocultural heritage are also closely linked to the construction and confirmation of identities and social cohesion [25,41–45]. Biocultural heritage has been key in developing both local advocacy groups and legal frameworks focused on the protection and ownership of landscapes and resources by and for local communities [4,46–48]. The concept has also been incorporated into conservation biology and broadened to include deep-time landscape history [7,8,49,50].

While there are now several different conceptualisations of biocultural heritage (see, for instance, [4,5]), we draw on the framework developed by Lindholm and Ekblom [50] in defining biocultural heritage as consisting of four interactive elements, each operating at interlinked temporal and spatial scales, that can only be understood through integrated landscape analysis. *Ecosystem memories* (Figure 1) can be defined as practices and outcomes operating on larger or deep-time scales, where agricultural, grazing and/or fire management activities have reshaped landscapes with long lasting effects on both biological and landscape structures. *Landscape memories* represent smaller scale materialised human practices and ways of organising landscapes and their outcomes. These include changes in soils, geological formations, flora and fauna but also archaeological sites, built environments and living land-use practices. Local heritage practices and narratives are often interlinked with such land-use activities and play a vital part in maintaining them. *Place memories* are also defined by local narratives, place names and signs of earlier or continuing practices whose significance is under constant debate and re-negotiation both locally and with external actors. These memory elements will be exemplified in more detail below. The fourth element, *stewardship and change*, concerns the conceptualisation and transfer of knowledge pertaining to landscape management, collaborative innovation and self-determination. The fusion of biodiversity goals with social and economic goals founded on self-determination is essential for establishing ecologically sound and equitable landscape management practices, as we explain below. To identify and explore how and why these four elements intersect requires *integrated landscape analysis*. These inclusive methodological and conceptual approaches to knowledge and landscape management are essential to both documenting and researching biocultural heritage and applying the insights generated to future stewardship and adaptive practices.

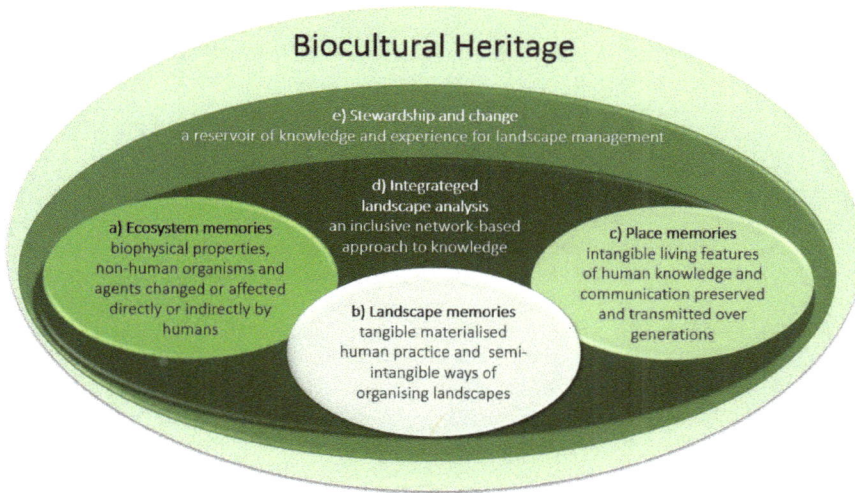

Figure 1. The four 'elements' of biocultural heritage (**a–d**) as defined by Lindholm and Ekblom [50].

More specifically, integrated landscape analysis is an interdisciplinary toolbox allowing us to trace elements of biocultural heritage and their internal relationships over time, incorporating contemporary botanical surveys, pollen analysis, archaeology, geographical information systems, cartography, historical research and interviews, and participant observation [50]. Studies spanning over both millennia long and shorter timescales have been critical in terms of understanding the biophysical and social ecological aspects of biocultural heritage. Such studies have also been important for re-evaluating degradation narratives that risk impeding effective biodiversity conservation and landscape management practices. A good early example of this is provided by the work of the People, Land and Time in Africa (PLATINA) research group at Stockholm University on deconstructing environmental narratives concerning the origins and drivers of severe soil erosion in Kondoa District, central Tanzania, that had informed colonial and post-colonial interventions for decades (see [51–53] for summaries, and additional references). A deeper understanding of how humans shape landscapes is an essential component in any plan for a sustainable future [7,54–62].

Below we review examples of land-use practices in sub-Saharan Africa, structured around the five interactive elements of biocultural heritage. We combine biological inventories and/or assessments by local residents on the ecological effects of land-use practices, with condensed summaries of archaeology, vegetation history, and interviews with local practitioners. We also complement these with our own and our students' field studies. In none of the cases presented below can the elements of bicultural heritage be understood from one single discipline nor from one single vantage point, whether that is the perspective of a local herder or farmer, researcher, conservationist, development worker or government official.

3. Biocultural Heritage in Space and Time

3.1. Ecosystem-Scale Memories

The concept of 'ecosystem memories' comes closest to UNESCO's [40] definition of biocultural heritage as ecosystems developing from human practices. Meadow pastures and wood pastures in Africa can be seen as continuing cultural landscapes, found in savannas and woodlands that are grazing- and fire-dependent [63–66]. Agroforest- and fire- managed landscapes carry structural and species level memories in terms of biological diversity. Interdisciplinary studies are still too few to allow us to assess the performance of these landscapes in terms of biodiversity, but broader scale studies on fire ecology in Africa have shown that landscapes that are fire-managed regularly tend to have both lower intensity and cooler fires that occur earlier in the dry season and which are more beneficial to sustaining biodiversity [65–67]. Fire also plays a crucial part in local landscape management in the semi-dry regions of southern and eastern Africa. In the savanna shrubland of the Chyulu area, Kenya, people burn to improve either hunting or pasture by, in the case of the latter, removing unpalatable grass and ticks. Grazing and dead wood collection are also used to create fire breaks [68]. Such systems are practiced in several regions of southern Africa, but the Chitemene system in Zambia has become particularly renowned for promoting pasture and fertilisation of farms, while also protecting individual trees, and thereby creating fire breaks [69]. In savannas, such mixed fire regimes and patch mosaic burning results in a heterogenous landscape structure comprising a range of post-fire ages, favourable to biodiversity, but managed fires also prevent damaging late season hot fires and uncontrollable wildfires that homogenise landscapes and eliminate fire-sensitive species [65,66,70].

In forested regions, agroforest landscapes create parkland and mosaic landscapes that are structurally diverse and high in agro-biodiversity [71]. The parkland mosaic landscape has a continuity over millennia. Previously, linkages were made between the extent of savannas and parklands and degradation from farming and fire management going back ca. 4000 years. However, palaeoecological studies now suggest climate dynamics have been more important in shaping the distribution of savannas in West Africa over the long term [72,73]. However, humans have also contributed to shaping mosaic landscapes, and fires are an important tool for maintaining landscape structure. In the Koulikoro district of southern Mali, fire is used to create landscape mosaics that increase micro- and edge-habitats, which are favourable for biodiversity. Here, fires and clearings create mosaic landscapes of semi-open areas, fields, fallows, and old growth forests [74,75]. Similarly, in the Kissidougou savanna region in Guinea, West Africa, forest patches and boundaries are continuously protected by households using methods that include mounding (to encourage plant growth), mulching, tilling, planting of crops beneath the trees and protection of tree species [76]. With sufficient fallow periods, such parkland management has positive impacts on tree biodiversity, as has been shown by studies in southwestern Burkina Faso [77]. In East Africa, diverse parkland landscapes are created through a variety of off- and on-farm management and forest protection, resulting in a diversity of trees [78,79]. Apart from shaping landscape structure and biology, fire management and shifting agriculture leave memories in terms of soil nutrients. The combination of burning and mulching of soils leads to the formation of black earths [76,80,81]. These black earths are conducive to both agricultural and biological diversity, as they are higher in organic carbon, pH, and plant-available nutrients than other local soils, and are also less prone to nutrient leaching or acidification [81].

Another similar example of long-term 'soil memory' important for ecosystems and agrobiodiversity comes from the East African savannas. Historic occupations have resulted in the enrichment of soils from dung and the formation of grassy glades [82–86]. In Kenya, pastoralists recognise these glades as marking former settlement sites and value them for the nutrient-rich grasses, especially *Cynodon* spp., that recolonize these former 'human' spaces [87,88]. Interviews and archaeological data presented by Shoemaker [89] covering the last ca. 150 years show how these places have been resettled over generations and are preferred sites for settlement. These 'anthropogenic' glades, described by Veblen [90,91] as biological 'hotspots', produce good pasture grasses and are functionally important for a range of taxa, including wild megaherbivores [91], but also birds [92,93], geckos, and arthropods (e.g., [94], see also [82,85]). As pastoral communities have been highly mobile over time and soil nutrient compositions can last for millennia [87,88,95–97], the total surface area of such glades is important for overall ecosystem functioning in East African savannas.

3.2. Landscape Memories

Landscape memories can be understood as forms of materialised human practice, such as built environments and archaeological sites, including settlement systems and land-use systems linked to user and property rights—what Widgren [98,99] has called 'landesque capital'. Across the continent, there are many examples of irrigation or terracing system landscapes that are relict (see for instance Engaruka in Tanzania [100–102], Nyanga in Zimbabwe [103] and Mpumulanga in South Africa [104]).

In addition to these discontinued terrace systems, there are also landscapes where precolonial irrigation practices are extant (see, for instance, the Mbulu Highlands, Tanzania [105,106]; and the Cheranagni escarpment, Kenya [107]). We will here expand on one example, the furrow irrigation found on the southern slopes of Mount Kilimanjaro, termed *mfongo* by Chagga-speaking people (Figure 2). The earliest firmly dated irrigation features on Kilimanjaro were built in the 18th century, though oral traditions, historical references, and linguistic evidence indicate that irrigation schemes were present on the mountain by the 17th century at the latest [108,109]. This system, built on customary land tenure, supports multi-layered agroforest gardens high in biodiversity (500 species of which 400 are non-cultivated) [110,111]. Despite their productivity and longevity, government policies have long been remiss in terms of promoting and maintaining *mfongo* practices [108,112,113]. Sunday [114] has explored the continuity and legacy of the *mfongo* through interviews with 200 households. The practice of making and maintaining water channels is less common today than it was in the remembered past. When asked why *mfongo* practices are discontinued, local residents replied that increasing droughts and lessened water runoff from Mount Kilimanjaro were the primary causes, although state policies were also mentioned for the wider contextual setting (see also [113]). However, local residents still value the *mfongo* system for ensuring water access and for maintaining crop yields and diversity. The water channels are also embedded in local heritage and identity. In the interviews conducted by Sunday [114] (p. 48), one elderly woman expressed her worry that this knowledge was now being lost:

Figure 2. (a) a traditional water furrow (*mfongo*) cutting across a hillside on Mount Kilimanjaro; (b) a water furrow leading up to a Chagga home garden (photos published by Sunday [114] and reproduced with his permission @ Sabbath Sunday).

"Times have changed very fast! Our children who go to school, when they live in cities for long and God blesses them with children: those children will not produce a generation that will revive culture practices of *mfongo* and rituals".

For that woman, and also many other respondents interviewed by Sunday, the making of water channels is more than an agricultural practice, it is also a 'spiritual obligation' and a living knowledge that must be actively maintained.

3.3. Place-Based Memories

Place-based memories refer to intangible living features of human knowledge and communication: know-how, place names, orature, arts, ideas and culture, and also biological heritage. A good example is the practice of the protection of 'sacred areas', often ancient and/or community burial areas, or old settlement sites that are surrounded by very strong rules of community protection, some of which have become havens for old growth forests. Though together they may constitute a landscape memory, the foundation of protection is very much placed-based. Sacred sites/forests occur across the African continent, but their potential in terms of biodiversity protection has only been recognised in the last few decades [115–122]. A study in northern Tanzania (the North Pare mountains) located 290 sacred sites. Though small in size, together they covered a total area of 370 ha [123]. Similarly, on Zanzibar there are a great number of sacred areas that are important as reservoirs for the endemic Zanzibar-Inhambane forest phytochoria, which are known to be high in species diversity [124]. Another well-known example is the World Heritage sacred Mijkenda Kaya forests in Kenya. This heritage landscape consists of 11 separated forest islands (growing on abandoned occupation settlements) containing as many as 307 species that are listed as endangered [125,126] (Figure 3). These are sanctuaries for forest-adapted species, both plants and animals, and have been estimated to comprise 4.2–5.6% of the entire Zanzibar-Inhambane forests mosaic [127]. These sacred areas and the knowledge systems surrounding the activities that brought them into being are as key to the transmission of traditional practices as to the innovation of new ones [128,129].

In the Muzarabani Communal Lands of Zimbabwe (Zambezi valley), forest loss has been shown to be dramatically less in areas considered sacred, or under protection of traditional custodians [130]. Similar observations have also been made in Mozambique in the Licuati forest in the south [131] and the Chôa Highlands in Manica Province. In Chôa, sacred areas have greater species diversity, more complex forest structure, and higher incidence of fire-sensitive species [132]. Palaeoecological techniques have been used in southern Mozambique [133] to study the long-term history of littoral forests, suggesting the existence of mosaic landscapes for 1400 years. Though more studies are needed, it has been suggested that existing forest patches should possibly be re-assessed as having been actively protected through long-term (i.e., over centennial scales) management [133]. Though not supported by palaeoecological data, long-term protection of forest patches has also been suggested for East Africa [134]. In Madagascar, there are also positive examples of customary protection of sacred forest. One example is Ankodida, where customary protection builds on customary rules surrounding resource use and local custodianship [135]. Such culturally protected small forest islands have been shown to be essential for maintaining ecosystem services [136].

Sacred areas are also common in West Africa. In Benin, 2940 sacred areas have been documented ranging in size from 0.1 ha to 1600 ha, covering a total of 18,360 ha [137,138]. Apart from being heritage places, these areas are highly important as biological refugia and function as seed-banks and genetic reservoirs [139]. In numerous regions of Benin, sacred areas can exhibit higher tree species diversity than state-established conservation spaces [140]. Estimates in Ghana suggest there are 2000–3200 such sacred sites [141]. Studies within sacred groves on the Accra Plains found that the biomass and diversity of small mammals often exceed that of surrounding biomes [141]; similar correlations have also been made more recently regarding butterfly populations [142]. In the Loma area in Liberia, dominated by the Upper Guinea forests, sacred areas are often the sites of old towns or graves and the long history and mobility of people and settlements has created a dense network of such sacred areas [143]. Socially proscribed systems of protection surrounding sacred areas allow for the maturation of old growth trees. At the same time, economically important shade-tolerant tree crops (e.g., Kola, cocoa and coffee) can be planted amongst the trees, thriving in the nutrient-dense soils that often result from anthropogenic inputs associated with settlements, such as charcoal, animal dung and food refuse. A comparison between the sacred areas and unmanaged fallow areas shows that vegetation in sacred areas is more heterogeneous in the basal layers and higher in the upper layers of the canopy [143].

Figure 3. Mijkenda Kaya forests in Kenya (picture from Wikimedia commons, photo by Victor Ochieng @Victor Ochieng).

4. Discussion: Stewardship and Change

The memories exemplified above act on different spatial and temporal scales shaping the structure of the ecosystem or landscape. However, the practices that shape them are usually place bound. As physically discernible memories (shown through landforms, vegetation and archaeological and/or heritage sites), they inform practices in the present. Biocultural heritage is closely connected with identity, social cohesion and practice but also social and political negotiation locally [25,42,43]. Heritage places, in particular, are often the arena for community meetings where this negotiation is played out. Such events frequently feature the transmission of old and new knowledge [127–129]. Biocultural heritage may then be considered to build on local practices and initiatives, and is therefore key in promoting stewardship, innovation and change [4–6,9,10].

As exemplified in the case of the *mfongo* system, but also more broadly in the case of fire management [20], state and development policies have tended to work against local practices. After nearly a century of state fire suppression polices across the continent, there is now an expanding body of research calling for better integration of fire management and conservation goals and hence closer collaboration between conservation workers and local farmers and pastoralists [67,74,144]. For instance, in the Chyulu Hills area of Kenya reviewed above, local residents lament the fact that they are not allowed to pick dead wood or instigate controlled burns around park boundaries, which increases the likelihood of uncontrolled fires within the park, and both local residents and conservationists are stressing the need for collaboration [68]. The Caprivi Integrated Fire Management Programme in Namibia is a positive example here: through collaboration, fire management policies have changed from a centralised fire suppression policy to community-managed fire areas where early season burns are used as a fire management tool by local communities [145].

There is also now a growing awareness of the potential for customary practices and heritage sites to promote forest conservation [119,121,127]. A positive example is the Ankodida forest in Madagascar where the World Wild Fund for Nature (WWF) worked with local residents to incorporate sacred groves into the newly forming protected area network. As communities were concerned that erosion of traditional values was leading to threats to the groves, the formal conservation status was seen as reinforcing cultural taboos promoting the protection of the forest [135]. The Mijkenda Kaya forests in Kenya, discussed above, provide another positive example of stewardship based on local practices of protection and heritage values, where local communities and authorities have collaborated to proclaim the area as a World Heritage Site [127], though as discussed below, there are also problems here with inclusivity and access.

One of the biggest challenges since the emergence of community-based resource management (CBRM) has remained the definition of community itself [36,146,147]. Individual community representatives may become a shorthand for 'community', which risks entrenching local power structures that are less than equitable, in particular when it comes to issues related to heritage, often in control of male elders and/or particular lineages. Continuing with the case of the Mijkenda forests, male elders lament the loss of respect for traditions as well as the loss of traditions themselves, processes they believe will ultimately threaten the protection of the sacred forests [127,128]. Meanwhile, based on interviews carried out by Groh [129], youths and women report experiencing issues surrounding access to sacred forest areas, and also the transfer of knowledge and innovations occurring within them during community meetings and ceremonies. For both community leaders and for other local people not so clearly in positions of power, this feeling of exclusion is then a double loss, as crucial knowledge regarding agrobiodiversity and resource use is not being transferred to younger generations (nor women) and the traditions and values associated with forest protection are thus not prioritised by younger community members. Biocultural heritage, while featuring existing (and customary) local practices of land-use and heritage management must therefore also build on principles of inclusiveness and transparency, otherwise it will simply not be accepted by the wider community.

When integrated with principles of equity, justice and representation, while also building on local practices and innovation, biocultural heritage has the promise of combining the goals and aspirations of local residents with the national and international goals of sustainability and biodiversity protection [148]. Concrete collaborations between local communities and conservation in biocultural heritage are still too few and far between, but as argued here, there are ample opportunities to learn from existing local practices of biodiversity stewardship but also from local processes of agrobiodiversity innovation and change. To conclude, in this paper we have discussed a series of cases illustrating how a conceptual framework of biocultural heritage allows for new approaches to heritage, nature conservation, landscape planning and development goals. Biocultural heritage assets, we argue, provide the means to negotiate management goals in these areas, and in certain cases, also to combine them.

Author Contributions: Main author: A.E. Author contributions are as follows: Conceptualization: A.E. and K.-J.L.; Writing—Original Draft Preparation: A.E., A.S.; L.G., P.L.; K.-J.L.; Writing—Review and Editing: P.L., A.S. and L.G.; Visualization: A.E. and K.-J.L.

Funding: The paper has been partly produced within the Adaptation & Resilience to Climate Change (ARCC) in Eastern Africa project funded by the Swedish Research Council (Vetenskapsrådet), Formas and the Swedish

International Development and Cooperation Agency (SIDA), grant number 2016-06355, awarded to Paul Lane and Anneli Ekblom. Anna Shoemaker's work was supported as part of the European Commission Marie Skłodowska-Curie Initial Training Network titled "Resilience in East African Landscapes (REAL)" (FP7-PEOPLE-2013-ITN project number 606879, awarded to Paul Lane).

Acknowledgments: This article is a contribution to the Integrated History and Future of People on Earth (IHOPE) initiative (http://ihopenet.org/), and to the Pages LandCover6K and LandUse6K Africa working group activities.

Conflicts of Interest: The authors declare no conflict of interest.

References

1. Reid, W.V.; Berkes, F.; Wilbanks, T.; Capistriano, D. Introduction. In *Bridging Scales and Local Knowledge in Assessments*; Reid, W.V., Berkes, F., Wilbanks, T., Capistriano, D., Eds.; Island Press: Washington, DC, USA, 2006; pp. 1–18.
2. Chazdon, R.L.; Harvey, C.A.; Komar, O.; Griffith, D.M.; Ferguson, B.G.; Ramos, M.M.; Morales, H.; Nigh, R.; Soto-Pinto, L.; van Breugel, M.; et al. Beyond reserves: A research agenda for conserving biodiversity in human-modified tropical landscapes. *Biotropica* **2009**, *41*, 142–153. [CrossRef]
3. Maffi, L. *On Biocultural Diversity Linking Language, Knowledge, and the Environment*; Smithsonian Institution Press: Washington, DC, USA, 2001; pp. 1–578.
4. Davidson-Hunt, I.J.; Turner, K.L.; Te Pareake Mead, A.; Cabrera-Lopez, J.; Bolton, R.; Idrobo, C.J.; Miretski, I.; Morrison, A.; Robson, J.P. Biocultural design: A new conceptual framework for sustainable development in rural indigenous and local communities. *Sapiens* **2012**, *5*, 33–45.
5. Maffi, L. *Biocultural Diversity Conservation: A Global Sourcebook*, 2nd ed.; Taylor and Francis: Abingdon, UK, 2012; pp. 3–23.
6. Barthel, S.; Crumley, C.L.; Svedin, U. Bio-cultural refugia—Safeguarding diversity of practices for food security and biodiversity. *Glob. Environ. Chang.* **2013**, *23*, 1142–1152. [CrossRef]
7. Cevasco, R.; Moreno, D.; Hearn, R. Biodiversification as an historical process: An appeal for the application of historical ecology to bio-cultural diversity research. *Biodivers. Conserv.* **2015**, *24*, 3167–3183. [CrossRef]
8. Gavin, M.C.; McCarter, J.; Mead, A.; Berkes, F.; Stepp, J.R.; Peterson, D.; Tang, R. Defining biocultural approaches to conservation. *Trends Ecol. Evol.* **2015**, *303*, 140–145. [CrossRef] [PubMed]
9. Swiderska, K.; Argumedo, A.; Song, Y.; Rastogi, A.; Gurung, N.; Wekesa, C. Biocultural Innovation: The Key to Global Food Security? *IIED Briefing*, 2018. Available online: http://pubs.iied.org/17465IIED/ (accessed on 29 November 2018).
10. Poole, A.K. Where is Goal 18? The Need for Biocultural Heritage in the Sustainable Development Goals. *Environ. Value* **2018**, *27*, 55–80. [CrossRef]
11. IUCN. Cultural Landscapes and Protected Areas. Unfolding the Linkages and Synergies. *World Heritage Magazine*, 2013. Available online: https://www.iucn.org/content/cultural-landscapes-and-protected-areas-unfolding-linkages-and-synergies (accessed on 30 November 2018).
12. Agnoletti, M. Introduction: Framing the issue—A transdisciplinary reflection on cultural landscapes. In *Conservation of Cultural Landscapes*; Agnoletti, M., Ed.; CABI Publishing: Wallingford, UK, 2006; pp. xi–xix.
13. Agnoletti, M. Rural landscape, nature conservation and culture: Some notes on research trends and management approaches from a (southern) European perspective. *Landsc. Urban Plan.* **2014**, *126*, 66–73. [CrossRef]
14. Angelstam, P. Maintaining cultural and natural biodiversity in Europe's economic centre and periphery. In *Conservation of Cultural Landscapes*; Agnoletti, M., Ed.; CABI Publishing: Wallingford, UK, 2006; pp. 125–143.
15. De Chazal, J.; Rounsevell, M.D.A. Land-use and climate change within assessments of biodiversity change: A review. *Glob. Environ. Chang.* **2009**, *19*, 306–315. [CrossRef]
16. Barthel, S.; Crumley, C.L.; Svedin, U. Biocultural refugia: Combating the erosion of diversity in landscapes of food production. *Ecol. Soc.* **2013**, *18*, 71. [CrossRef]
17. Richards, P. *Indigenous Agricultural Revolution: Ecology and Food Production in West Africa*; Hutchinson: London, UK, 1985; pp. 1–192.
18. Richards, P. *Coping with Hunger: Hazard and Experiment in an African Rice-Farming System*; Allen and Unwin: London, UK, 1986; pp. 1–76.

19. Fairhead, J.; Leach, M. *Misreading the African Landscape: Society and Ecology in a Forest-Savanna Mosaic;* Cambridge University Press: Cambridge, UK, 1996; pp. 1–354.

20. Fairhead, J.; Leach, M. Enriching the landscape: Social history and the management of transition ecology in the forest savanna mosaic of the republic of Guinea, Africa. *J. Int. Afr. Inst.* **1996**, *66*, 14–36. [CrossRef]

21. Fairhead, J.; Leach, M. *Science, Society and Power: Environmental Knowledge and Policy in West Africa and the Caribbean;* Cambridge University Press: Cambridge, UK, 2003; pp. 1–272.

22. Ekblom, A. Archaeology, historical sciences and environmental conservation. In *Oxford Handbook of Historical Ecology and Applied Archaeology;* Stump, D., Isendahl, C., Eds.; Oxford University Press: Oxford, UK, 2015. Available online: http://www.oxfordhandbooks.com/view/10.1093/oxfordhb/9780199672691.001.0001/oxfordhb-9780199672691-e-9 (accessed on 29 November 2018).

23. Veldman, J.W.; Buisson, E.; Durigan, G.; Wilson Fernandes, G.; Le Stradic, S.; Mahy, G.; Negreiros, D.; Overbeck, G.E.; Veldman, R.G.; Zaloumis, N.P.; et al. Toward an old-growth concept for grasslands, savannas, and woodlands. *Front. Ecol. Environ.* **2015**, *13*, 154–162. [CrossRef]

24. Bond, W.; Zaloumis, N.P. The deforestation story: Testing for anthropogenic origins of Africa's flammable grassy biomes. *Phil. Trans. R. Soc.* **2016**, *B371*, 20150170. [CrossRef] [PubMed]

25. Berkes, F.; Colding, J.; Folke, C. Rediscovery of traditional ecological knowledge as adaptive management. *Ecol. Appl.* **2000**, *10*, 1251–1262. [CrossRef]

26. Kotir, J.H. Climate change and variability in Sub-Saharan Africa: A review of current and future trends and impacts on agriculture and food security. *Environ. Dev. Sustain.* **2011**, *13*, 587–605. [CrossRef]

27. Niang, I.; Ruppel, O.C.; Abdrabo, M.A.; Essel, A.; Lennard, C.; Padgham, J.; Urquhart, P. Africa. In *Climate Change 2014: Impacts, Adaptation and Vulnerability. Contribution of Working Group II to the Fifth Assessment Report of the Intergovernmental Panel on Climate Change;* Field, C.B., Ed.; Cambridge University Press: Cambridge, UK, 2014; pp. 1199–1265.

28. Serdeczby, O.; Adams, S.; Baarsch, F.; Coumou, D.; Robinson, A.; Hare, W.; Shaeffler, M.; Perrette, M.; Reinhart, J. Climate change impacts in Sub-Saharan Africa: From physical changes to their social repercussions. *Reg. Environ. Chang.* **2016**, *17*, 1585–1600. [CrossRef]

29. Hutton, J.; Adams, W.; Murombedzi, J. Back to the barriers? Changing narratives in biodiversity conservation. *For. Dev. Stud.* **2005**, *2*, 341–370. [CrossRef]

30. Brockington, D.; Igoe, J. Eviction for conservation: A global overview. *Conserv. Soc.* **2006**, *4*, 424–470.

31. Fairhead, J.; Leach, M.; Scoones, I. Green grabbing: A new appropriation of nature? *J. Peasant Stud.* **2012**, *392*, 237–261. [CrossRef]

32. Anseeuw, W. The rush for land in Africa: Resource grabbing or green revolution? *S. Afr. J. Int. Aff.* **2013**, *201*, 159–177. [CrossRef]

33. Romeu-Dalmau, C.; Gasparatos, A.; von Maltitz, D.; Graham, A.; Almagro-Garcia, J.; Wilebore, B.; Willis, K.J. Impacts of land use change due to biofuel crops on climate regulation services: Five case studies in Malawi, Mozambique and Swaziland. *Biomass Bioenergy* **2018**, *114*, 30–40. [CrossRef]

34. Maffi, L. Cultural, and biological diversity. *Ann. Rev. Anthropol.* **2005**, *34*, 599–617. [CrossRef]

35. Pretty, J.; Adams, B.; Berkes, F.; de Athayde, S.F.; Dudley, N.; Hunn, E.; Maffi, L.; Milton, K.; Rapport, D.; Robbins, P.; et al. The intersections of biological diversity and cultural diversity: Towards integration. *Conserv. Soc.* **2009**, *72*, 100–112.

36. Agrawal, A.; Gibson, C.C. Enchantment and disenchantment: The role of community in natural resource conservation. *World Dev.* **1999**, *27*, 629–650. [CrossRef]

37. Adams, W.; Hulme, D. Changing narratives, policies & practices in African conservation. In *African Wildlife & Livelihoods: The Promise & Performance of Community Conservation;* Hulme, D., Murphree, M., Eds.; James Currey: London, UK, 2001; pp. 9–23.

38. Dove, M.D. Indigenous People and Environmental Politics. *Ann. Rev. Anthropol.* **2006**, *35*, 191–208. [CrossRef]

39. West, P.; Igoe, J.; Brockington, D. Parks and peoples: The social impact of protected areas. *Ann. Rev. Anthropol.* **2006**, *25*, 251–277. [CrossRef]

40. UNESCO. *Operational Guidelines for the Implementation of the World Heritage Convention;* UNESCO World Heritage Centre: Paris, France, 2008; pp. 1–163.

41. Berkes, F. Rethinking community-based conservation. *Conserv. Biol.* **2004**, *18*, 621–630. [CrossRef]

42. Olsson, P.; Folke, C.; Berkes, F. Adaptative comanagement for building resilience in social-ecological systems. *Environ. Manag.* **2004**, *34*, 75–90. [CrossRef] [PubMed]

43. Ruiz-Mallén, I.; Corbera, E. Community-based conservation and traditional ecological knowledge: Implications for social-ecological resilience. *Ecol. Soc.* **2013**, *18*, 12. [CrossRef]

44. Gómez-Baggethun, E.; Corbera, E.; Reyes-García, V. Traditional ecological knowledge and Global Environemtal Change: Research findings and policy implications. *Ecol. Soc.* **2013**, *184*, 72. [CrossRef] [PubMed]

45. Nazarea, D.V. Local knowledge and memory in biodiversity conservation. *Ann. Rev. Anthropol.* **2006**, *35*, 317–335. [CrossRef]

46. Swiderska, K. *Banishing the Biopirates: A New Approach to Protecting Traditional Knowledge*; IIED: London, UK, 2016; pp. 1–24.

47. Chen, C.; Gilmore, M. Biocultural rights: A new paradigm for protecting natural and cultural resources of indigenous communities. *Int. Indig. Policy J.* **2015**, *6*, 3. [CrossRef]

48. Bavikatte, K.S.; Bennet, T. Community stewardship: The foundation of biocultural rights. *J. Hum. Rights Environ.* **2015**, *6*, 7–29. [CrossRef]

49. Eriksson, O. Historical and current niche construction in an Anthropogenic biome: Old cultural landscapes in southern Scandinavia. *Land* **2016**, *54*, 42. [CrossRef]

50. Lindholm, K.-J.; Ekblom, A. A framework for exploring and managing biocultural heritage. *Anthropocene* **2019**, accepted.

51. Payton, R.W.; Christiansson, C.; Shishira, E.K.; Yanda, P.; Eriksson, M.G. Landform, soils and erosion in the north-eastern Irangi Hills, Kondoa, Tanzania. *Geogr. Ann. Ser. A Phys. Geogr.* **1992**, *74*, 65–79. [CrossRef]

52. Östberg, W. *The Kondoa Transformation: Coming to Grips with Soil Erosion in Central Tanzania*; Nordic Africa Institute: Uppsala, Sweden, 1986; pp. 1–99.

53. Lane, P. Environmental narratives and the history of soil erosion in Kondoa District, Tanzania: An archaeological perspective. *Int. J. Afr. Hist. Stud.* **2009**, *42*, 457–483.

54. Crumley, C.L. Historical ecology: A multidimensional ecological orientation. In *Historical Ecology: Cultural Knowledge and Changing Landscapes*; Crumley, C., Ed.; School of American Research: Santa Fe, NM, USA, 1994; pp. 1–16.

55. Crumley, C.L. Historical ecology: Integrated thinking at multiple temporal and spatial scales. In *The World System and the Earth System-Global Socioenvironmental Change and Sustainability Since the Neolithic*; Hornborg, A., Crumley, C.L., Eds.; Left Coast Press: Walnut Creek, CA, USA, 2007; pp. 15–28.

56. Antrop, M. Why landscapes of the past are important for the future. *Landsc. Urban Plan.* **2005**, *701*, 21–34. [CrossRef]

57. Foster, D.R. Conservation issues and approaches for dynamic cultural landscapes. *J. Biogeogr.* **2002**, *29*, 1533–1535. [CrossRef]

58. Erickson, C.; Baleé, W. The historical ecology of a complex landscape in Bolivia. In *Time and Complexity in Historical Ecology: Studies in the Neotropical Lowlands*; Baleé, W., Erickson, C., Eds.; Columbia University Press: New York, NY, USA, 2006; pp. 187–233.

59. Chambers, F.; Daniell, J. Conservation and habitat restoration of moorland and bog in the United Kingdom uplands, a regional, paleoecological perspective. *Pages Newslett.* **2010**, *19*, 45–47. [CrossRef]

60. Gillson, L. *Biodiversity Conservation and Environmental Change: Using Paleoecology to Manage Dynamic Landscapes in the Anthropocene*; Oxford University Press: Oxford, UK, 2015; pp. 87–114.

61. Grove, A.T.; Rackham, O. *The Nature of Mediterranean Europe. An Ecological History*; Yale University Press: New Haven, VT, USA, 2001; pp. 1–384.

62. Krzywinski, K. Unity in diversity: The concept and significance of cultural landscape for the heritages of Europe. In *Cultural Landscapes of Europe. Fields of Demeter, Haunts of Pan*; Krzywinski, K., O'Connell, M., Küster, H.J., Eds.; Aschenbeck Media: Bremen, Germany, 2009; pp. 9–22.

63. Scholes, R.J.; Archer, R.S. Tree–grass interactions in savannas. *Ann. Rev. Ecol. Syst.* **1997**, *28*, 517–544. [CrossRef]

64. Sankaran, M.; Hanan, N.P.; Scholes, R.J.; Ratnam, J.; Augustine, D.J.; Cade, B.S.; Gignoux, J.; Higgins, S.I.; Le Roux, X.; Ludwig, F.; et al. Determinants of woody cover in African savannas. *Nature* **2005**, *438*, 846–849. [CrossRef]

65. Archibald, S.; Staver, A.C.; Levin, S.A. Evolution of human-driven fire regimes in Africa. *Proc. Natl. Acad. Sci. USA* **2012**, *109*, 847–852. [CrossRef] [PubMed]

66. Archibald, S.; Lehmann, C.E.; Gómez-Dans, J.L.; Bradstock, R.A. Defining pyromes and global syndromes of fire regimes. *Proc. Natl. Acad. Sci. USA* **2013**, *110*, 6442–6447. [CrossRef] [PubMed]

67. Laris, P.; Dembele, F. Humanizing savanna models: Integrating natural factors and anthropogenic disturbance regimes to determine tree–grass dynamics in savannas. *J. Land Use Sci.* **2012**, *7*, 459–482. [CrossRef]

68. Kamau, P.N.; Medley, K.E. Anthropogenic fires and local livelihoods at Chyulu Hills, Kenya. *Landsc. Urban Plan.* **2014**, *124*, 76–84. [CrossRef]

69. Eriksen, C. Why do they burn the bush? Fire, rural livelihoods and conservation in Zambia. *Geogr. J.* **2007**, *173*, 242–256. [CrossRef]

70. Van Wilgen, B.W.; Govender, N.; Biggs, H.C.; Ntsala, D.; Funda, X.N. Response of savanna fire regimes to changing fire management policies in a large African National Park. *Conserv. Biol.* **2004**, *18*, 1533–1540. [CrossRef]

71. Boffa, J.-M. *Agroforestry Parklands in Sub-Saharan Africa*; FAO: Rome, Italy, 1999. Available online: http://www.fao.org/docrep/005/x3940e/X3940E00.htm (accessed on 29 November 2018).

72. Vincens, A.; Schwartz, D.; Elenga, H.; Reynaud-Farrera, I.; Alexander, A.; Bertaux, J.; Marlotti, A.; Martin, L.; Meunier, J.-M.; Nguetsop, F.; et al. Forest response to climate changes in Atlantic Equatorial Africa during the last 4000 years BP and inheritance on the modern landscapes. *J. Biogeogr.* **1999**, *264*, 879–885. [CrossRef]

73. Bonnefille, R. Rainforest response to past climatic changes in tropical Africa. In *Tropical Rainforest Responses to Climate Change*; Bush, M.B., Flenley, J.R., Eds.; Springer: New York, NY, USA, 2007; pp. 117–163.

74. Laris, P. Burning the seasonal mosaic: Preventive burning strategies in the wooded savanna of southern Mali. *Hum. Ecol.* **2002**, *30*, 155–186. [CrossRef]

75. Kull, C.A.; Laris, P. Fire ecology and fire politics in Mali and Madagascar. In *Tropical Fire Ecology: Climate Change, Land Use, and Ecosystem Dynamics*; Cochrane, M.A., Ed.; Springer: New York, NY, USA, 2009; pp. 171–226.

76. Fairhead, J.; Leach, M. Amazonian Dark Earths in Africa? In *Amazonian Dark Earths: Wim Sombroek's Vision*; Woods, W.I., Teixeira, W.G., Lehmann, J., Steiner, C., WinklerPrins, A., Rebellato, L., Eds.; Springer: New York, NY, USA, 2009; pp. 265–278.

77. Augusseau, X.; Nikiema, P.; Torquebiau, E. Tree biodiversity, land dynamic and farmers strategies on the agricultural frontier of southwestern Burkina Faso. *Biodivers Conserv.* **2006**, *15*, 613–630. [CrossRef]

78. Tiffen, M.; Mortimore, M.; GichUnited Kingdomi, F. *More People, Less Erosion: Environmental Recovery in Kenya*; ACTS Press: Nairobi, Kenya, 1994; pp. 1–309.

79. Backen, M.M. The role of indigenous trees for the conservation of biocultural diversity in traditional agroforestry land use systems: The Bungoma case study. In-situ conservation of indigenous tree species. *Agroforest. Syst.* **2001**, *52*, 119–132. [CrossRef]

80. Frausin, V.; Fraser, J.; Narmah, W.; Lahai, M.; Winnebah, T.A.; Fairhead, J.; Leach, M. "God made the soil but we made it fertile": Gender, knowledge, and practice in the formation and use of African Dark Earths in Liberia and Sierra Leone. *Hum. Ecol.* **2014**, *42*, 695–710. [CrossRef]

81. Solomon, D.; Lehmann, J.; Fraser, J.A.; Leach, M.; Amanor, K.; Frausin, V.; Kristiansen, S.M.; Millimouno, D.; Fairhead, J. Indigenous African soil enrichment as climate-smart sustainable agriculture alternative. *Front. Ecol. Environ.* **2016**, *142*, 71–76. [CrossRef]

82. Blackmore, A.C.; Mentis, M.T.; Scholes, R.J. The origin and extent of nutrient-enriched patches within a nutrient-poor savanna in South Africa. *J. Biogeogr.* **1990**, *17*, 463–470. [CrossRef]

83. Augustine, D.J. Long-term, livestock-mediated redistribution of nitrogen and phosphorus in an East African savanna. *J. Appl. Ecol.* **2003**, *49*, 137–149. [CrossRef]

84. Augustine, D.J. Influence of cattle management on habitat selection by impala on central Kenyan rangeland. *J. Wildl. Manag.* **2004**, *68*, 916–923. [CrossRef]

85. Muchiru, A.N.; Western, D.J.; Reid, R.S. The role of abandoned pastoral settlements in the dynamics of African large herbivore communities. *J. Arid Environ.* **2008**, *72*, 940–952. [CrossRef]

86. Muchiru, A.N.; Western, D.J.; Reid, R.S. The impact of abandoned pastoral settlements on plant and nutrient succession in an African savanna ecosystem. *J. Arid Environ.* **2009**, *73*, 322–331. [CrossRef]

87. Boles, O.J.C.; Lane, P.J. The Green, Green Grass of Home: An archaeo-ecological approach to pastoralist settlement in central Kenya. *Azania* **2016**, *51*, 507–530. [CrossRef]

88. Petek, N. *Archaeological Perspectives on Risk and Community Resilience in the Baringo Lowlands, Kenya*; Department of Archaeology and Ancient History, Uppsala University: Uppsala, Sweden, 2018; pp. 1–294.

89. Shoemaker, A. *Pastoral Pasts in the Amboseli Landscape: An Archaeological Exploration of the Amboseli Ecosystem from the Later Holocene to the Colonial Period*; Department of Archaeology and Ancient History, Uppsala University: Uppsala, Sweden, 2018; pp. 1–320.

90. Veblen, K.E. Savanna glade hotspots: Plant community development and synergy with large herbivores. *J. Arid Environ.* **2012**, *78*, 119–127. [CrossRef]

91. Veblen, K.E. Impacts of traditional livestock corrals on woody plant communities in an East African savanna. *Rangel. J.* **2013**, *35*, 349–353. [CrossRef]

92. Söderström, B.; Reid, R.S. Abandoned pastoral settlements provide concentrations of resources for savannah birds. *Acta Oecol.* **2010**, *36*, 184–196. [CrossRef]

93. Morris, D.L.; Western, D.; Maitumo, D. Pastoralists' livestock and settlements influence game bird diversity and abundance in a savanna ecosystem of southern Kenya. *Afr. J. Ecol.* **2009**, *47*, 48–55. [CrossRef]

94. Donihue, C.M.; Porensky, L.M.; Foufopoulos, J.; Riginos, C.; Pringle, R.M. Glade cascades: Indirect legacy effects of pastoralism enhance the abundance and spatial structuring of arboreal fauna. *Ecology* **2013**, *94*, 827–837. [CrossRef]

95. Lane, P.J. Archaeology in the age of the Anthropocene: A critical assessment of its scope and societal contributions. *J. Field Archaeol.* **2015**, *40*, 485–498. [CrossRef]

96. Shahack-Gross, R.; Marshall, F.; Ryan, K.; Weiner, S. Reconstruction of spatial organization in abandoned Maasai settlements: Implications for site structure in the Pastoral Neolithic of East Africa. *J. Archaeol. Sci.* **2004**, *31*, 1395–1411. [CrossRef]

97. Marshall, F.; Reid, R.E.B.; Goldstein, S.; Storozum, M.; Wreschnig, A.; Hu, L.; Kiura, P.; Shahack-Gross, R.; Ambrose, S.H. Ancient herders enriched and restructured African grasslands. *Nature* **2018**, *561*, 387–390. [CrossRef]

98. Widgren, M. Towards a historical geography of intensive farming in eastern Africa. In *Islands of Intensive Agriculture in Eastern Africa*; Widgren, M., Sutton, J.E.G., Eds.; James Currey: Oxford, UK, 2004; pp. 1–18.

99. Widgren, M.; Hakansson, T. Introduction. Landesque capital: What is the concept good for? In *Landesque Capital: The Historical Ecology of Enduring Landscape Modifications*; Hakansson, T., Widgren, M., Eds.; Routledge: Abingdon, UK, 2007.

100. Sutton, J.E. Irrigation and soil-conservation in African agricultural history: With a reconsideration of the Inyanga terracing Zimbabwe and Engaruka irrigation works Tanzania. *J. Afr. Hist.* **1984**, *251*, 25–41. [CrossRef]

101. Stump, D. The development and expansion of the field and irrigation systems at Engaruka, Tanzania. *Azania* **2006**, *41*, 69–94. [CrossRef]

102. Lang, C.; Stump, D. Geoarchaeological evidence for the construction, irrigation, cultivation, and resilience of 15th–18th century AD terraced landscape at Engaruka, Tanzania. *Quat. Res.* **2017**, *883*, 382–399. [CrossRef]

103. Soper, R.C. *Nyanga: Ancient Fields, Settlements and Agricultural History in Zimbabwe*; British Institute in Eastern Africa: London, UK, 2002; pp. 1–277.

104. Widgren, M. Precolonial agricultural terracing in Bokoni, South Africa. *J. Afr. Archaeol.* **2016**, *14*, 33–53. [CrossRef]

105. Börjensson, L. *A History under Siege: Intensive Agriculture in the Mbulu Highlands, Tanzania, 19th Century to the Present*; Stockholm University: Stockholm, Sweden, 2004; pp. 1–187.

106. Börjesson, L. Boserup Backwards? Agricultural intensification as 'its own driving force' in the Mbulu Highlands, Tanzania. *Geogr. Ann. Ser. B Hum. Geogr.* **2007**, *89*, 249–267. [CrossRef]

107. Davies, M.I.J. The temporality of landesque capital: Cultivation and the routines of Pokot life. In *Landesque Capital: The Historical Ecology of Enduring Landscape Modifications*; Håkansson, N.T., Widgren, M., Eds.; Routledge: London, UK, 2004; pp. 172–196.

108. Stump, D.; Tagseth, M. The history of precolonial and early colonial agriculture on Kilimanjaro: A review. In *Culture, History Identity: Landscapes of Inhabitation in the Mount Kilimanjaro Area, Tanzania*; Clack, T.A.R., Ed.; BAR International Series 1966; Archaeopress: Oxford, UK, 2009; pp. 107–124.

109. Stump, D. The archaeology of agricultural intensification in Africa. In *Oxford Handbook of African Archaeology*; Mitchell, P., Lane, P., Eds.; Oxford University Press: Oxford, UK, 2013. [CrossRef]

110. Hemp, A. The banana forests of Kilimanjaro: Biodiversity conservation of the agroforestry system of the Chagga home gardens. *Biodivers. Conserv.* **2006**, *154*, 1193–1221. [CrossRef]

111. Hemp, C.; Hemp, A. The Chagga Home Gardens on Kilimanjaro: Diversity and Refuge Function for Indigenous Fauna and Flora in Anthropogenically Influenced Habitats in Tropical Regions under Global Change on Kilimanjaro, Tanzania, Africa. 2008. Available online: https://www.ihdp.unu.edu/file/get/7728 (accessed on 29 November 2018).

112. Chuhila, M.J. Coming down the Mountain: History of Land Use Change in Kilimanjaro, ca. 1920 to 2000s. Ph.D. Thesis, University of Warwick, Warwick, UK, 2016; pp. 1–349.

113. De Bont, C. The continuous quest for control by African irrigation planners in the face of farmer-led irrigation development: The case of the Lower Moshi Area, Tanzania 1935–2017. *Water Altern.* **2018**, *113*, 893–915.

114. Sunday, S. Adaption, Resilience and Transformability A Historical Ecology of a Traditional Furrow Irrigation System on the Slopes of Kilimanjaro. Master's Thesis, Department of Archaeology and Ancient History, Uppsala University, Uppsala, Sweden, 20 December 2015.

115. Decher, J. Conservation, small mammals, and the future of sacred groves in West Africa. *Biodivers. Conserv.* **1997**, *6*, 1007–1026. [CrossRef]

116. UNESCO. *Conserving Cultural and Biological Diversity: The Role of Sacred Natural Sites and Cultural Landscape*; UNESCO: Paris, France, 2005; pp. 1–340.

117. Mgumia, F.H.; Oba, G. Potential role of sacred groves in biodiversity conservation in Tanzania. *Environ. Conserv.* **2003**, *30*, 259–265. [CrossRef]

118. Bhagwat, S.A.; Rutte, C. Sacred groves: Potential for biodiversity management. *Front. Ecol. Environ.* **2006**, *410*, 519–524. [CrossRef]

119. Sheridan, M.J.; Nyamweru, C. *African Sacred Groves: Ecological Dynamics and Social Change*; James Currey: London, UK, 2008; pp. 1–230.

120. Metcalfe, K.; French-Constant, R.; Gordon, I. Sacred sites as hotspots for biodiversity: The Three Sisters Cave complex in coastal Kenya. *Oryx* **2010**, *44*, 118–123. [CrossRef]

121. Dudley, N.; Bhagwat, S.; Higgins-Zogib, L.; Lassen, B.; Verschuuren, B.; Wild, R. Conservation of biodiversity in sacred natural sites in Asia and Africa: A review of the scientific literature. In *Sacred Natural Sites: Conserving Nature and Culture*; Verschuuren, B., Wild, R., McNeely, J., Oviedo, G., Eds.; Earthscan: London, UK, 2010; pp. 19–32.

122. Bhagwat, S.A.; Dudley, N.; Harrop, S.R. Religious following in biodiversity hotspots: Challenges and opportunities for conservation and development. *Conserv. Lett.* **2014**, *4*, 234–240. [CrossRef]

123. Akida, A.; Blomley, R. *Trends in Forest Ownership, Forest Resources Tenure and Institutional Arrangements: Are They Contributing to Better Forest Management and Poverty Reduction? Case Study from the United Republic of Tanzania*; FAO: Dar es Salaam, Tanzania, 2006; pp. 1–26.

124. Madewaya, K.H.; Oka, H.; Matsumoto, M. Sustainable management of sacred forests and their potential for eco-tourism in Zanzibar. *Bull. FFPRI* **2004**, *3*, 33–48.

125. Schipper, J.; Burgess, N. Tropical and Subtropical Moist Broadleaf Forests: Eastern Africa: Coastal Areas of Kenya, Somalia, and Tanzania. WWF. Available online: http://www.worldwildlife.org/ecoregions/at0125 (accessed on 1 December 2015).

126. Kibet, S. Plant communities, species diversity, richness, and regeneration of a traditionally managed coastal forest, Kenya. *For. Ecol. Manag.* **2011**, *261*, 949–957. [CrossRef]

127. Shepheard-Walwyn, E. *Culture and Conservation in the Sacred Sites of Coastal Kenya*; University of Kent: Canterbury, UK, 2014.

128. Ongugo, P.; Wekesa, C.; Ongugo, R.; Abdallah, A.; Akinyi, L.; Pakia, M. Smallholder Innovation for Resilience SIFOR Qualitative Baseline Study, Mijikenda Community, Kenyan Coast. *IIED* **2014**, 1–35. Available online: http://pubs.iied.org/G03830/ (accessed on 29 November 2018).

129. Groh, M.A. Community Based Adaptations to Climate Change: Experiences of the Mijikenda Community in Coastal Kenya. Master's Thesis, University of Lisbon, Lisbon, Portugal, 30 September 2016.

130. Byers, B.A.B.; Cuncliffe, R.N.; Hudak, A.T. Linking the conservation of culture and nature: A case study of sacred forests in Zimbabwe. *Hum. Ecol.* **2001**, *29*, 187–218. [CrossRef]

131. Izidine, S.A.; Siebert, S.; van Wyk, A.E.; Zobolo, A.M. Taboo and political authority in conservation policy: A case study of the Licuati Forest in Maputaland, Mozambique. *J. Study Relig. Nat. Cult.* **2008**, *23*. [CrossRef]

132. Virtanen, P. The role of customary institutions in the conservation of biodiversity: Sacred forests in Mozambique. *Environ. Value* **2002**, *11*, 227–241. [CrossRef]

133. Ekblom, A. Forest-savanna dynamics in the coastal lowland of southern Mozambique since 400 AD. *Holocene* **2008**, *18*, 1247–1257. [CrossRef]

134. Sheridan, M.J. The dynamics of Africa Sacred grooves: Ecological, social and symbolic processes. In *African Sacred Groves: Ecological Dynamics & Social Change*; Sheridan, M.J., Nyamweru, C., Eds.; James Currey: Oxford, UK, 2007; pp. 9–41.

135. Virah-Sawmy, M.; Gardner, C.J. The Durban Vision in practice: Experiences in the participatory governance of Madagascar's new protected areas. In *Conservation and Environment Management in Madagascar*; Scales, I.R., Ed.; Routledge: London, UK, 2014; pp. 240–276.

136. Bodin, Ö.; Tengö, M.; Norman, A.; Lundberg, J.; Elmqvist, T. The value of small size: Loss of forest patches and ecological thresholds in southern Madagascar. *Ecol. Appl.* **2006**, *16*, 440–451. [CrossRef]

137. Boussou, B. Sustainable Management of Sacred Forests in Republic of Benin. CBD International, 2011. Available online: http://www.cbd.int/traditional/doc/8jcu-01/2011-06-03-bossou-en.pdf (accessed on 31 March 2018).

138. Alohou, E.C.; Gbemavo, D.S.J.C.; Ouinsavi, C.; Sokpon, N. Local perceptions and importance of endogenous beliefs on sacred groves conservation in South Benin. *Int. J. Biodivers. Conserv.* **2016**, *5*, 105–112. [CrossRef]

139. Lokossou, A.O.; Boussou, B. Benin's Experience in the Management of Sacred Forests for Biodiversity Conservation. Available online: http://satoyama-initiative.org/benins-experience-in-the-management-of-sacred-forests-for-biodiversity-conservation/ (accessed on 31 March 2018).

140. Ceperley, N.; Montagnini, F.; Natta, A. Significance of sacred sites for riparian forest conservation in Central Benin. *Bois Forêts Troipques* **2010**, *303*, 5–23. [CrossRef]

141. Gordon, C. Sacred groves and conservation in Ghana. In *Newsletter of the IUCN SSC African Reptile and Amphibian Specialist Group*; IUCN Species Survival Commission: Pretoria, South Africa, 1992; Volume 1, pp. 3–4.

142. Bossart, J.L.; Antwi, J.B. Limited erosion of genetic and species diversity from small forest patches: Sacred forest groves in an Afrotropical biodiversity hotspot have high conservation value for butterflies. *Biol. Conserv.* **2016**, *198*, 122–134. [CrossRef]

143. Fraser, J.A.; Diabaté, M.; Narmah, W.; Beavogui, P.; Guilavogui, K.; de Foresta, H.; Junqueira, A.B. Cultural valuation and biodiversity conservation in the Upper Guinea forest, West Africa. *Ecol. Soc.* **2016**, *213*, 36. [CrossRef]

144. Donaldson, J.E.; Archibald, S.; Govender, N.; Pollard, D.; Luhdo, Z.; Parr, C.L. Ecological engineering through fire-herbivory feedbacks drives the formation of savanna grazing lawns. *J. Appl. Ecol.* **2018**, *55*, 225–235. [CrossRef]

145. Beatty, R. Annexes–CBFiM Case Studies: Annex 1: CBFiM in Namibia: The Caprivi Integrated Fire Management Programme. In *Community-Based Fire Management: A Review*; FAO: Rome, Italy, 2011; pp. 41–47.

146. Ribot, J.C. *Democratic Decentralization of Natural Resources: Institutionalizing Popular Participation*; World Resources Institute: Washington, DC, USA, 2002; pp. 1–30.

147. Virtanen, P. Community-based natural resource management in Mozambique: A critical review of the concept's applicability at the local level. *Sustain. Dev.* **2005**, *13*, 1–12. [CrossRef]

148. Berkes, F. Community-based conservation in a globalized world. *Proc. Natl. Acad. Sci. USA* **2007**, *104*, 15188–15193. [CrossRef] [PubMed]

MDPI

St. Alban-Anlage 66

4052 Basel

Switzerland

Tel. +41 61 683 77 34

Fax +41 61 302 89 18

www.mdpi.com

Land Editorial Office

E-mail: land@mdpi.com

www.mdpi.com/journal/land

www.ingramcontent.com/pod-product-compliance
Lightning Source LLC
Chambersburg PA
CBHW051852210326
41597CB00033B/5867